2nd Materials Research Society of Indonesia Meeting (MRS-Id 2016)

IOP Conference Series: Materials Science and Engineering
Volume 214

Bandung, Indonesia
24 - 26 October 2016

Editors:

Khairurrijal Khairurrijal
Kilkuo Okuyama
Ferry Iskandar
Satria Zulkarnaen Bisri

ISBN: 978-1-5108-4416-2
ISSN: 1757-8981

Printed from e-media with permission by:

Curran Associates, Inc.
57 Morehouse Lane
Red Hook, NY 12571

CURRAN ASSOCIATES INC.
proceedings.com

Some format issues inherent in the e-media version may also appear in this print version.

Printed by Curran Associates, Inc. (2017)

For permission requests, please contact the Institute of Physics at the address below.

Institute of Physics
Dirac House, Temple Back
Bristol BS1 6BE UK

Phone: 44 1 17 929 7481
Fax: 44 1 17 920 0979

techtracking@iop.org

Additional copies of this publication are available from:

Curran Associates, Inc.
57 Morehouse Lane
Red Hook, NY 12571 USA
Phone: 845-758-0400
Fax: 845-758-2633
Email: curran@proceedings.com
Web: www.proceedings.com

Table of contents

Volume 214

2nd Materials Research Society of Indonesia Meeting (MRS-Id 2016)24–26 October 2016, Bandung, West Java, Indonesia

Accepted papers received: 22 June 2017 Published online: 6 July 2017

Preface

Papers

IOP Conf. Series: Materials Science and Engineering **214** (2017) 011001 doi:10.1088/1757-899X/214/1/011001

PREFACE

The Materials Research Society of Indonesia (MRS-Id) Meeting is a bi-annual society meeting of the MRS-Id or Himpunan Riset Material Indonesia (HRMI). The 2nd MRS-Id Meeting was held in Bandung from 24th to 26th October 2016 while the first one was in Bali in 2014. It was co-organized by Physics of Electronic Materials Research Division, Faculty of Mathematics and Natural Sciences of Institut Teknologi Bandung (ITB).

The aim of this event was to deliver the focus of MRS Indonesia that is to promote materials related with science and technology in Indonesia, regional, and global. We were very pleased that this event could bring together scientists, researchers, and engineers in Indonesia, regional, and global to share their newest scientific works and discoveries in the field of materials science and engineering.

Approximately 160 participants have attended the meeting. Among the participants, there were 6 plenary speakers, 26 invited speakers, and 121 contributed speakers. Based on the country where the participants came from, they were from 10 (ten) countries namely France, Germany, Indonesia, Japan, Malaysia, Saudi Arabia, Singapore, Sweden, UK, and USA. This MRS-Id Meeting consisted of 10 (ten) symposia that cover the broad spectra of materials science and engineering.

The speakers of the meeting have submitted their full papers for possible publications in a proceeding of IOP Publishing (IOP Conference Series: Materials Science and Engineering) or a journal of IOP Publishing (Materials Research Express). After peer reviewed by experts in related fields, finally, this proceeding contains 36 accepted papers. In addition, there are 20 papers published in the journal after an additional peer-review process of the journal.

The Materials Research Society of Indonesia (MRS-Id) acknowledges the enthusiasm of the participants and numerous sponsors, all members of the committees, the international advisory board, Faculty of Mathematics and Natural Sciences of ITB, and all those who have contributed to the success of the 2nd MRS-Id Meeting.

The next MRS-Id Meeting will be held in 2018. We are looking forward to welcoming you to the 3rd MRS-Id Meeting!

Prof. Khairurrijal Khairurrijal
Chairman of MRS-Id Meeting 2016 / Current President of MRS-Id

Dr. Ferry Iskandar
Co-Chairman of MRS-Id Meeting 2016

Prof. Khairurrijal Khairurrijal (ITB, Indonesia), Prof. Kikuo Okuyama (Hiroshima University, Japan), Dr. Ferry Iskandar (ITB, Indonesia), and Dr. Satria Zulkarnaen Bisri (RIKEN, Japan)
Editorial Board

Peer review statement

All papers published in this volume of *IOP Conference Series: Materials Science and Engineering* have been peer reviewed through processes administered by the proceedings Editors. Reviews were conducted by expert referees to the professional and scientific standards expected of a proceedings journal published by IOP Publishing.

IOP Conf. Series: Materials Science and Engineering **214** (2017) 012001 doi:10.1088/1757-899X/214/1/012001

Preparation of Zinc Oxide (ZnO) Thin Film as Transparent Conductive Oxide (TCO) from Zinc Complex Compound on Thin Film Solar Cells: A Study of O_2 Effect on Annealing Process

E Y Muslih[1,2] and K H Kim[3]

[1]Physical Metallurgy Laboratory, Mechanical Engineering Department, Faculty of Industrial Technology, Trisakti University, Jalan Kyai Tapa No. 1, Grogol, Jakarta 11440, Indonesia.
[2]Nano Center Indonesia, Jalan Raya Bogor KM 47, Cibinong, Bogor 16912, Indonesia.
[3]School of Material Science and Engineering, Yeungnam University, 214-1 Daedong, Gyeongsan, Gyeongbuk 712-749, Republic of Korea.

Email: ersan@nano.or.id

Abstract. Zinc oxide (ZnO) thin film as a transparent conductive oxide (TCO) for thin film solar cell application was successfully prepared through two step preparations which consisted of deposition by spin coating at 2000 rpm for 10 second and followed by annealing at 500 °C for 2 hours under O_2 and ambient atmosphere. Zinc acetate dehydrate was used as a precursor which dissolved in ethanol and acetone (1:1 mol) mixture in order to make a zinc complex compound. In this work, we reported the O_2 effect, reaction mechanism, structure, morphology, optical and electrical properties. ZnO thin film in this work shows a single phase of wurtzite, with *n*-type semiconductor and has band gap, carrier concentration, mobility, and resistivity as 3.18 eV, $1.21 \times 10^{-19} cm^3$, 11 cm^2/Vs, 2.35×10^{-3} Ωcm respectively which is suitable for TCO at thin film solar cell.

1. Introduction

Zinc oxide (ZnO) thin film is the II-IV thin film semiconductor which is widely used in many important applications such as sensors, photodetectors, LEDs, transistors and also as a conductive layer in the solar cells [1,2,3]. As a conductive layer, ZnO thin film has band gap and resistivity of 3.35 eV and 4.4×10^{-3} Ωcm respectively [4]. ZnO can be fabricated by vacuum and non-vacuum processes and both methods have their own advantages. However to fabricate a simple, low cost and easy scalability of ZnO thin film, non-vacuum process is more preferable especially chemical or wet process [5]. Commonly, to fabricate thin film by chemical or wet process, organic solvents is used such as 2-propanol, methanol, ethanol, oleylamine, aminoalcohols, hexamethyltetraamine [6-10].

In this work, ZnO thin film successfully synthesized from zinc acetate dehydrate as starting material and using a mixture of acetylacetone and ethanol as solvents. Spin coating method is used to deposit the precursor solution and continued by annealing to get ZnO thin film layer. However, this work focuses on O_2 effects to optical and electrical properties of ZnO thin film layer to be applied as a transparent conductive oxide (TCO) in thin film solar cell.

IOP Conf. Series: Materials Science and Engineering **214** (2017) 012001 doi:10.1088/1757-899X/214/1/012001

2. Experimental

2.1. Materials and Methods

ZnO thin film layer was prepared from 0,1 M of zinc acetate dihydrate from Aldrich as starting material which dissolved in the mixture of ethanol and acetylacetone (both from Aldrich) in 1:1 mol ratio. This solution defines as a precursor solution and it was a clear and homogeneous solution and deposited on soda lime glass (SLG) which cleaned with acetone, ethanol, deionized water using ultrasonic for 15 minutes respectively and followed by nitrogen gas drying in order to eliminate further contaminants. Moreover, 0.2 mL of precursor solution was deposited by a homemade spin coating apparatus at 2000 rpm for 10 seconds and dried at room temperature for 3 minutes. This process was repeated for 5 times and continued by annealing at 500 ^{o}C for 2 hours under O_2 and ambient atmosphere using a tube furnace from Best Made In Korea (BMIK, South Korea).

2.2. Characterization of ZnO Thin Film

The morphology of the thin film was characterized by scanning electron microscope (SEM; Hitachi S-4800, Japan) with an operating voltage 15 kV. The chemical composition of the thin films was analyzed by energy dispersive spectrometry (EDS; Horiba, Japan) attached to the SEM with an acceleration voltage, working distance and emission current of 15 kV, 15 mm, and 10 μA respectively. The phase of crystallographic information was obtained from the x-ray diffraction (XRD; PANanalitycal, Netherlands). The spectra absorption was recorded by UV-Vis-NIR Spectrophotometer in the wavelength range of 300-1500 nm (Carry 5000, Varian, USA). Electrical properties were determined using Hall-Effect measurement at 0.01 μA (HMS-300, Ecopia, USA).

3. Results and Discussion

In the preparation, the mixture of acetylacetone and ethanol with the same mol ratio could form a hemiketal compound which could dissolve and form a zinc complex compound with zinc acetate dihydrate simultaneously. Then, the zinc complex solution was deposited on the substrate and allowed to evaporate under room temperature. Moreover, ZnO thin film was fabricated by annealing the deposited zinc complex on the substrate under O_2 atmosphere at 500 ^{o}C for 2 hours, the zinc complex compound was decomposed into ZnO and deposited on the substrate, meanwhile the organic compounds were decomposed into carbon dioxide and water vapor.

$$C_2H_6O\ (l) + C_5H_8O_2\ (l) \rightarrow C_7H_{14}O_3\ (aq)$$

$$C_7H_{14}O_3\ (aq) + Zn^{2+}\ (aq) \rightarrow Zn[C_7H_{14}O_3]_x\ (s)$$

$$Zn[C_7H_{14}O_3]_x\ (aq) + 11O_2\ (g) \rightarrow ZnO\ (s) + 7CO_2\ (g) + 7H_2O\ (g)$$

After annealing at 500 ^{o}C under O_2 atmosphere for 120 minutes, ZnO thin film had been formed with a single phase of hexagonal wurtzite. X-ray diffraction (XRD) pattern, showed identical XRD peaks with ZnO standard (ICDD: 01-089-7102). Meanwhile, ZnO thin film annealed under ambient atmosphere showed the existence of carbon residue as amorphous which identified as a broadened peak from 20 to 50°. However, although ZnO thin film also could be formed by annealing under ambient atmosphere, but there was carbon remained as a residue due less oxygen contained in the air. This condition makes the organic compounds were not decomposed completely and formed carbon compounds as residue which trapped inside the thin film.

$$Zn[C_7H_{14}O_3]_x\ (aq) + 3O_2\ (aq) \rightarrow ZnO\ (s) + CO\ (g) + 6C\ (s) + 7H_2O\ (g)$$

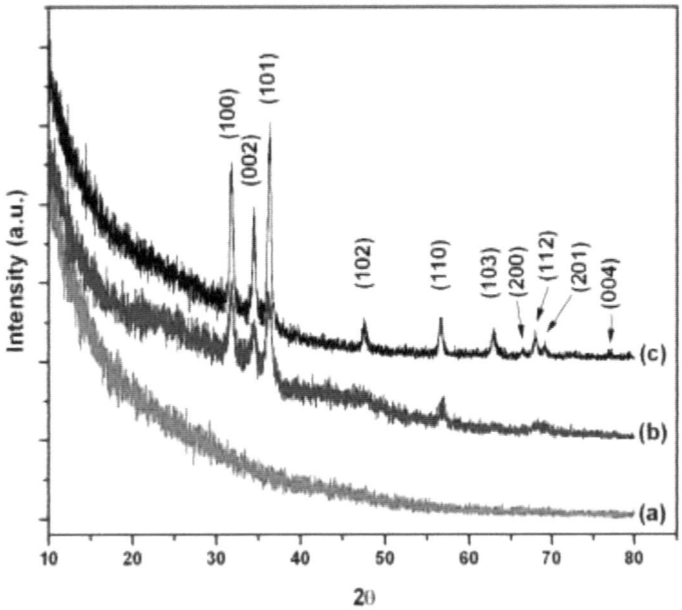

Figure 1. XRD pattern of substrate before annealing (a), ZnO thin film after annealed under ambient atmosphere (b), and O_2 gas atmosphere (c).

From scanning electron microscope (SEM), ZnO thin film annealed under O_2 gas atmosphere showed a similar shape with solid ZnO [11] with uniform granular with grain size around 60 nm and from cross-sectional image could be seen a dense thin film with 507 nm of thickness. On the other hand, ZnO thin film annealed under ambient atmosphere showed carbon as residue surrounded by small granular of ZnO. Meanwhile, carbon in the thin film came from organic compounds left and trapped in the thin film due to every incomplete combustion reaction from organic compounds would leave carbon behind as residue was detected by energy dispersive spectrometry (EDS).

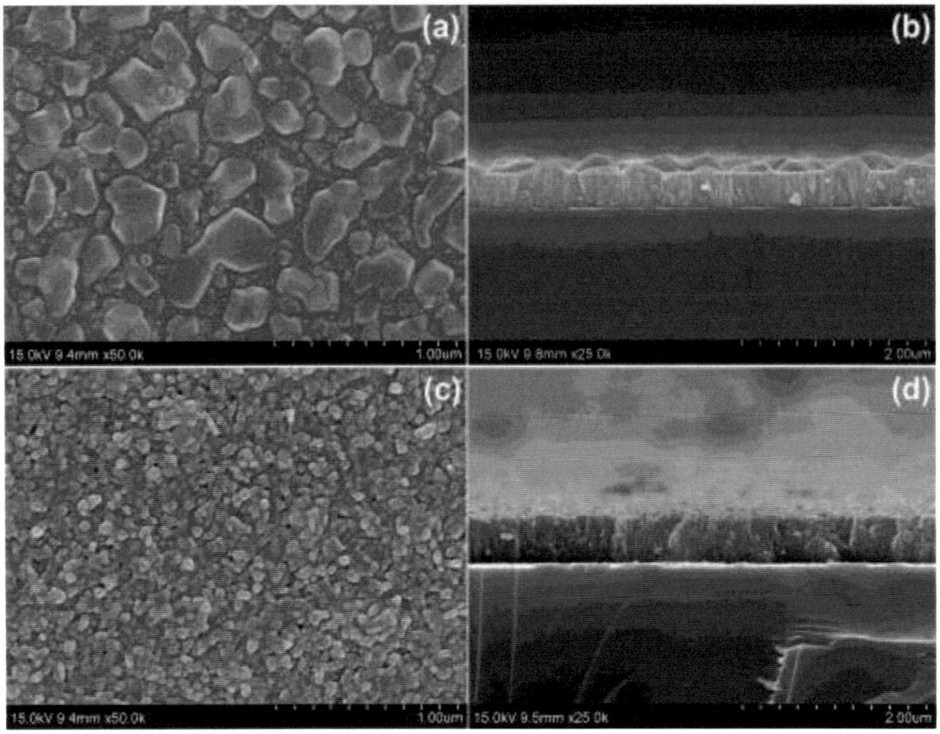

Figure 2. SEM images of ZnO thin film annealed under ambient atmosphere (a) and O_2 (c), and their cross-sectional image respectively (b and d).

From the EDS spectrum, it was clearly shown that ZnO thin film annealed under ambient atmosphere still contained carbon as residue whereas ZnO thin film annealed under O_2 gas atmosphere showed that no carbon spectrum was detected due to complete combustion reaction between O_2 with organic compounds in the thin film. Besides, O_2 from the atmosphere also helped oxidation of Zn became ZnO. However, Si and Ca from substrate (soda lime glass) containing Si and Ca were detected due to dispersion spectrum of Si and Ca could penetrate the ZnO thin film and were detected by detector as shown in Figure 3b.

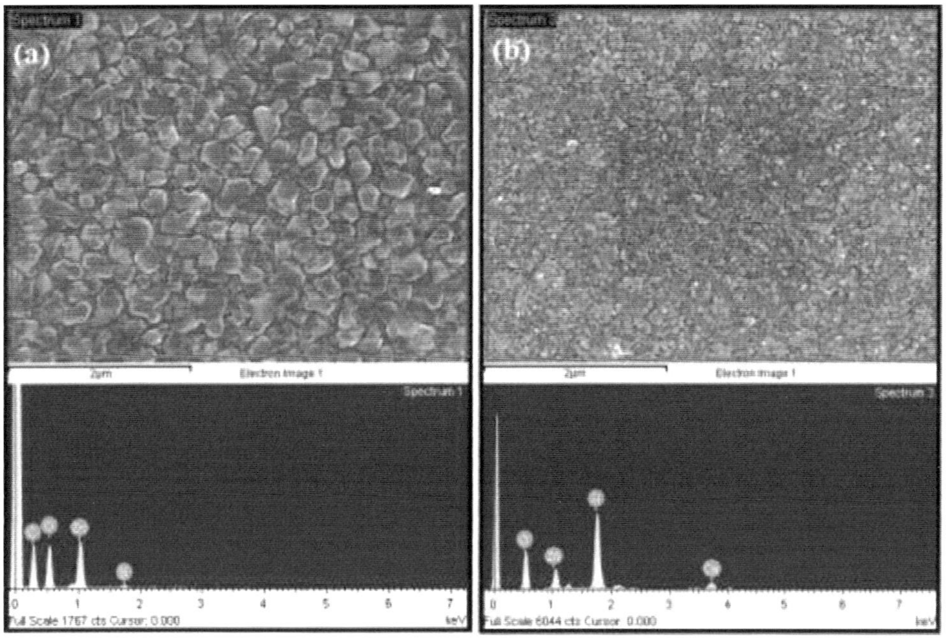

Figure 3. EDS images and spectrums of ZnO thin film annealed under ambient atmosphere (a) and O_2 (b).

The optical properties of ZnO thin film annealed either under O_2 atmosphere or ambient atmospheres showed low transmittance at 300 nm to 400 nm and high transmittance at 400 nm to 1500 nm. However, the ZnO thin film annealed under O_2 atmosphere had higher transmittance and band gap energy due to no carbon residue which could absorb the light. The band gap energy of ZnO thin film annealed under air and O_2 atmosphere was 3.07 and 3.18 eV respectively. ZnO annealed under ambient atmosphere absorbed less light than annealed under O_2 atmosphere due to the existence of carbon as residue. However, ZnO annealed under O_2 atmosphere had similar transmittance and band gap energy with the reference [12].

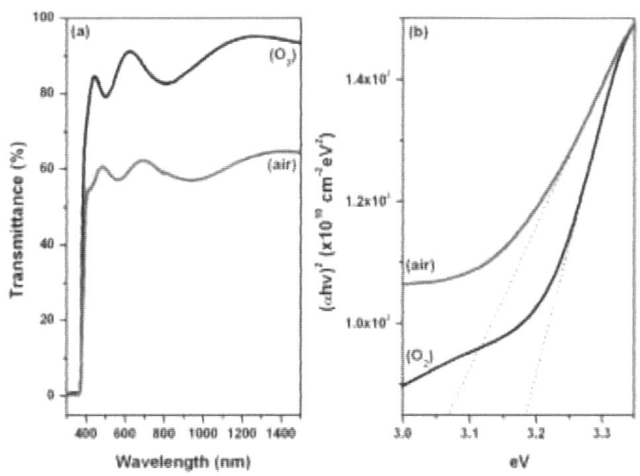

Figure 4. Optical transmittance of ZnO annealed under O_2 and ambient atmosphere (a) and their energy band gap (b).

The electrical properties of ZnO thin films both showed *n*-type semiconductor with electrical properties such as carrier concentration and mobility as shown in Table 1. Carrier concentration and mobility of ZnO annealed under O_2 and ambient atmosphere showed appropriate by reference. However, carrier concentration and mobility of ZnO annealed under an ambient atmosphere were lower than annealed under O_2 atmosphere. This phenomenon was also caused by carbon which left behind the thin film as a residue that could restrict the electrons movement in the ZnO thin film.

Table 1. Electrical properties of ZnO wurtzite thin film prepared under several conditions

Electrical Properties	Under O_2 atmosphere	Under ambient atmosphere	Reference [13]
Carrier Concentration	$1.21 x 10^{-19}$ cm^3	$2.35 x 10^{-18}$ cm^3	10^{-17} - 10^{-20} cm^3
Mobility	11 cm^2/Vs	9 cm^2/Vs	5 - 50 cm^2/Vs

4. Conclusion

The ZnO thin film could be fabricated trough zinc complex compound pathway which is made from zinc acetate dihydrate and hemiketal without pre heat treatment process. The amount of O_2 during annealing of ZnO thin film is critical due to acts as a reagent for eliminating the organic compounds and as an oxidant simultaneously. The ZnO thin film annealed under O_2 atmosphere shows better morphology, optical and electrical properties due to sufficient of O_2 during reactions. ZnO thin film has suitable properties as TCO for thin film solar cell application.

5. References

[1] Musić, S and Šarić A 2012 *Ceramics International* **7** 6047
[2] Aranovich J A, Golmayo D, Fahrenbruch A L and Bube R H 1980 *Journal of Applied Physics* **8** 4260
[3] Kołodziejczak-Radzimska A and Jesionowski T 2014 *Materials* **4** 2833
[4] Kim K H, Wibowo R A and Munir B 2006.*Materials Letters* **15** 1931
[5] Thongsuriwong K, Amornpitoksuk P and Suwanboon S 2010 *Journal of Physics and Chemistry of Solids* **5** 730
[6] Musić S, Dragčević Đ, Popović S and Ivanda M 2005 *Materials letters* **19** 2388
[7] Natsume Y and Sakata H 2000 *Thin solid films* **1** 30
[8] Smirnov M, Baban C and Rusu G I 2010 *Applied Surface Science* **8** 2405
[9] Salavati-Niasari M, Davar F and Mazaheri M 2008 *Materials Letters* **12** 1890
[10] Ismail A A, El-Midany A, Abdel-Aal E A and El-Shall H 2005 *Materials Letters* **14** 1924
[11] Thongsuriwong K, Amornpitoksuk P and Suwanboon S 2013 *Advanced Powder Technology* **1** 275
[12] Bedia A, Bedia F Z, Aillerie M, Maloufi N and Benyoucef B 2015 *Energy Procedia* **74** 529
[13] Pearton S J, Norton D P, Ip K, Heo Y W and Steiner T 2005 *Progress in materials science* **3** 293

Acknowledgment

This study was supported by the Basic Science Research Program (2013R1A1A2013408) and Center for Inorganic Photovoltaic Materials (No.2012-0001170) through the National Research Foundation of Korea (NRF) funded by the Ministry of Science, ICT and Future Planning.

IOP Conf. Series: Materials Science and Engineering **214** (2017) 012002 doi:10.1088/1757-899X/214/1/012002

Natural Rubber Nanocomposite with Human-Tissue-Like Mechanical Characteristic

Riri Murniati, Nanda Novita, Sutisna, Edy Wibowo, Ferry Iskandar and Mikrajuddin Abdullah*

Material Electronics Research Group, Department of Physics, Faculty of Mathematics and Natural Sciences, Institut Teknologi Bandung Jalan Ganesa 10 Bandung 40132 Indonesia

*din@fi.itb.ac.id

Abstract. The blends of synthetic rubber and natural rubber with nanosilica were prepared using a blending technique in presence of different filler volume fraction. The effect of filler on morphological and mechanical characteristics was studied. Utilization of human cadaver in means of medical study has been commonly used primarily as tools of medical teaching and training such as surgery. Nonetheless, human cadaver brought inevitable problems. So it is necessary to find a substitute material that can be used to replace cadavers. In orthopaedics, the materials that resemble in mechanical properties to biological tissues are elastomers such as natural rubber (latex) and synthetic rubber (polyurethanes, silicones). This substitution material needs to consider the potential of Indonesia to help the development of the nation. Indonesia is the second largest country producer of natural rubber in the world. This paper aims to contribute to adjusting the mechanical properties of tissue-mimicking materials (TMMs) to the recommended range of biological tissue value and thus allow the development of phantoms with greater stability and similarity to human tissues. Repeatability for the phantom fabrication process was also explored. Characteristics were then compared to the control and mechanical characteristics of different human body part tissue. Nanosilica is the best filler to produce the best nanocomposite similarities with human tissue. We produced composites that approaching the properties of human internal tissues.

1. Introduction

In medical, cadavers had been used as an object of study and research, especially for medical surgical training. However arise some problems behind the use of this cadaver such as durability, costly, availability, and issues of morality. The cadavers that will be studied by medical students come from a range of ages and conditions were completely random [1]. As medical science has progressed, it has become increasingly important to provide non-human interactive formats for teaching patient care. While it is desirable to train medical personnel in patient care protocols before allowing contact with real patients. On the other hand, allowing inexperienced students to perform medical procedures on actual patients that would allow for the hands-on practice cannot be considered a viable alternative because of the inherent risk to the patient. Non-human interactive devices and systems can be used to teach the skills needed to successfully identify and treat various patient conditions without putting actual patients at risk [2].

Such training devices and systems can be used by medical personnel and medical students to teach and assess competencies such as patient care, medical knowledge, practice-based learning and systems based practice. The training devices and systems can also be used by patients to learn the proper way to perform self-examinations. In particular, structures for simulating body tissue for use in practicing surgical and/or clinical techniques. In the medical field, it is necessary for students, doctors, and surgeons to be able to practice surgical and clinical techniques. Especially, they need to practice incisions, the removal of various kinds of complaint (such as cysts, and melanomas), access to venous structures below the epidermis and the insertion of sutures. As an alternative to providing an actual body or part of an actual body for practicing such techniques, there is a need for artificial means whereby they can be practiced. Known structures providing simulations of body tissue surfer from the disadvantage that they are not sufficiently analogous, either visually or physically, to actual body tissue [3].

We use a substitute material to create a mimicking human cadaver through this study. This is due to the urgency of material selection in accordance with the potential of Indonesia. Indonesia is a country with a long history in agriculture and plantations. So in the manufacture of replacement cadaveric material need to choose a material that has become a mainstay product Indonesia for a long time. Rubber production in Indonesia also have to be a concern, still not fully used by the industry in Indonesia and the price drops dramatically. Indonesia is the second largest rubber producer in the world. Excess production of rubber will then be exported without being processed or in a raw state. It is very worrying that should be considered a solution to steer this product to a more promising alternative industries.

Natural rubber (NR) is known to exhibit numerous outstanding properties such as good oil resistance, low gas permeability, improved wet grip and rolling resistance, coupled with high strength; having properties resembling those of synthetic rubbers. Natural rubber coming from latex is mostly polymerized isoprene with a small percentage of impurities in it. This will limit the range of properties available to it, although addition of sulfur and vulcanization are used to improve the properties. Synthetic rubber is any type of artificially made polymer material which acts as an elastomer. Synthetic rubber serves as a substitute for natural rubber in many cases, especially when improved material properties are needed. A wide variety of particulate fillers are used in the rubber industry for various purposes, of which the most important are reinforcement, reduction in material costs and improvements in processing [4]. Reinforcement is primarily the enhancement of strength and strengthrelated properties, abrasion resistance, hardness and modulus [5,6]. The idea of blending synthetic rubbers with natural rubber is certainly not a new one, but it is only now that this can be shown to be possible with consistently positive results, by the use of new techniques developed over the last five years. These compounds are capable of forming a chemical link between these dissimilar rubbers to produce a technologically compatible blend. The blend vulcanizates thus produced exhibit enhanced physical properties by judicious selection of the NR: HSR ratio [7]. The use of carbon black is synonymous with the history of tyres. Although it has lost some ground to other reinforcing fillers such as silica, by virtue of its unrivalled performance, it is still the most popular and widely used reinforcing filler. However, the primary properties of carbon blacks are normally controlled by particle size, surface area, structure, surface activity and they are in most cases interrelated [8].

Silicone-based tissue-mimicking phantom is widely used as a surrogate of tissue for clinical simulators, allowing clinicians to practice medical procedures and researchers to study the performance of medical devices [9]. Polydimethylsiloxane (PDMS) tissue simulating phantoms with tuneable optical properties to be used for optical system calibration and performance testing in visible and near infrared domain also had been studied. Compared to liquid phantoms, cured PDMS phantoms are easier to transport and use, and have a longer usable life than gelatine based phantoms [10].

Matrix materials typically are water, gelatine, agar, polyester or epoxy and polyurethane resin, room-temperature vulcanizing (RTV) silicone, or polyvinyl alcohol gels. The water and hydrogel materials provide a soft medium that is biologically and biochemically compatible with the addition of organic molecules and are optimal for scientific laboratory studies. Polyester, polyurethane, and silicone phantoms are essentially permanent matrix compositions that are suitable for routine calibration and testing of established systems. The most common three choices for scatters have been: lipid-based emulsions, titanium or aluminium oxide powders, and polymer microspheres [11].

2. Methods

The novelty of this study uses natural rubber-based tissue mimicking human tissue. This is due to the urgency of material selection in accordance with the potential of Indonesia. So in the manufacture of replacement cadaveric material need to choose a material that has become a mainstay product Indonesia for a long time. The product is natural rubber. This study investigates using some formula of materials to create the desired mechanical properties and surgery characteristics of a tissue-mimicking tissue. The substitute materials have been determined as well as the composite material to be mixed into the rubber material. To be able to have the cadaver characteristics, we choose three types of filler material; caolin, calcium carbonate, and nanosilica. This is to see which is most effective filler in modifying natural rubber character resembles a human cadaver, and nanosilica is the one. Then the use of filler will be further optimized by using a variation of the volume fraction. It also conducted a matrix constituent mix between natural rubber and silicone rubber. Then it will do a comparison between the two materials. Silicone itself has a price that is much more expensive than natural rubber. So the comparison also serves to optimization apart from evoking the characteristics of a human cadaver in natural rubber by using a mixture of silicone rubber.

Basically, this experiment is divided into several phases: Phase design and manufacture of hot press mold and incision test equipment, the synthesis of samples, material characterization and retrieval of data, and analysis phase. Molds needed with special specifications, good material, size, and performance. A molding used is made of stainless steel, with a length of 10 cm, width 7 cm, and height 3 cm. That is not only as container scored but also serves as a heater that will enhance curing maturation. The precursors will be made by the optimization of the type of base material used rubber (liquid latex and silicone rubber). Samples were synthesized by mixing several chemicals (Table 1), the formula that is used depends on the type of materials used, the order of input materials, stirring time and the amount of material used. All formulas were found to be tested and repeated continuously until it was the most appropriate formula to produce nanocomposite most similar in structure mechanics with human flesh.

Table 1. Formula of the Materials

Materials	phr (part per hundred rubber)
Latex-silicone rubber	100
Filler	1.5
TiO_2	1
Minarex oil	7.5
ZnO	0.5
Stearat acid	0.5
Parafin wax	0.5
Food coloring	1
Texapan	1
CMC	0.5
Sulphur	0.5
Gelling agent	0.5

Do mixing precursor liquid latex with a foaming agent and a curing agent for a certain time (according to the formula), then added with a stabilizer and gelling agent. Meanwhile, another sample was mixed using filler then mixed with a stabilizer and gelling agent. Then the precursors and the material is dried and burned [12]. For a variety of rubber, natural rubber is mixed in advance with silicone rubber before being added foaming agent, gelling agent, curing agent and stabilizer. Also use softener and activator to facilitate the mixing process of the composite, and then put into the oven for 2 hours at a temperature of 150 °C, and then inserted into the mold molding and curing hot press for 1 hour at 100 °C. The sample was tested with the incision test equipment that had been designed specifically to testing of meat slicing (surgery-like) using tools that created by our team. Testing of the control samples was also performed on samples made from synthetic rubber (silicone) and beef. The material crosslink densities measured by swelling using the Flory–Rehner theory and the filler volume fractions are calculated with the component densities. The control samples also tested and compared. Data obtained from the results of this characterization then compared with reference to mechanical strength and morphology meat/human cadaver to see the similarity of the material and what could be improved for future research.

Mechanical tests were performed using an Universal Testing Machine, the sample needs to be formed into the shape of a small dumbbell. The length of each sample was homogeny to facilitate the processing of data. In addition, the material needs to be in a solid phase to perform this mechanical test. Actually for mechanical tests, not only the stress-strain testing needs to be done, but also the viscoelastic characteristics [13]. The data obtained from this test in the form of stress-strain curve then we can calculate the Young's modulus of composites. In an undeformed thermoplastic polymer tensile sample [14-15], the polymer chains are randomly oriented. When a stress is applied, a neck develops as chains become aligned locally. The neck continues to grow until the chains in the entire gage length have aligned. Elastic modulus is the mathematical description of an object or substance's tendency to be deformed elastically (i.e., nonpermanently) when a force is applied to it. The elastic modulus of an object is defined as the slope of its stress-strain curve in the elastic deformation region.

3. Results and Discussion

We conducted a study of variation filler, the filler used in this experiment is kaolin, $CaCO_3$, and nanosilica. The sample with nanosilica filler shows experimental results that more constant and material strength is better than other fillers.The nanosilica filler mass then optimized then we get the sample with 12 grams nanosilica filler is the best one to be added to the composite before the gelling agent added, with filler volume fraction variation can be obtained a variety of composite strength obtained by difference filler [16]. The properties of rubber will determine which kind of rubber is best-suited for any sort of application, or perhaps whether or not a combination of the two kinds of rubbers is necessary. Such an occurrence happens with automobile tires, which are constructed out of a mixture of natural gum rubber and synthetic rubber. On the other hand, the design of aircraft tires necessitates that they be made completely out of natural rubber. Clearly then, there are things which distinguish the two products, as no matter how great science and technology is today, it cannot perfectly replicate natural rubber.

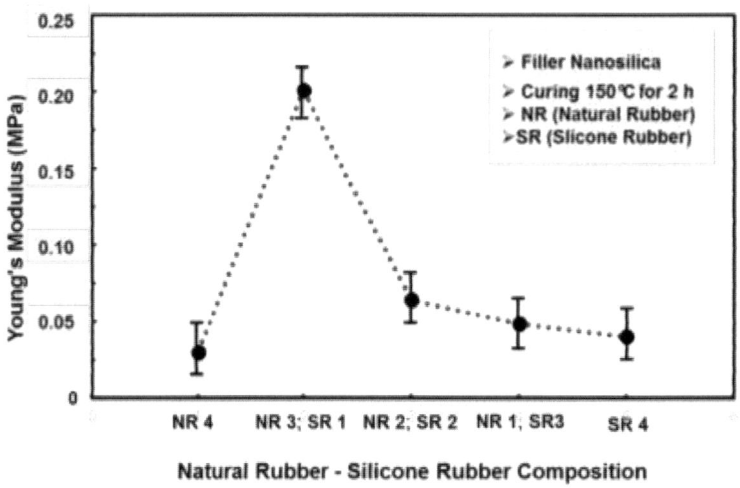

Figure 1. Curve of Young’s Modulus of Natural-Silicone Rubber Composition.

In the Figure 1 shown the area elastic force curve of the change in length along the slope dF / dl are then used to calculate the Young's modulus. Elastic behavior is a reversible behavior that often shows a linear relation between stress and strain. In an undeformed thermoplastic polymer tensile sample, the polymer chains are randomly oriented. When a stress is applied, a neck develops as chains become aligned locally. The neck continues to grow until the chains in the entire gage length have aligned. Then the strength of the polymer is increased. The data will then be compared to the literature of cadavers or human flesh characteristics.

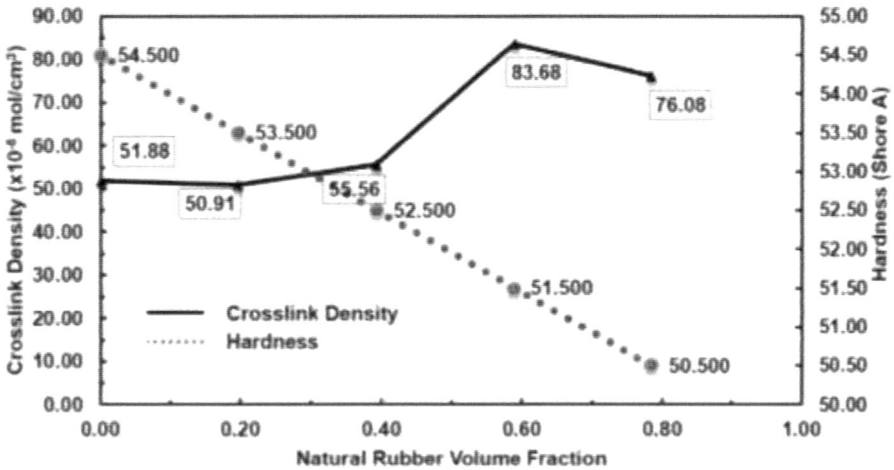

Figure 2. Hardness and Crosslink Density of Natural Rubber Volume Fraction.

Each material that added give hardness contribution in compound, specially for fillers and the elastomers. Kaolin and Calium Carbonate give hardness contribution + 0.25 shore A / 1 phr, nanosilica give hardness contribution + 0.5 shore A / 1 phr, natural rubber (latex) + 40 shore A / 1 phr, and silicone rubber + 44 shore A / 1 phr. We can see the hardness calculation results as seen in Figure 2 and Figure 3, hardness of the composite is linear with nanosilica filler and silicone rubber volume fraction, reverse with natural rubber volume fraction, and relative with natural-silicone rubber combination.

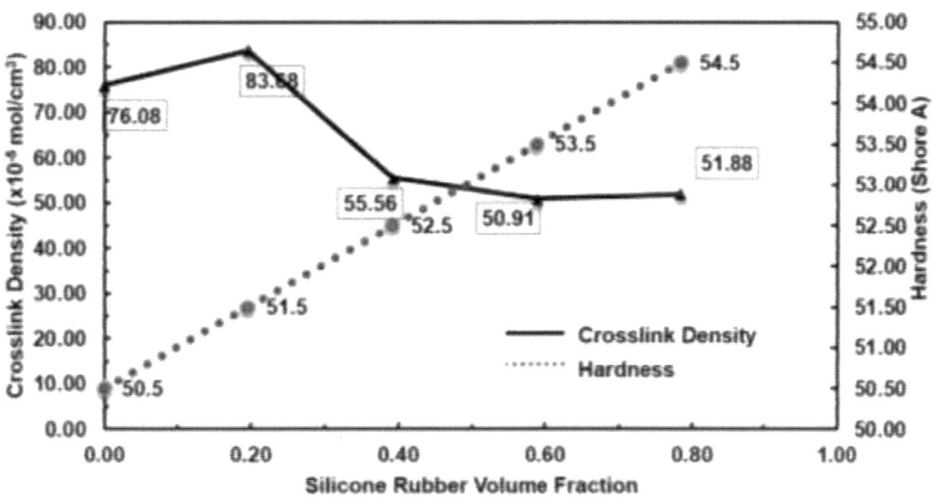

Figure 3. Hardness and Crosslink Density of Silicone Rubber Volume Fraction.

The cross-link density was determined by immersing a small amount (known mass) of sample in 100 ml toluene to attain equilibrium swelling. After this the sample was taken out from the toluene and the solvent was blotted from the surface of the sample and weighed immediately. This sample was then dried out at 80 °C to constant weight. Then the chemical cross-link density was calculated by the Flory–Rehner equation[17]. Higher crosslink density is observed in Figure 3 for the compounds containing 100 phr elastomers with combination of 75% natural rubber and 25 % silicone rubber, or containing 58.8 % natural rubber volume fraction and 19.6 % silicone rubber volume fraction of all materials in the composite. Natural rubber has a quality known as "tack," meaning that it adheres well to itself and to other materials. Silicone rubber has chemical resistance, it cannot be readily changed when it comes into contact with other materials. Silicone and natural rubber combination properties is non-reactive: both are chemically resistant to many fluids including water and weak acids. Excellent as an adhesive or a protective coating. Both can undergo vulcanization, meaning that chemical bonds are strengthened by sulfur. The backbone of silicone contains silicon and oxygen, while the backbones of most of the other rubbers contain carbon-carbon bonds. Silicone can be distinguished from rubbers by the atomic structure.

Calculation results of Young's modulus of this experiment then compared to references that can be seen in Table 2. Referring to this table, it can be seen that composite with natural-silicone rubber composition with nanosilica filler shows experimental results that relatively constant close to the machanical characteristic of some human body part. So that using nanosilica filler, it will be easier to manipulate characteristics, then the experiment will be conducted the filler mass variation of nanosilica [18].

Table 2. Mechanical Characteristic of Human Body

Body	Gender	Age (year)	Modulus Young (Mpa)	Ref.
Bicep Muscle	Man	65	49.02	[19]
Bicep Muscle	Man	85	59.15	[19]
Bicep Muscle *	Man	75	48.74	[19]
Leather Back		30	200	[20]
Leather Back		9	0.42	[20]
Cartilage in the knee	Man	16	0.13	[20]
Cartilage in the knee	Woman	65	1.91	[20]
Shoulder Muscle (*Anterior*)			16.5	[22]
Shoulder Muscle (*Middle*)			6	[22]
Shoulder Muscle (*Posterior*)			4.1	[22]
Esophagus			0.077	[22]
Gut			0.0356	[22]
Gut*			0.0359	[22]
Pericardium			2.51	[22]
Achilles tendon			819	[23]
Patellar tendon			643	[24]
Ligament (cruciate ligament)			345	[24]
Semitendinosus tendon		26	362.2	[25]
Ligament (ligamentum flavum)		young	98	[26]
Ligament (ligamentum flavum)		old	20	[26]
Ankle ligament (anterior talofibular ligament)	Woman	27	86	[27]
Cartilage (HWA of femoral condyle, superficial zone)			5	[28]
Cartilage (HWA of femoral condyle, middle zone)			3.1	[28]
Cartilage (LWA of femoral condyle, superficial zone)			10.1	[28]
Cartilage (LWA of femoral condyle, middle zone)			5.4	[28]

The increase in young's modulus in Figure 6 may be due to the fact that an increasing volume fraction of nanosilica caused the vulcanization reaction to increase and create more active cross-link sites in the rubber compound. From both frosslink density and young's modulus results, the compounds with 7.84 % volume fraction of nanosilica filler have given higher result. From these studies, a balancing correlation has obtained for the same compounds under standard abraders.

4. Conclusion

Nanosilica is used in filler has nano-sized particles while the caolin and CaCO3 that used in powder size. Visually, hardness and density of material obtained that a composite rubber calcium carbonate and kaolin is not too much different from the rubber control. While the nanosilica composite rubber is much denser and Young's modulus values obtained from the composite rubber nanosilica also greater than calcium carbonate and kaolin composite rubber. Young's modulus data compared between the experimental results and the human body characteristics, it can be concluded that the rubber composite of 25% natural rubber and 75 % silicone rubber (0.07 MPa), and 100 % silicone rubber (0.08 MPa) has mechanical characteristics that resemble human esophagus (0.077 MPa). Rubber composite of 50% natural rubber and 50 % silicone rubber has characteristics resembling cartilage in the knee man teens (0.13MPa). While the composite rubber of 75% natural rubber and 25 % silicone rubber (0.22 MPa) has characteristics resembling children leather back (0.42 MPa). Nanosilica volume fraction shows the best result at 58.8 % natural rubber volume fraction and 19.6 % silicone rubber volume fraction with 12 gram nanosilica filler (10 per hundred rubber), the highest young's modulus and crosslink density are performed.

5. References

[1] Kaku M 2011 *Physics of the Future: How Science Will Shape Human Destiny and Our Daily Lives by the Year 2100* (New York: Doubleday).
[2] Lowe S 2015 *Surgical simulation models, materials, and methods* Patent US9123261 B.
[3] Cooper CM, Sunderland J 1998 *Method of Making A Surgical And/ Or Clinical Apparatus*. Patent US005775916A.
[4] Tinker AJ, Jones KP 1998 *Blends of natural rubber novel techniques for blending with speciality polymers* (London: Chapman and Hall).
[5] Choi SS, Nah C, Lee SG, Joo CW Effect of filler-filler interaction on rheological behaviour of natural rubber compounds filled with both carbon black and silica 2003 *Polym. Int* **52** (1) 23.
[6] Heinrich G, Kluppel M, Vilgis TA. Reinforcement of elastomers 2002 *Curr Opin Sol. State Mater Sci* **6** (3)195–203.
[7] Naskar N, Debnath SC, Basu DK 2001 Novel method for preparation of carboxylated nitrile rubber-natural rubber blends using bis(diisopropyl)thiophosphoryl polysulfides *J. Appl. Polym. Sci* **80** (10) 1725–36.
[8] Nasir M, Choo CH 1989 Chemical modification of natural rubber latex with peracetic acid *Polym. J* **25**(4) 355–9.
[9] Wang Y, Tai BL, Yu H, Shih AJ 2014 Silicone-Based Tissue-Mimicking Phantom for Needle Insertion Simulation *J. Medical Devices - Asme* **8** 117-123.
[10] Ayers F, et al. 2008 Fabrication and characterization of silicone-based tissue phantoms with tunable optical properties in the visible and near infrared domain *Proc. of SPIE* **6870** 1-9.
[11] Pogue BW, Patterson MS 2006 Review of tissue simulating phantoms for optical spectroscopy, imaging and dosimetry *J. Biomed. Opt* **11** 041102.
[12] ASM International 2004 *Tensile Testing* Second Edition.
[13] ISO 37 2011 *Methods of test for vulcanized rubber, Part 1 determination of tensile stress-strain properties (* Third Revision).
[14] Sakrani S 2009 Mechanical Properties and Testing, Open course ware (University Teknologi Malaysia).
[15] Gedney R 2002 Guide To Testing Metals Under Tension *Advan. Mat. Process.* pp. 29–31.
[16] Kaku M 1982 *Hand book of composites* (Van Nostarnd: New York).
[17] Manik SP, Banerjee S 1979 Determination of chemical cross-links in rubbers *Macromol. Mater. Eng* **6** (71) 171–8.

[18] Cassagnau P 2003 Payne Effect and Shear Elasticity Of Silica-Filled Polymers In Concentrated Solutions and in Molten State *Polym* **44** (8) 2455-2462.

[19] Egorov VI, et al. 2002 Mechanical Properties of The Human Gastrointestinal Tract *J. of Biomech* **35** (10) 1417-1425.

[20] Agache PG, et al. 1980 Mechanical Properties And Young's Modulus Of Human Skin In Vivo, *Arch. Dermatol. Res* **269.3** 221-232.

[21] Armstrong CG, Mow VC 1982 Variations In The Intrinsic Mechanical Properties Of Human Articular Cartilage With Age, Degeneration, And Water Content *J. of Bone and Joint Surgery* 87-94.

[22] Itoi E, et al. 1996 Tensile Properties of The Supraspinatus Tendon *J.of Shoulder and Elbow Surgery* **5** (2) S28.

[23] Wren TAL, Yerby SA, Beaupre GS and Carter DR 2001 Mechanical properties of the human Achilles tendon *Clinical Biomech* **16** 245–251.

[24] Butler DL, Kay MD and Stouffer DC 1986 Comparison of material properties in fascicle-bone units from human patellar tendon and knee ligaments *J. of Biomech* **19** 425–432.

[25] Butler DL, Grood ES, Noyes FR, Zernicke RF, Brackett K 1984 Effects of structure and strain measurement technique on the material properties of young human tendons and fascia, *J.of Biomech* **17** 579-596.

[26] Nachemson AL, Evans JH 1968 Some mechanical properties of the third human lumbar interlaminar ligament (ligamentum flavum) *J. of Biomech* **1** 211–220.

[27] Pierre RKS, Rosen J, Whitesides TE, Szczukowski M, Fleming LL, Hutton WC 1983 The tensile strength of the anterior talofibular ligament *Foot Ankle* **4** 83-85.

[28] Akizuki S, et al. 1986 Tensile properties of human knee joint cartilage: I. Influence of ionic conditions, weight bearing, and fibrillation on the tensile modulus *J. of Orthop. Research* **4** 379–392.

[29] Edwards M, et al. 2005 Mechanical Testing Of Human Cardiac Tissue: Some Implications For MRI Safety *J. of Cardiovas. Mag.Reson* **7** (5) 835-840.

Acknowledgements

This work was financially supported in part by the “PDU” program in collaboration with BPKLN of the Ministry of Higher Education of Republic of Indonesia, and a Research Grant (no. 310y/I1.C01/PL/2015) from the Ministry of Research and Higher Education, Republic of Indonesia, 2015–2017.

IOP Conf. Series: Materials Science and Engineering **214** (2017) 012003 doi:10.1088/1757-899X/214/1/012003

Theoretical Investigation on Particle Brownian Motion on Micro-air-bubble Characteristic in H_2O Solvent

Irana Eka Putri[1] and Grace Gita Redhyka

Technical Implementation Unit of Instrumentation Development, Indonesia Institute of Sciences, Bandung, 40135, Indonesia

E-mail: iran003@lipi.go.id

Abstract. Micro-air-bubble has a high potential contribution in waste water, farming, and fishery treatment. In this research, submicron scale of micro-air-bubble was observed to determine its stability in H_2O solvent. By increasing its stability, it can be used for several applications, such as bio-preservative for medical and food transport. The micro-air-bubble was assumed in spherical shape that in incompressible gas boundary condition. So, the random motion of particle (Brownian motion) can be solved by using Stokes-Einstein approximation. But, Hadamard and Rybczynski equation is promoted to solve for larger bubble (micro scale). While, the effect of physical properties (e.g. diffusion coefficient, density, and flow rate) have taken important role in its characteristics in water. According to the theoretical investigation that have been done, decreasing of bubble velocity indicates that the bubble dissolves away or shrinking to the surface. To obtain longevity bubble in pure water medium, it is recomended to apply some surfactant molecules (e.g. NaCl) in micro-air-bubble medium.

1. Introduction

Several researches have addressed benefits of micro-air-bubble in some area, such as in fishery, waste treatment, agricultural industry, and mining [1-3]. Oh et. al. had been proven that micro-gas bubble can be found in gasoline fuel to improve engine efficiency and emission performance [4]. In other hand, it can also be found in solid-liquid phase, such as Pt-nanoelectrode, graphite and polymer surface [5-7]. Physically, micro-air-bubble has diameter size about 4 to 15 μm at constant periodic rate [8,9]. It is segmented into three phases, e.g., gas, aqueous liquid, and shell phase. Actually, measured size of micro-air-bubble is equal to hydrodynamic size of particle (main particle and Debye length constant).

According to fluid dynamics properties, micro-air-bubble has a small rising speed in high viscosity. Similarly, according to Stokes's law, a smaller bubble also has a smaller rising speed and so tends the bubble stay longer in some specific fluid [10]. Its longevity can be increase over a few months by the addition of surfactant molecules (salt, hydrocarbons, or hydrochloride acid) that decrease the pressure difference and lower the surface tension between inner bubble and surroundings in fluid medium [11-13]. High interaction of van der Waals bonding and longevity in several solvent are specific characteristics of micro-air-bubble that has been observed [14,15]. A smaller micro-air-bubble than 5 μm in diameters does not rise and its physical stability does not dissolve away [16].

Micro-air-bubble may maintain in constant small size at low and medium gas flow rates, but increases dramatically at high gas flow rates. The bubble is much smaller in salt water medium and its radius less than 0.01 cm because the varying drag forces may be applied. So, the bubble density may also take an effect of its behavior [17]. According to Hadamard and Rybczynski equation, first solution of a slowly moving sphere micro-air-bubble through an ambient fluid is determined by the different viscosity and density of inside and outside bubble [18].

In this letter we consider micro-air-bubble in perfect spherical shape assumption is taken to analyze its characteristics in pure water medium. So, the effects of some parameters (bubble density, water flow rate, and capillarity dimensional) are included to determine its characteristics.

2. Theoretical and experimental model

2.1. Single-bubble Formation

The single Nano-gas-bubble has the chemical balance reaction according to yield hydrogen gas that described in the following equation,

$$2H^{+}{}_{(aq)} + 2e^{-} \Leftrightarrow H_{2(g)} \tag{1}$$

According to the reaction, the generation of micro-air-bubble is done through electrolysis process, which is combining process of hydrogen ions (H^+) with electrons (e^-) through in H_2O interface.

$$O_{3(g)} + H_2O_{(\ell)} \Leftrightarrow OH_{(aq)} + OH^{2-}{}_{(aq)} \tag{2}$$

Single bubble growth occurs when the bubble passes the maximum pressure and it grows quickly, then separates from its capillarity and the new bubble is formed. When the capillarity is immersed into h (depth of liquid medium), so the hydrostatic pressure for bubble maximum pressure (P_{max}) can be determine by using the following equation.

$$P_{max} = 2\gamma/r + \Delta\rho g h \tag{3}$$

The effect of surface tension (γ) and internal hydrostatic pressure of liquid ($\Delta\rho gh$) are included to determine the hydrostatic pressure of bubble. Hydrostatic pressure regulation allows for the quantitative control of growth, shrinkage, or the obtained bubble size stability [19]. Another characteristics like bubble size and rise velocity distribution are depended on its value.

2.2. Velocity and Displacement of Bubble

The bubble motion near a surface is greatly influenced by Brownian motion between the bubble and the surface, as well as Van der Waals force. When a particle or bubble moves in certain medium, it is not attached to a surface, its movement can be determined by using electrostatic and Van der Waals force [20,21]. The Brownian motion theory is about the random motion of very small particle that immerse in a fluid medium. Rise of temperature surrounds the particle can affect the Brownian motion.

In modern theoretical of Brownian motion based on kinetic theory that has been developed by George G Stokes can determine the spherical small particle size by using the particle motion (velocity) that can be affected by viscosity and density of fluid medium. A relation between D and atomic properties of matter are found and can be expressed by using Stokes's law equation. For larger bubble (> 200 μm) is appropriate to use Hadamard and Rybczynski equation (see equation (4)) by including the effect of viscosity (η) and density (ρ) of fluid medium and bubble [22,23].

$$v_b = \frac{2}{3}\frac{d^2 g(\rho_b - \rho_f)}{\eta_f}\left[\frac{\eta_f + \eta_b}{2\eta_f + 3\eta_b}\right] \tag{4}$$

Where d, g, ρ_b, ρ_f, η_b and η_f are bubble diameter (μm), gravity acceleration (m/s^2), bubble density, fluid medium density, viscosity of bubble, and viscosity of fluid medium, respectively. The velocity of bubble movement is calculated by depending on the rise velocity when viscosity of fluid is taken place and surface tension effects are significant in the system.

3. Result and discussion

In this part, the effect of capillarity on bubble quantities was studied in theoretical predictions. The results are shown in figure 1. It indicates that a wider capillarity can increase the bubble quantities rapidly. A bubble generated system is built by using single capillarity (pipe) with length l and diameter outlet d_p are about 29.3 mm and 1 to 8 mm, respectively. According to [5], bubble diameter d under 5 μm did not rise to the surface and it could maintain floating in water medium. So, variations of bubble diameter (1 to 4 μm) were determined. The bubble quantities q that can be generated from single capillarity are about,

$$q(d_p) = \frac{\frac{2}{3} Q^2 \rho_g d^3}{d_p^2}\left[1 - \frac{1}{24}\pi d_p^2 d\right] \tag{5}$$

Where Q is the flow rate of water (ml/s) and ρ_g is the density of bubble (0.00143 gram/cm^3). The nucleation of a new bubble can determine the bubble's lifetime [24].

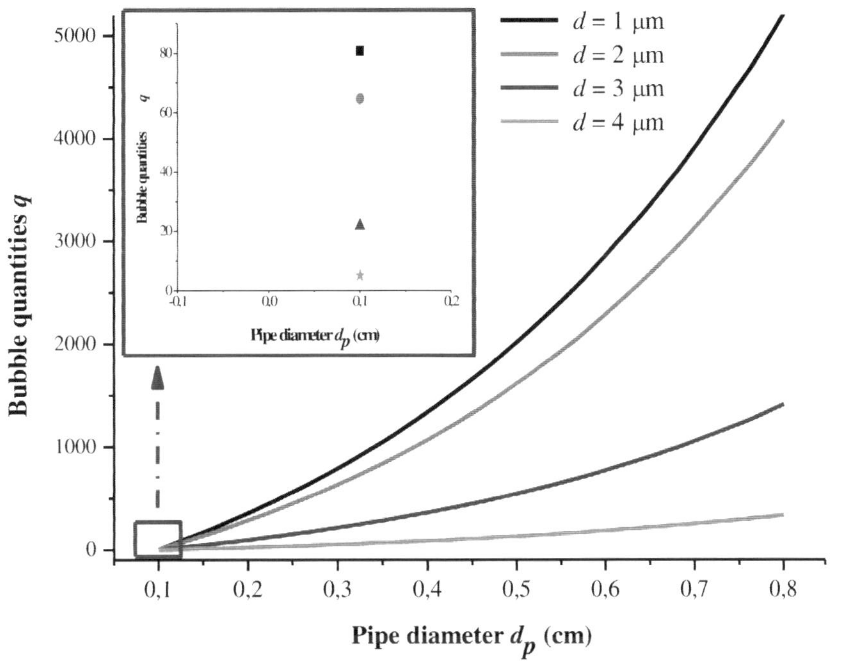

Figure 1. The effect of pipe (capillarity) diameter d_p on bubble quantities q with single-bubble diameter d difference.

Figure 2 expresses the theoretical prediction of the bubble growth rate R (figure 2a) and the bubble velocity (figure 2b) from t_0 to t_n according its Brownian movement. At the initial time, the bubble moves rapidly, it indicates that the bubble does not dissolve. In the opposite, the bubble dissolves away when it moves in slow velocity. The decreasing of bubble velocity shows that its movement reaches saturated point, it means the bubble is dissolved in water or shrinking. The terminal velocity was particularly sensitive to the presence of contaminant molecules [25].

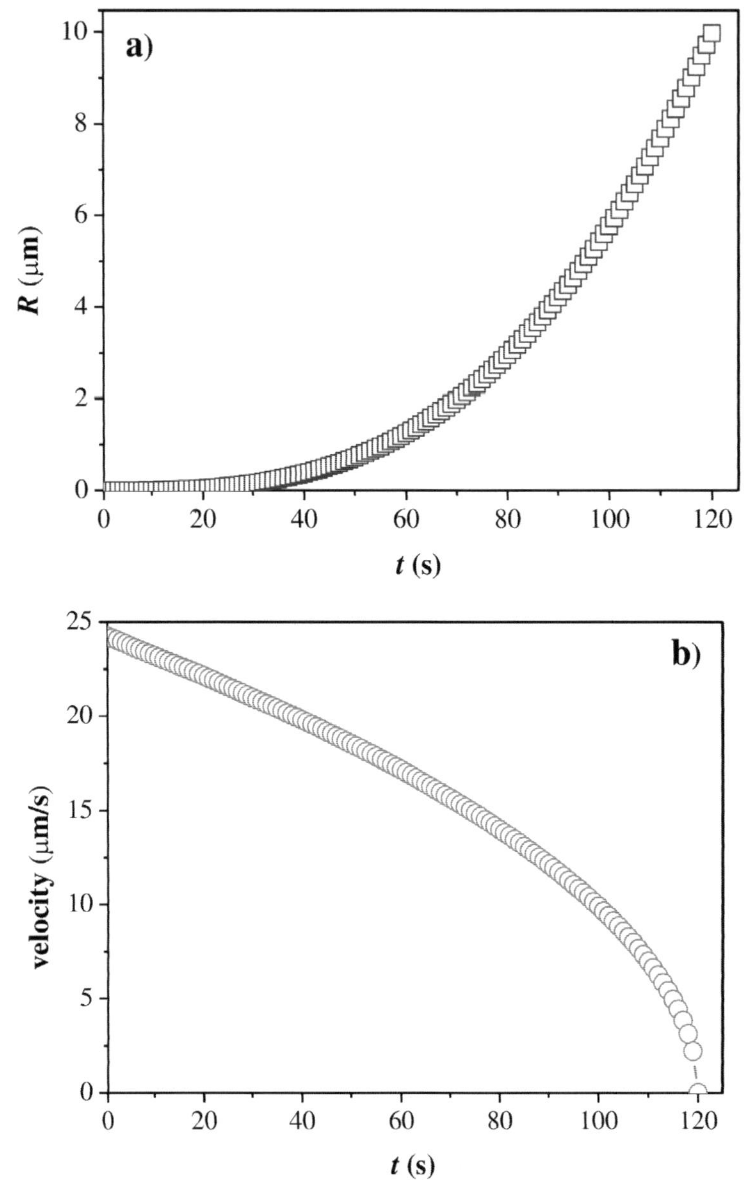

Figure 2. a) The theoretical predictions of bubble growth *R* in *t* seconds, and **b)** The velocity of bubble movement.

4. Conclusions

The theoretical study of bubble movement in pure water medium has been formulated by using some parameters. The bubble quantities are increase above 5000 bubbles per second when bubble particle is about 1 μm. While, its size grows rapidly above 5 μm in around 90 seconds, then the velocity is decrease slowly.

5. References

[1] Ebina K, Shi K, Hirao M, Hashimoto J, Kawato Y, Kaneshiro S, Morimoto T, Koizumi K and Yoshikawa H 2013 *Plos One* **8** e65339.
[2] Oshita S and Liu S 2013 *Proc. Int. Symp. on Agri-Foods for Health and Wealth (Bangkok)* **2013** 23.
[3] Agarwal A, Ng W J and Liu Y 2011 *Chemosphere* **84** 1175.
[4] Oh S H, Han J G and Kim J M 2015 *Fuel* **158** 399.
[5] Nishiyama T, Yamada Y, Ikuta T, Takahashi K and Tanaka Y 2015 *Langmuir* **31** 982
[6] Nishiyama T, Takahashi K, Ikuta T, Yamada Y and Tanaka Y 2016 *Chem. Phys. Chem* **17** 1500.
[7] Chen Q, Wiedenroth H S, German S R and White H S 2015 *J. Am. Chem. Soc* **137** 12064.
[8] Makuta T, Takemura F, Hihara E, Matsumoto Y and Shoji M 2006 *J. Fluid Mech* **548** 113
[9] Makuta T and Takemura F 2006 *Physics of Fluids* **18** 108102.
[10] Cordero J R V, Sámano D, Yue P, Feng J J and Zenit R 2011 *J. Non-Newtonian Fluid Mech* **166** 118.
[11] Dressaire E, Bee R, Bell D C, Lips A and Stone H A 2008 *Science* **320** 1198.
[12] Bonsu K O, Shokri N and Grassia P 2015 *Colloids and Surface A: Physicochemical and Engineering Aspects* **481** 514.
[13] Uchida T, Liu S, Enari M, Oshita S, Yamazaki K and Gohara K 2016 *Nanomaterials* **6** 31.
[14] Soga Y, Imanaka H, Imamura K and Ishida N 2015 *Adv. Powder Technol* **26** 1729.
[15] Lijuan Z, Hao C, Zhaoxia L, Haiping F and Jun H 2008 *Sci. China Ser. G-Phys. Mech. Astron* **51** 219.
[16] Sirsi S, Feshitan J, Kwan J, Homma S and Borden M 2010 *Ultrasound Med. Biol* **36** 935.
[17] Bournival G, Ata S and Wanless E J 2016 *Langmuir* **32** 6226.
[18] Tripathi M K, Sahu K C and Govindarajan R 2014 *Sci. Rep* **4** 4771.
[19] Volk A, Rossi M, Kähler C J, Hilgenfeldt S and Marin A 2015 *Lab. Chip* **15** 4607.
[20] Ng D and Liang H 2008 *J. Tribol* **130** 021603.
[21] Mei H, Zhao D, Wang T, Cheng J and Lu X 2014 *Sci. China Tech Sci* **57** 2144.
[22] Celata G P, D'Annibale F, Marco P D, Memoli G and Akio T 2007 *Exp. Thermal and Fluid Sci.* **31** 609.
[23] Parkinson L, Sedev R, Fornasiero D and Ralston J 2008 *J. of Coll. and Interface Sci* **322** 168.
[24] German S R, Chen Q, Edwards M A and White H S 2016 *J. Electrochem. Soc* **163** 3160.
[25] Rodríguea S B, Corona A A and Soria A 2012 *Revista Mexicana de Ingeniería Química* **11** 269.

Acknowledgement

This research was supported by the thematic program and the flagship program through Bandung Management Unit for Instrumentation Development (Deputy for Scientific Services) funded by Indonesia Institute of Sciences, Indonesia.

IOP Conf. Series: Materials Science and Engineering **214** (2017) 012004 doi:10.1088/1757-899X/214/1/012004

Density Functional Theory (DFT) Study of Molecularly Imprinted Polymer (MIP) Methacrylic Acid (MAA) with D-Glucose

T D K Wungu[1,2], S E Marsha[1], Widayani[1] and Suprijadi[1,2]

[1] Department of Physics, Faculty of Mathematics and Natural Sciences, Institut Teknologi Bandung, Jl. Ganesa 10 Bandung 40132, Indonesia
[2] Research Center for Nanosciences and Nanotechnology, Institut Teknologi Bandung, Jl. Ganesa 10 Bandung 40132, Indonesia

Corresponding author: triati@fi.itb.ac.id

Abstract. In order to find an alternative biosensor material which enables to detect the glucose level, therefore in this study, the interaction between Methacrylic Acid (MAA) based Molecularly Imprinted Polymer (MIP) with D-Glucose is investigated using the Density Functional Theory (DFT). The aim of this study is to determine whether a molecule of the MAA can be functioned as a bio-sensing of glucose. In this calculation, the Gaussian 09 with B3LYP and 631+G(d) basis sets is used to calculate all electronic properties. It is found that the interaction between a molecule of MAA and a molecule of D-Glucose was observed through the shortened distance between the two molecules. The binding energy of MAA/D-glucose and the Mulliken population analysis are investigated for checking possible interaction. From analysis, the MAA based MIP can be used as a bio-sensing material.

1. Introduction

Nowadays, diabetes becomes a major public health problem that can strike young ages people in the world. Diabetes is a disease which is caused when the system in the body cannot produce enough insulin. One of the factors of diabetes is high glucose level in blood which can be caused by heredity or unbalanced diet. In order to find an alternative biosensor material enabling to detect the glucose level, therefore in this study the interaction between Methacrylic Acid (MAA) based Molecularly Imprinted Polymer (MIP) with D-Glucose is investigated.

MIP is a polymer that has cavities in polymer matrix which is obtained through the molecularly imprinted technique. MIP technique is known as the very useful technique to design the receptors that will mimic some other molecules and recognize them [1,2]. The MIP purposed to make the template that will be able to recognize the target molecules. The mechanism of MIP technique is shown in Figure 1.

Figure 1 shows the process of MIP technique step by step. At first, template molecule and functional monomers are prepared to create functional monomer complex mediated by covalent and/or non covalent interaction during the polymerization process. During template extraction step, complementary recognition site are left the polymer and becomes the polymer matrix [1] [2] [4]. Thus, the target molecule will easily detach from the template and make MIP have high selectivity with other molecules [4-5].

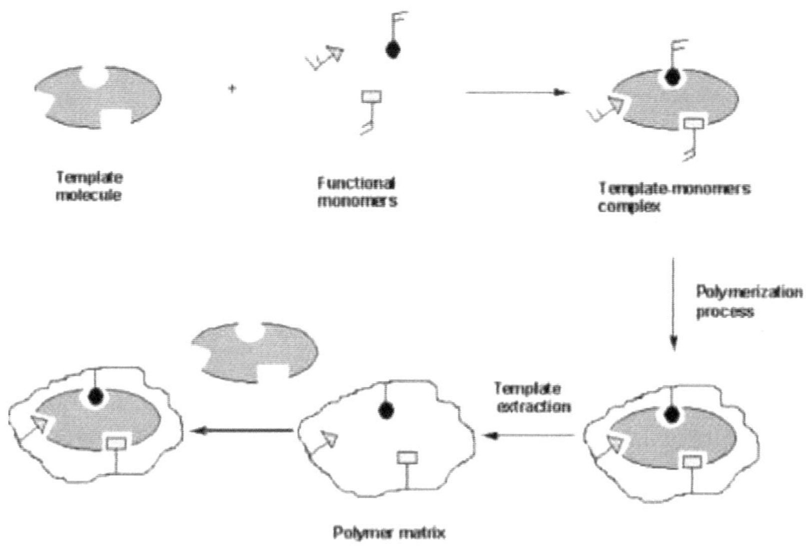

Figure 1. MIP technique scheme [3].

Methacrylic acid (MAA) is one of functional polymers that act as template molecule in MIP process. The MAA have been used due to its ability to take part in ion-ion, ion-dipole, dipole-dipole, Hydrogen, and Van Deer Waals interactions [6]. Furthermore, MAA is easy to have interaction with alkali molecules, hence, MAA can be used as template molecule and can have interaction with glucose. Here, glucose acts as a functional monomer.

2. Model and Computational Method

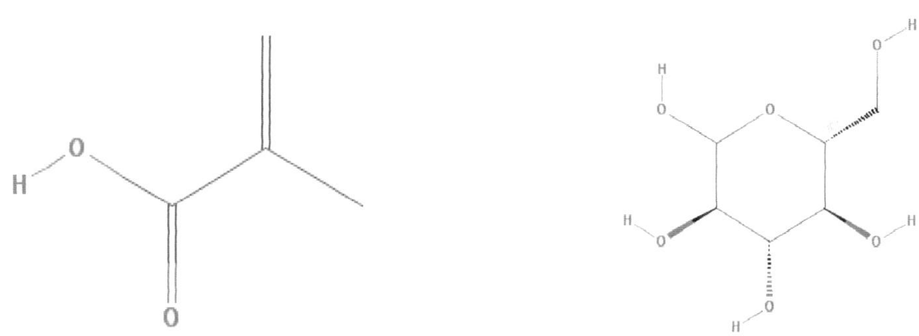

(a) (b)

Figure 2. (a) The MAA structure [7]; (b) the D-Glucose Structure [8].

Figure 2, shows the structure of both MAA and D-Glucose. MAA is a clear colourless liquid or known as low-melting solid, vapour heavier than air, and may polymerize exothermically molecule [7]. The D-glucose is used as primary source of energy for organisms and naturally occurs in fruits and other parts of plant [8]. However, if the D-glucose or glucose is over consumed it will affect the human body health which is known as diabetes. Therefore, research related to MIP with functional monomer computationally and experimentally has been conducted. A computational approach using the Density Functional Theory (DFT) calculation and semi-empirical simulation have done to screening monomer for preparing MIP and to obtain the best monomer able to strongly interact with required functional monomer [9-11]. Also, our previous work related to synthesis and characterization of MAA-based MIP with D-glucose template was done to seek the interaction of template and D-Glucose [12]. Therefore, to support our previous work [12] we present simulation on the interaction of MAA with D-glucose using DFT calculation.

The DFT methods have been carried out using the Gaussian 09 Program to calculate all electronic properties. For the molecular properties of both the Methacrylic Acid (MAA) and the D-Glucose molecules, the B3LYP/6-311+G(d) basis set is used. The cut-off energy was chosen based on the high accuracy precision mode applied in the calculation and the prediction change in energy was 2.55×10^{-8}. The energy of the system computed at RB+HF-LYP was -306.56 Hartree. The maximum force, the RMS force, the maximum displacement, and the RMS displacement was converged to reach a threshold value of 44×10^{-6}, 15×10^{-6}, 15.5×10^{-6} and 49.2×10^{-6}, respectively. The electron correlation is taken into account by using hybrid functional (B3LYP), that use the approximation from exact exchange from Hartree-Fock theory with exchange and correlation expressed through Kohn-Sham. Figure 3 shows the design of MAA and D-glucose molecule. A molecule structure of MMA (Figure 3a) consists of 12 atoms while a molecule structure of D-Glucose (Figure 3b) consists of 22 atoms. The optimization calculation has been done by completely relaxing all atoms without any symmetry constrain.

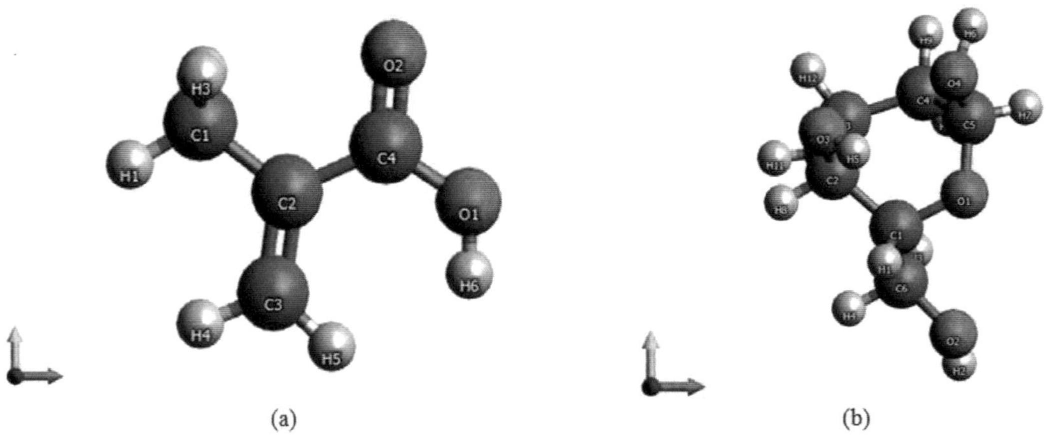

Figure 3. Design structure of (a) MAA (b) D-glucose.

3. Results and Discussion

In this section we present the results of the optimized geometry and the parameters related to the electronic structure of MAA and D-glucose. The optimization calculation was done by using B3LYP basis set, however for single molecule of both MAA and D-glucose the 6-311+G(d) basis set was used. The Optimization parameters are listed in Table 1.

Table 1. Optimization result parameters for MAA and D-glucose

Parameter	Maximum force	RMS Force	Maximum Displacement	RMS Displacement
MAA	0.000044	0.000015	0.001515	0.000492
D-Glucose	0.000044	0.000015	0.001515	0.000492
Threshold	0.00045	0.0003	0.0018	0.0012

IOP Conf. Series: Materials Science and Engineering **214** (2017) 012004 doi:10.1088/1757-899X/214/1/012004

Table 2. Total Mulliken charge between MAA and D-glucose after optimization calculation

Atom Label	Mulliken Charge	Atom Label	Mulliken charge
C1	-0.11717	O19	-0.00586
C2	0.294533	O20	0.004427
H3	0.007057	H21	0.026242
H4	0.029586	H22	-0.00368
H5	0.00355	O23	-0.00176
C6	-0.05226	C24	-0.0164
H7	0.008824	H25	0.003561
H8	0.001928	C26	-0.09627
C9	-0.1692	C27	0.062832
O10	0.032911	H28	-0.00717
H11	0.000272	H29	-0.00205
O12	-0.02275	O30	0.02089
O13	0.022566	C31	-0.00514
H14	-0.01351	H32	-0.0063
H15	-0.00699	H33	0.004014
H16	-0.00647	H34	0.01065
C17	-0.00881	H35	-0.00872
C18	0.016667		

From Table 1, we acquire that the convergence criteria are already met since the value of each parameters (maximum force, RMS force, maximum displacement, and RMS displacement) are lower than threshold. Hence, we can say that the MAA and D-Glucose molecule are already optimized by using B3LYP/6-311+G(d) basis set. After the optimization process, the structure of MAA and D-Glucose does not change; this can be seen from Figure 4. The shape of MAA structure is still planar after optimization (Figure 4a). The bonding within D-Glucose seems very tight because it has double bond within its structure and hence the shape of D-Glucose does not change as well.

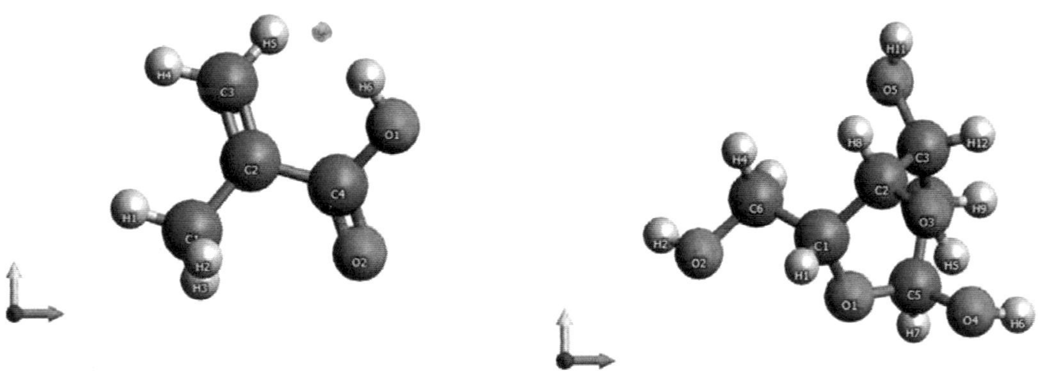

(a) (b)

Figure 4. The optimized structure of (a) MAA and (b) D-glucose.

Next is the calculation of MAA and D-glucose interaction. The initial design of MAA and D-Glucose is depicted in Figure 5. Interaction between MAA and D-Glucose can be checked from their distance. At this point, our reference distance between 2 molecules of MAA and D-glucose is taken from C2-C6 initial distance that is 5.65 Å (see Figure 5). The C2 atom belongs to MAA while C6 atom belongs to D-glucose. The selection of mentioned atoms is based on the center of each molecule and to make the measurement easier to do.

After the optimization process, the distance between MAA and D-Glucose increases by 0.85 Å that is change from 5.65 Å to 6.41 Å. The change in distance shows that there are some forces, called as the non covalent interaction, that make MAA and D-Glucose repels each other. However, if we look into Figures 5 and 6, the distance between O2 in MAA and O4 in D-glucose are getting closer. The O2-O4 distance is decreased by 3.815 Å that is change from initially 6.62 Å to 2.81 Å. It is shown that the non covalent interaction occurred. In addition, the hydrogen bonding between O2 and O4 observed, meaning that the MAA and D-glucose interaction occurred.

The interaction between MAA and D-Glucose can also be observed from the total Mulliken charge. Once we have optimized structure, we can not only depend on position but also the electronic density. The fastest and easiest way to see the density is through the Mulliken charges population. Table 2 shows the total Mulliken charge of MAA- D glucose structure. The sum of Mulliken charge is not 0, but they still have the charge between them. Despite the Mulliken charge which is the easiest way to analyse the interaction, it very depends on the basis set. We cannot, therefore, depend only on Mulliken population analysis. The other way to see the interaction between 2 molecules is through binding energy (ΔE_{bind}).
The formulation of binding energy used is $\Delta E_{bind} = E_{complex} - (E_{MAA} + E_{D\text{-}glucose})$. Using 6-311+G(d) basis set, the ΔE_{bind} between MAA and D-glucose is strong that is -2143.927 a.u.

From the 3 (three) ways of checking the interaction between MAA-D glucose: bonding interaction, Mulliken population analysis, and binding energy, we can see that MAA interacts with D-glucose. Hence, MAA based MIP can be used as a sensor to sense the D-Glucose. The D-Glucose will fill the cavity in the template because MAA will attract them to fill the cavity. From the chemical reaction point of view, the interaction between MAA and D-Glucose does not form chemical reaction, so it can be easily removed from the template.

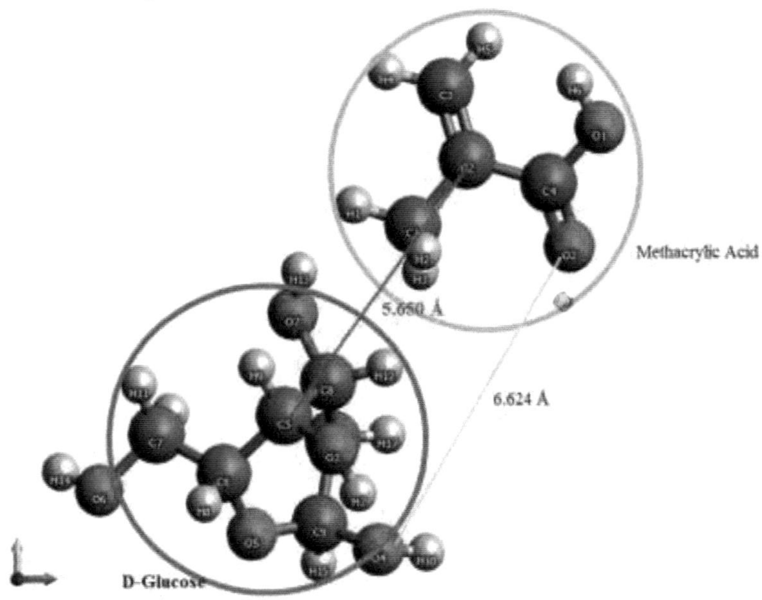

Figure 5. The initial structure of MAA-D glucose before optimization calculation.

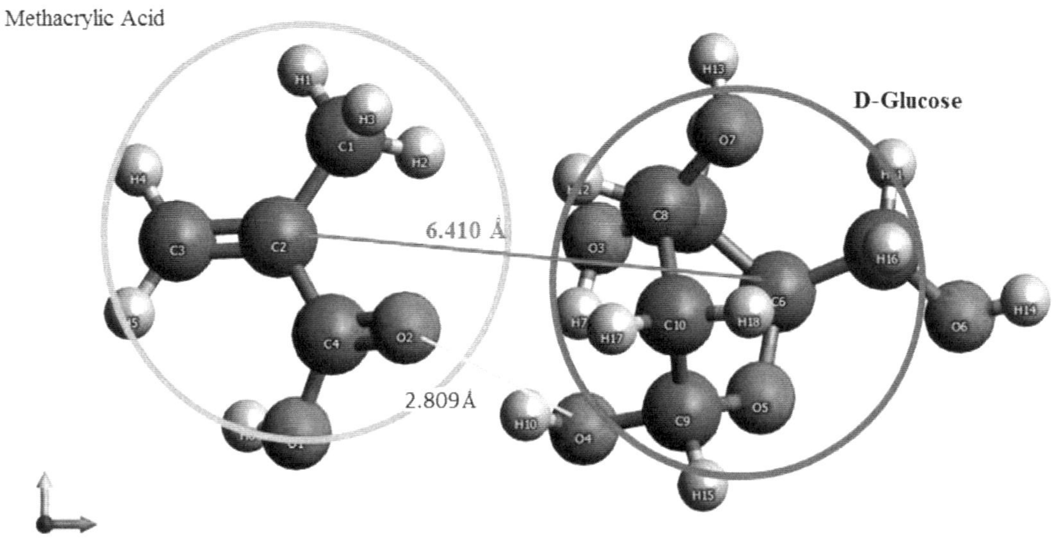

Figure 6. The optimized structure of MAA - D glucose.

4. Summary

The calculation of MAA based MIP interactions with D-glucose using the density functional theory (DFT) was done to study the structure and electronic properties of MAA/D-glucose for diabetes application. The interaction of a molecule of MAA and a molecule of D-glucose is non covalent interaction. The Mulliken population analysis and binding energy calculation also shows that both MAA and D-glucose are having interaction one another though a single MAA molecule is presented, there still interaction can be observed. From this study, MAA based MIP is possible to be used as a glucose sensor.

References

[1] Haupt K 2003 *Anal. Chem* **126** (6) p 747.
[2] Haupt K and Mosbach K 2000 *Chem. Rev* **100** (7) p 2495.
[3] Vasapollo G , Roberta S D, Merg L, Scorrano S and Mele G 2011 *Int. J. Mol. Sci* 5908.
[4] Shariatinia Z et. al. 2011 *Struct. Chem* **22** 1347.
[5] Wulff G and Schönfeld R 1998 *Adv. Mater* **10** p 957.
[6] Ramström O, Yei L and Klaus M 1996 *Chem. Bio.l* **3** no 6 p 471.
[7] N. C. f. B. Information, "PubChem Compound Database; CID=4093," [Online]. Available: https://pubchem.ncbi.nlm.nih.gov/compound/4093. [Accessed 23 June 2016].
[8] N. C. f. B. Information, "PubChem Compound Database; CID=5793," [Online]. Available: https://pubchem.ncbi.nlm.nih.gov/compound/5793. [Accessed 23 June 2016].
[9] Dong Q, Yan M, Zhang M, Liu Z and Li Y 2005 *Anal. Chim. Acta* **542** 186–192.
[10] Huang Y and Zhu Q 2015 *J. Chem* **2015** 1.
[11] Marestonia J D, Wonga A, Felicianoa G T, Marchia M R R, Tarleyc C R T, Sotomayora M D P T 2016 *J. Braz. Chem. Soc* **27**(1) 199.
[12] Yanti, Nurhayati T, Royani I, Widayani and Khairurrijal 2016 *J. Phys. Conf. Series* **739** 012143.

Optimization of Microencapsulation Composition of Menthol, Vanillin, and Benzyl Acetate inside Polyvinyl Alcohol with Coacervation Method for Application in Perfumery

Muhamad Sahlan[1], Mohammad Raihani Rahman[2,*]

[1]Research Leader and Lecturer, Chemical Engineering, Faculty of Engineering, Universitas Indonesia
[2]Master Degree Student, Chemical Engineering, Faculty of Engineering, Universitas Indonesia

*m.raihani.rahman@gmail.com

Abstract. One of many applications of essential oils is as fragrance in perfumery. Menthol, benzyl acetate, and vanillin, each represents olfactive characteristic of peppermint leaves, jasmine flowers, and vanilla beans, are commonly used in perfumery. These components are highly volatile, hence the fragrance components will quickly evaporate resulting in short-lasting scent and low shelf life. In this research, said components have been successfully encapsulated simultaneously inside Polyvinyl Alcohol (PVA) using simple coacervation method to increase its shelf life. Optimization has been done using Central Composite Diagram with 4 independent variables, i.e. composition of menthol, benzyl acetate, vanillin, and tergitol 15-S-9 (as emulsifier). Encapsulation efficiency, loading capacity, and microcapsule size have been measured. In optimized composition of menthol (13.98 %w/w), benzyl acetate (14.75 %w/w), vanillin (17.84 %w/w), and tergitol 15-S-9 (13.4 %w/w) encapsulation efficiency of 97,34% and loading capacity of 46,46% have been achieved. Mean diameter of microcapsule is 20,24 μm and within range of 2,011-36,24 μm. Final product was achieved in the form of cross linked polyvinyl alcohol with hydrogel consistency and orange to yellow in color.

1. Introduction

Indonesia with its tropical climate has been a place for various plants that yield essential oils. Essential oils is widely used as main components in perfumery. Every perfume consists of 3 main components, top notes which are the most volatile, middle notes, and base notes which are the least volatile[1]. Its nature as volatile component makes essential oils can not release its fragrant aroma for relatively long period[2].

Encapsulation is one of many method to retain essential oils scent. In general, there are 3 methods of utilizing encapsulation technology, there are chemical (e.g. interfacial polycondensation, in situ polymerization), physico-chemical (coacervation, sol-gel encapsulation), and mechanical (spray drying, solvent evaporation)[3]. Coacervation is the most suitable encapsulation method to apply in food and beverage, pharmacy, cosmetics, and fragrances and flavors[4].

Limonene has been encapsulated inside chitosan with complex coacervation method. Lutensol as surfactant has been reported to decrease particle size[5]. Martins encapsulated thyme oil inside polylactide using 3 kind of surfactants with various hydrophilic-lypophilic balance (HLB), Tween 20 (HLB 16,4), Tween 80 (HLB 15,5) and Tergitol 15-S-9 (HLB 13,3). Encapsulation using Tergitol 15-S-9 yields the best efficiency (65%)[6]. Polyvinyl alcohol (PVA) has been used to encapsulate santosol oil with simple coacervation method using glutaraldehyde as crosslinking agent. Higher concentration of sodium sulphate as coacervating agent has been reported to increase microcapsule wall thickness, hence slower release of santosol oil[7].

In this research, our group has recently encapsulated menthol, benzyl acetate, and vanillin simultaneously inside PVA with aqueous phase simple coacervation method and Tergitol 15-S-9 as surfactant. Composition of menthol, benzyl acetate, vanillin, and Tergitol 15-S-9 was optimized to achieve optimum encapsulation efficiency and loading capacity. Optimization has been done using Response Surface Method (RSM) with Central Composite Diagram (CCD) experiment design. RSM is equipped with statistical tools to determine the significance of a factor over a response. The evaluation of factors using the RSM uses experimental design in order to distribute the selected variables within the boundaries of the design[8].

2. Materials and Methods

2.1. Materials

Sodium sulfate (ACS reagent, ≥99.0%, anhydrous, powder) as coacervating agent, glutaraldehyde solution (grade II, 25% in H_2O) as crosslinking agent, Tergitol 15-S-9, l-menthol (≥99%, FCC, FG) as emulsifier, benzyl acetate (≥99%, FCC, FG), vanillin (≥97%, FCC, FG), and Mowiol 4-98 (Mw 27.000) were all shipped from Sigma Aldrich Singapore.

2.2. Methods

2.2.1 Preparation of Microcapsule

PVA solution was made by adding 6g of Mowiol 4-98 into 100 ml aqua dm at 60^0C and stirred at 250 rpm for 30 minutes. Sodium sulfate solution was made by adding 6g of sodium sulfate into 50 ml aqua dm and then added to PVA solution. A mixture of menthol, benzyl acetate, vanillin, and tergitol was added into solution and then stirred at 750 rpm for 5 minutes. After 5 minutes, the temperature of solution was raised to 70^0C to induce coacervation. Then, 6 ml of glutaraldehyde solution is added and then stirred at 500 rpm and heated at 70^0C for 60 minutes. After that, solution was cooled and sedimented for 24 hours. After 24 hours period, the solution separated into PVA hydrogel and aqueous phase. Aqueous phase was then analyzed using GC FID.

2.2.2 Particle size measurement

Particle size was measured using Beckman Coulter LS 100 Q. PVA microcapsules suspension was collected before it formed hydrogel. Microcapsule suspension was dispersed in aqua dm and then analyzed.

2.2.3 Gas chromatography analysis

GC analysis was done using Shimadzu GC-FID, with DB-5 column, oven temperature at 200^0C, injector temperature at 110-160^0C, detector temperature at 200^0C, N_2 50kPa, H_2 70 kPa, and air 50kPa. 600 ppm sample was prepared by diluting aqueous phase with chloroform and vaporizing with turboevaporator-nitrogen.

3. Results and Discussion

3.1. Optimized composition of menthol, benzyl acetate, vanillin, and tergitol 15-S-9

Numerical optimization has been done using Design Expert 7.0, a statistical software from StateEase, to reach optimum encapsulation efficiency and loading capacity by adjusting composition of menthol, benzyl acetate, vanillin, and tergitol 15-S-9 with CCD experimental design. The low value is was to 1, intermediate value was set to 1.5, high value is set to 2, and total run was 30 runs. Encapsulation efficiency was set to 1 (as maximum value), and loading capacity was set between 0.32123 (as

IOP Conf. Series: Materials Science and Engineering **214** (2017) 012005 doi:10.1088/1757-899X/214/1/012005

minimum value achieved in experiment) and 0.5 (at value more than 0.5 , there will be encapsulated components on the surface of microcapsule)[9]. The result is shown in Table 3.1.

Table 3.1 Optimized value of menthol, benzyl acetate, vanillin, and tergitol

Menthol (%w/w)	Benzyl acetate (%w/w)	Vanillin (%w/w)	Tergitol (%w/w)	Encapsulation efficiency (%)	Loading capacity (%)
13.98	14.75	17.84	13.4	97.34	46.46

As shown above, vanillin was needed in much higher concentration compared to benzyl acetate and menthol. The reason for this is because vanillin is the most water-soluble components (1 g/100 ml) compared to benzyl acetate (<0.1 g/100 ml) and menthol (<0.046 g/100 ml). In aqueous phase coacervation, the more a component solubilized in water, the less it will be encapsulated inside the polymer. In this case, higher vanillin concentration was needed to increase vanillin that can be encapsulated inside polyvinyl alcohol.

3.2 Microcapsule particle size

Particle size analysis has been done using different concentration of tergitol as surfactant. This surfactant was used to modified the size of microcapsules. Tabel 3.2 shows the result.

Tabel 3.2 Microcapsule particle size

Surfactant (%w/w)	Mean diameter (μm)
19.28	17,35
13.4	20,24
9.64	22,63

As shown at Table 3.1, higher concentration of surfactant yields smaller particles. The reason is with higher concentration of surfactant, surface tension will decrease and wider oil-water interface is achieved, hence smaller particle size[10].

3.3 Microcapsule hydrogel analysis with ATR-FTIR

(a)

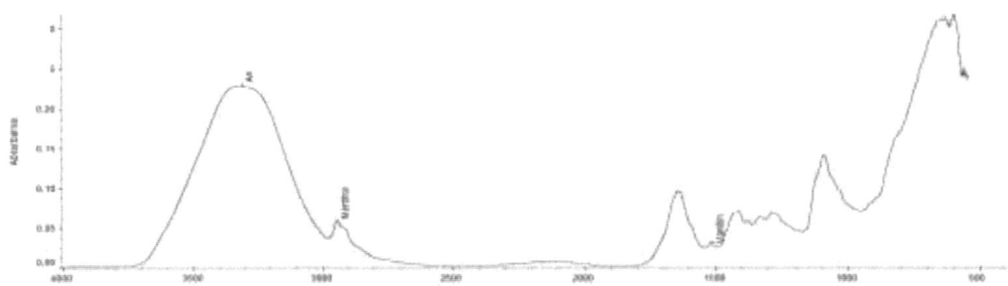

(b)

(c)

Figure 3.1 (a) Microcapsule sample, (b) Hollow microcapsule sample, (c) Substraction of a and b.

If we compare between Figure 3.1 (a) and (b), the graphs look almost identical except for the existence of vanilla peak between wavelength 1500-1600 cm^{-1} on figure 3.1 (a). Figure 3.1 (c) is the substarction form (a) and (b) and so the peak for menthol and vanillin become more visible. The peak for benzyl acetate can not be detected and the existence of benxyl acetate is confirmed with GC analysis in the following section.

3.4 Release profile of menthol

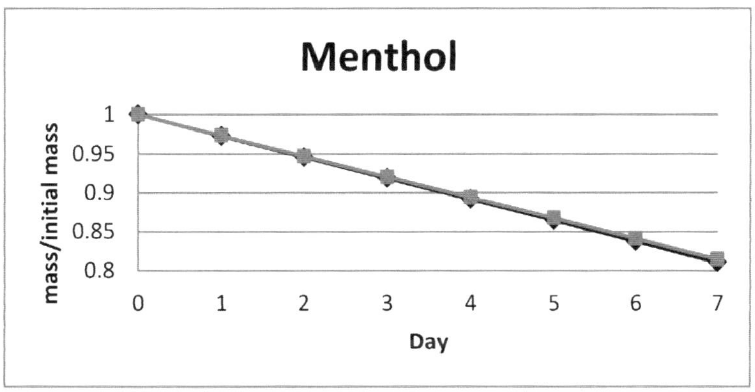

Figure 3.2 Release profile of menthol.
—— Standard menthol
—— Encapsulated menthol

Standard menthol is unencapsulated pure menthol and release data is collected by analytical balance periodically. Release data for encapsulated menthol is collected by GC Headspace analysis. The gradient for encapsulated menthol is -0.0265 and this is almost identical to the gradient for standard menthol which is -0.0271. This shows that encapsulation for menthol is not slowing down the release rate significantly. It may be because the evaporation rate of pure menthol is identical to the diffusion rate of menthol through PVA matrix. Further research on those two topics is expected to be done.

3.5 Release profile of benzyl acetate

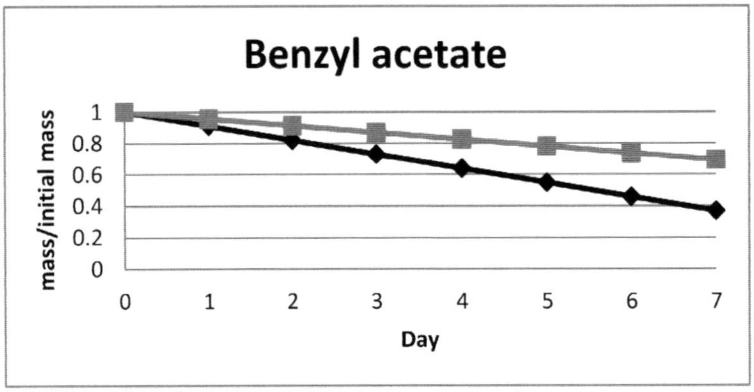

Figure 3.3 Release rate of benzyl acetate.
— Standard benzyl acetate
— Encapsulated benzyl acetate

Standard benzyl acetate is unencapsulated pure benzyl acetate and release data is collected by analytical balance periodically. Release data for encapsulated benzyl acetate is collected by GC Headspace analysis. The gradient for encapsulated benzyl acetate is -0.0439 and this is significantly lower compared to the gradient for standard benzyl acetate which is -0.0903. This shows that encapsulation for benzyl acetate is significantly slowing down the release rate. It may be because the evaporation rate of pure benzyl acetate is much more slower compared to the diffusion rate of benzyl acetate through PVA matrix. Further research on those two topics is expected to be done.

3.6 Release profile of vanillin

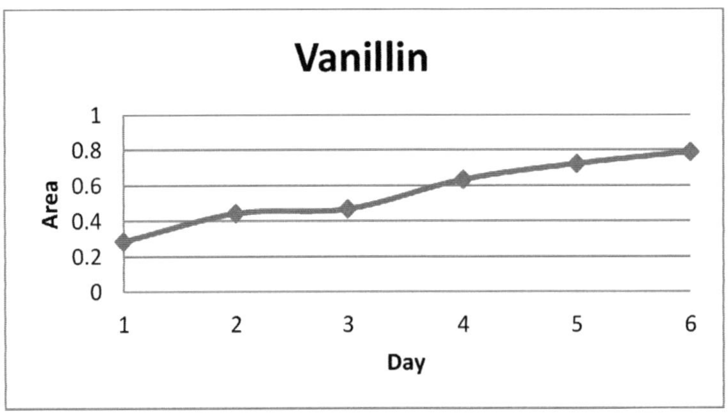

Figure 3.4 Change of area of vanillin FTIR spectra.

Release rate of vanillinis done using FTIR. Figure 3.4 shows the change of vanillin area over six days. Instead of getting lower, the absorbance of vanillin is getting higher. This shows that the release rate of vanillin is much more slower compared to the release rate of menthol and benzyl acetate, hence the increasing of vanillin composition in microcapsule over the period of six days.

3.7 Release rate of water

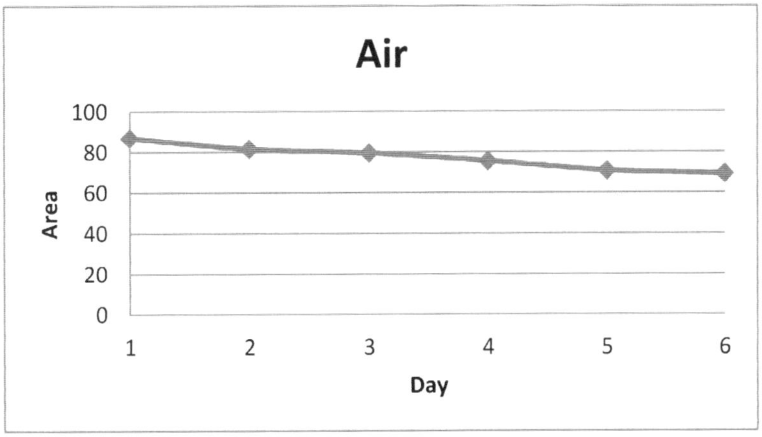

Figure 3.5 Change of area of water FTIR.

Qualitative analysis of released water is done by FTIR. If figure 3.5 compared to figure 3.4, it is clear that hydrogel microcapsule contains considerable amount of water, i.e. area for water is ranged around 60-90, while area for vanillin is ranged around 0.2-0.8. From day 5 to day 6, the change of area becomes significantly constant compared to the previous days and it shows that water content within the hydrogel microcapsule has reached its limit to be released in ambeient temperature.

3.8 Morphology of hydrogel microcapsule

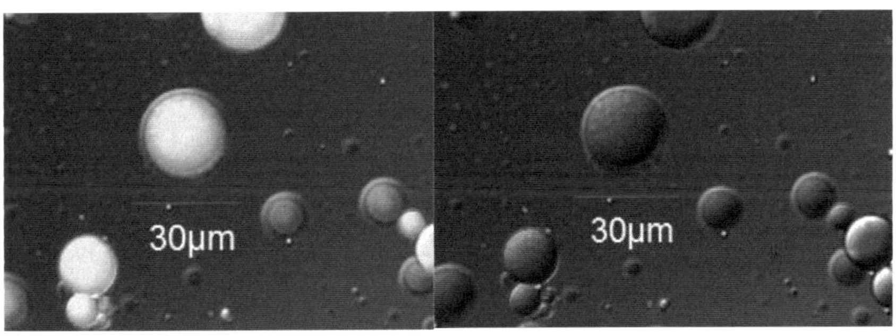

Figure 3.6 Microcapsules morphology.
Menthol
Benzyl acetate

Figure 3.6 is taken with confocal laser scanning microscope. It is shown that menthol and benzyl acetate can be encapsulated simultaneously inside one microcapsule. Another thing to note is that both menthol and benzyl acetate are homogenously distributed inside the microcapsules. As for vanillin, we can not detect its existence with this microscope. Existence of vanillin in microcapsules has been proven with FTIR analysis.

4. Conclusion

Our research group has successfully encapsulated menthol, benzyl acetate, and vanillin inside polyvinyl alcohol with encapsulation efficiency as high as 97,34% and loading capacity as high as

46,46%. These results is higher compared to several previous research, namely vanillin inside polysulfonate (45%)[11] and menthol inside gum arabic and gelatin (93%)[12]. The mean diameter microcapsules size with optimized composition is 20,24 μm has been achieved. Encapsulation can signifcantly slow down hte release rate of benzyl acetate, but not for menthol. Benzyl acetate and menthol can be encapsulated simultaneously and homogenously distributed inside one microcapsule.

5. References

[1] Herman, Stephen J 2005 *Chemistry and Tehnology of Flavors and Fragrances.Chapter 13.* Oxford: Blackwell Publishing

[2] Tzhayik, Oshrat, Artur Cavaco-Paulo, Aharon Gedanken 2012 *Fragrance release profile from sonochemically prepared protein microsphere containers* Ultrasonics Sonochemistry **19** 858-863

[3] Ghosh, S.K. Functional coatings and microencapsulation: A general perspective. In *Functional Coatings,* Ghosh, S.K., (Ed.) Wiley-VCH, Weinheim, Germany, 2006

[4] Martins, Isabel M., et.al. 2014 *Microencapsulation of essential oils with biodegradable polymeric carriers for cosmetic applications*. Chemical Engineering Journal **245** 191-200

[5] Souza, Jefferson M. et.al. 2014. *Properties and Controlled Release of Chitosan Encapsulated Limonene Oil*. Brazillian Journal of Pharmacognosy **24** 691-698

[6] Martins, Isabel M., et.al. 2014. *Microencapsulation of essential oils with biodegradable polymeric carriers for cosmetic applications*. Chemical Engineering Journal **245** 191-200

[7] Bachti, Aristi R., Costas Kiparissided. 1994. *Synthesis and Release Studies of Oil-Containing Poly(vinyl alcohol) Prepared by Coacervation.* Journal of Controlled Release **38** 49-58

[8] Fabian, Frank M. 2012. *Application of Response Surface Methodology and Central Composite Design for 5P12-RANTES Expression in the Pichia pastoris System.* Departemen of Chemical and Biological Engineering, University of Nebraska-Lincoln

[9] Oxley, James. 2016. Process Selection Criteria. In *Handbook of Encapsulation and Controlled Release*. Munmaya Ishra(Ed.) Florida: CRC Press 23-34

[10] Nass, R., D. Burgard, H. Schmidt. 1991. *Progress in Powder Synthesis by w/o-Microemulsion.* Eurogel **1991** 243-254

[11] Brisa, Pena, et.al. 2011. *Preparation and characterization of polysulfone microcapsules for perfume release.* Chemical Engineering Journal **179** 394-403

[12] Santos, Milla G., Debora A. Carpinteiro, Marcelo Thomazini, Glaucia A. Rocha-Selmi, Adriano G. da Cruz, Christiane E.C. Rodriguez, Carmen S. Favaro-Trindade. 2014. *Coencapsulation of Xylitol and Menthol by Double Emuslion Followed by Complex Coacervation and Microcapsule Application in Chewing Gum.* Food Research International **66** 454-462

IOP Conf. Series: Materials Science and Engineering **214** (2017) 012006 doi:10.1088/1757-899X/214/1/012006

Synthesis of Metal-organic Frameworks Based on Zr^{4+} and Benzene 1,3,5-Tricarboxylate Linker as Heterogeneous Catalyst in the Esterification Reaction of Palmitic Acid

Larasati, I D Winarni, F R Putri, Q A Hanif, W W Lestari*

Research Group of Porous Material for Sustainability, Chemistry Department, Faculty of Mathematics and Natural Sciences, Sebelas Maret University, Jl. Ir Sutami No.36A, Kentingan-Jebres, Surakarta, Central Java, Indonesia, 57126
Fax: +62 271 663375; Tel: +62 271 663375

*Email : witri@mipa.uns.ac.id

Abstract. The conversion of the biomass into biodiesels via catalytic esterification and trans-esterification became an interesting topic due to the depletion of fossil-based energy. Homogenous catalysts such as HCl, H_2SO_4 and NaOH commonly used as catalyst, however, the use of this kind of catalyst causes more problems, such as the difficulties on the separation from the product and the pollution effect on the environment. Heterogeneous catalysts, such as Metal-Organic Frameworks (MOFs) give an alternative promising way to substitute these limitations due to their strong catalytic site, porosity, high specific surface area, and easy-separation and reusable properties. Herein, we reported the synthesis of MOFs based on zirconium(IV) and H_3BTC linker (H_3BTC= benzene-1,3,5-tricarboxylic acid) by solvothermal and reflux method. Solvothermal reaction at 120 ºC was found to be the optimum method, that was indicated by most crystalline product compared to the simulated pattern in XRD analysis. The formation of the framework was characterized by FTIR analysis, which showed a significant shift from 1722 cm^{-1} to 1620 cm^{-1}. The synthesized Zr(IV)-BTC was thermally stable up to 322°C as shown by TG/DTA analysis. This high thermal stability was related to the high oxidation state of Zr(IV), which give a significant covalent character to the Zr-O bond.

1. Introduction

The development of alternative renewable energy source became an important and interesting issue to be evaluated and developed due to the depletion of fossil-based energy [1–3]. One of the interesting topics is the conversion of biomass into biodiesels via esterification and trans-esterification reaction. The production of biodiesel from biomass by esterification reaction frequently needs homogenous acid catalyst, such as H_2SO_4 and HCl due to its high selectivity and thermal stability [4, 5]. However, the use of this kind of catalyst causes more problems, such as the difficulties on the separation from the product and the pollution effect on the environment. [6]

Heterogeneous acid catalysts can be an alternative promising way to substitute the homogeneous catalyst and solve the separation and environmental problems [7,8]. One of the interesting heterogeneous catalysts is Metal-organic frameworks (MOFs). MOFs are porous crystalline materials constructed from metal cation or metal-oxide cluster as the node and organic ligand as the linker form an infinite three dimensional structure. Their high specific surface area, volume and active sites both of on the node and the linker enable them to be used as catalysts [9,10]. Their most adorable advantage is that their properties are tunable by modification of the node and the linker. The Lewis acid character of the MOFs come from the creation of vacant orbitals in the nodes by solvent removal, and this character can be raised by activation through heating [11, 12]. The only limitation of the use of MOFs in the catalysis is their low thermal stability.

Zirconium-based MOFs, with Zr^{4+} as the metal node can be an alternative to solve this problem with their chemical, mechanical and hydrothermal stabilities. They have the most stable construction with the contribution of the carboxylate-bridge ($Zr_6O_4(OH)_4(A)_6(CO_2)$ building units which raise the significant covalent character[13,14]. The first synthesized zirconium(IV) based MOFs was UiO-66, with BDC as the linkers by solvothermal method[15]. The structural variety and synthetic strategy of zirconium(IV) based MOFs are progressively developed by modifying the linkers such as aminoterephthalate[16], metallophophyrin[17] and BTC[18], and modifying the synthetic method such as microwave assisted synthesis[19] electrochemical deposition[20], reflux[11] and mechanochemical free solvent assembly[21,22], or by embedding metal cation like copper(II) and iron(III)[23].

The application of zirconium(IV) based MOFs as catalyst has been evaluated in many kinds of reactions, including cyanosilylation of carbonyl[24], Knoevenagel condensation[25], cyclization of citronellal[26] and biomimetic reaction[17]. Herein, we report the comparison synthetic method of MOFs based on zirconium(IV)- and H_3BTC linker through reflux and solvothermal reaction, their characterization as well as their catalytic properties on the esterification reaction of palmitic acid with ethanol.

2. Experimental Section

2.1. Materials

Anhydrous zirconium tetra chloride ($ZrCl_4$, Merck, 98.0 %), benzene 1,3,5- tricarboxylic acid (H_3BTC, Sigma Aldrich, 95%), acetic acid (CH_3COOH, Merck, 99%), ethanol (EtOH, Merck, 99.9 %), methanol (MeOH, Merck, 99.9 %), N,N'-dimethyl-formamide (DMF, Merck, 99.8 %), acetone were used without any further purification.

2.2. Synthesis of MOFs based on Zr^{4+}-BTC

2.2.1. Solvothermal Method

Two different trials were conducted using reported procedure [18]. Briefly, the first trial was synthesized by dissolving 0.1165 g (0.50 mmol) of $ZrCl_4$ and 0.0353 g (0.168 mmol) of 1,3,5-benzenetricarboxylic acid in 2.8 mL of acetic acid and 5 mL of N,N'-dimethylformamide in a closed teflon line. The mixture was stirred without heating by magnetic stirring for 15 minutes before being sealed and heated at 120 °C for 24 hours. The mixture was cooled to the room temperature in the oven and then the resulting materials were recovered by filtration and washed with 5 mL of N,N'-dimethyl formamide, 5 mL of acetone and 5 mL of methanol. Finally, the resulting precipitate was dried at room temperature and activated at 200 °C for 2 hours. The second trial batch was synthesized by simply changing the heating temperature from 120 °C to 135 °C with the same procedure.

2.2.2. Reflux Method

Synthesis was carried out using reported procedure [11], by dissolving 0.1 g of $ZrCl_4$ (1 mmol) and 0.09 g (1 mmol) 1,3,5-benzenetricarboxylic acid in 7.31 mL (220 mmol) of N,N'-dimethyl-formamide in a closed round bottom flask. The mixture was heated at 80 °C for 6 hours without stirring, and followed by heating at 100 °C for 24 hours with stirring. The resulting materials were recovered by filtration and washed with 5 mL of N,N'-dimethyl-formamide. The solids were then soaked in dichloromethane for 3 hours, followed by filtration and drying in air at room temperature.

2.3. Characterization

XRD diffractograms were obtained from JEOL JDX-3530 diffractometer using Cu–Kα radiation (λ = 1.5406 Å) to confirm the expected structure, phase purity and the crystallinity of MOFs based on Zr^{4+}-BTC. TG and DTA curves were recorded from STA Linseis PT-1600 over the temperature range 25

°C up to 800 °C at heating rate 10 °C min^{-1} under a flow of nitrogen to analyze the thermal stability. Scanning Electron Microscopy (SEM) were carried out on FEI type Inspect S50 to determine the morphology and surface area analysis were performed using a static volumetric apparatus Micrometrics, NOVA 1200e to provide the BET specific surface area.

2.4. Catalytic Test

The solvothermal synthesized Zr^{4+}-BTC MOFs at 120 °C was tested in a catalytic reaction on esterification reaction of palmitic acid with ethanol according to reported procedure [27]. The reaction was performed as follows: one mmol of palmitic acid and 60 mmol of ethanol were contacted with the MOF based on Zr^{4+}-BTC with variation 0; 0.6 and 1.8% mol of total weight) in a reflux reactor at 78 °C for 2 hours. At the end of reaction, the GC-MS analysis was carried out to determine the conversion of palmitic acid into palmitate ester as a model reaction in biodiesel production.

3. Results and Discussion

3.1. Material Characterization

The MOF based on Zr^{4+}-BTC was prepared through two different methods, the solvothermal and reflux method, to determine the optimum method and condition for the synthesis. The comparison of the diffraction pattern of the synthesized materials are presented in figure 1.

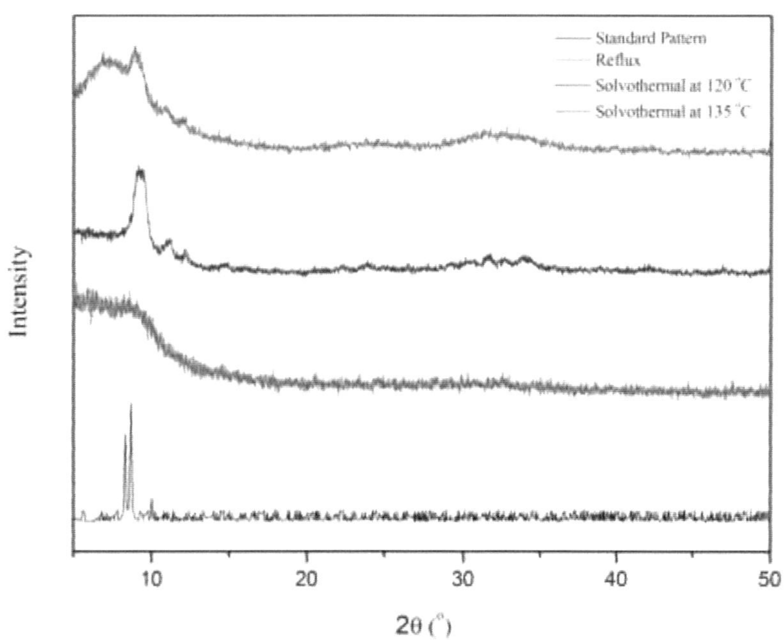

Figure 1. XRD diffractogram of synthesized MOFs based on Zr^{4+}-BTC compared to standard pattern.

The solvothermal method products show the higher crystallinity than the product synthesized by reflux method. Compared to the products obtained from reflux and solvothermal method at 135 °C, material prepared by solvothermal at 120 °C has the sharpest and similar peaks with the standard pattern with 2θ of the mean peak recorded on 8.6° and 8.3°.

Functional group analysis by FTIR (Fourier Transform Infra-Red) shows the significant absorption bands shifts of carbonyl (C=O) and hydroxyl (-OH) groups, as depicted in figure 2.

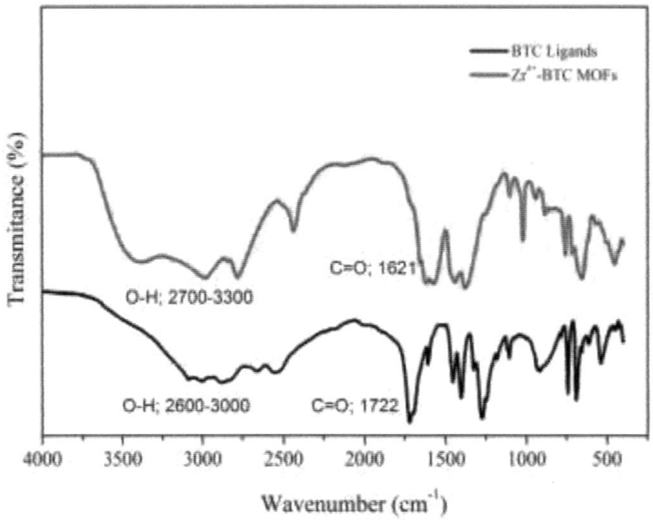

Figure 2. FTIR spectra of H_3BTC ligand and Zr^{4+}-BTC MOF synthesized under solvothermal reaction at 120 °C.

Significant shift was observed from 1722 to 1620 cm^{-1} corresponds to C=O stretching band, and peak at 1365 cm^{-1} is assigned as stretching vibration of C-O group, indicating the deprotonation of carboxylate group (from H_3BTC ligand) and coordination with zirconium(IV) metal ion to form $[Zr_6O_4(OH)_4(A)_6(BTC)_2]\cdot nH_2O$[18]. In addition, the absorptions bands of hydroxyl groups (O-H stretching) from the ligand clearly shifted from 2663; 2878 and 2984 cm^{-1} into 2779; 2983 and 3382 cm^{-1}. Broad band observed either in the ligand or in synthesized MOF correspond to the formation of hydrogen bonding either in the ligand or in MOFs. Moreover, broad peak observed in MOFs indicates that the zirconium ion was also coordinated with water molecules and associated with the formation of zirconium oxide cluster $(Zr_6O_4(OH)_4(A)_6(CO_2)$. The C–H aromatic vibration bands from the synthesized Zr-BTC MOF are shown at 3004 cm^{-1}. The peak at 1578 cm^{-1} is assigned with C=C bond in the aromatic ring, while the band at 658 cm^{-1} the characteristic absorption of Zr-O symmetric bond.

The thermal stability of the Zr^{4+}-BTC based MOF was measured with TG/DTA (fig. 3). According to thermogravimetric curves, there are three mass loss regions, which occurred at 37.8-147.7 °C; 147.7-321.7 °C and 321.7-541.1 °C. The first (53.3%) and the second (79.6%) mass losses were attributed to the removal of superficial solvents in the pore or surface of the material and the coordinated solvent in the material, respectively. The third mass loss occurred at 321.7 °C, where the framework started to decompose, and at 541.1 °C the frameworks was completely collapse. These mass loss confirmed by the DTA curves, which had endothermic peak, followed by sharp exothermic peak and the other endothermic peak. The high thermal stability of the Zr^{4+}-BTC MOFs was confirmed to be related to the strong Zr-O bonds. The high oxidation state and high charge density of zirconium(IV) polarize the oxygen atoms from the carboxylate groups, to result the high covalent character of Zr-O bonds [14].

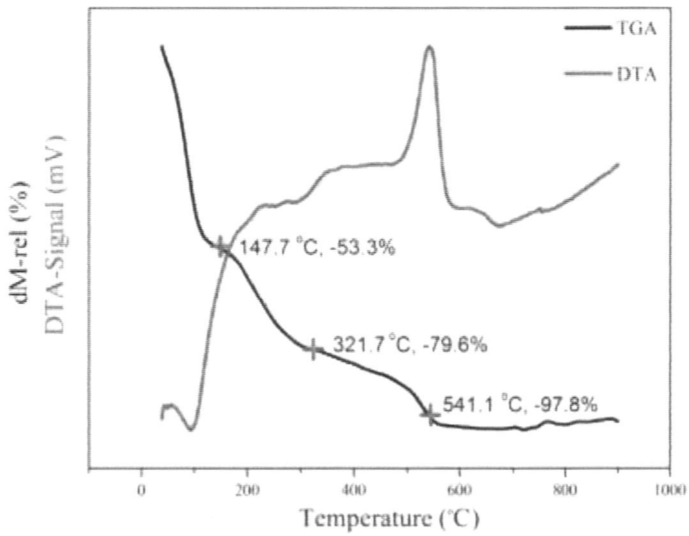

Figure 3. TG/DTA curves of synthesized Zr^{4+}-BTC MOFs.

(a) (b)

Figure 4. (a) SEM image and (b) particle size distribution of synthesized Zr^{4+}-BTC MOFs.

According to SEM analysis, the synthesized Zr^{4+}-BTC based MOF has irregular shape with the average particle size ca. 2.738 ± 1.632 µm (Figure 4). The bigger particles were assumed as the result of agglomeration from the smaller particles. The histogram (Figure 4b) shows the less homogenous of the grain, with size range 1-7 µm. Nitrogen sorption isotherm revealed this compound has the BET surface area of 223.766 m^2/g.

3.2. Catalytic Activity on Esterification Reaction

Catalytic activity of the Zr^{4+}-BTC based MOF was evaluated on esterification reaction of palmitic acid with ethanol. Concentration variation of the catalyst were 0, 0.6 and 1.8%. The resulting analysis carried out by GC-MS as given on Table 1 and shows that catalyst loading at 0.6% was the optimum condition for the esterification reaction.

Table 1. Esterification of palmitic acid and ethanol with percent variation of Zr^{4+}-BTC based MOF as catalyst

Condition	Conversion (%)
Without catalyst	38.61
0.6 % catalyst	69.20
1.8 % catalyst	54.82

4. Conclusions

The synthesis of Zr^{4+}-BTC based Metal-Organic Frameworks (MOFs) has been evaluated by various method. The solvothermal at 120 °C gives the most optimum resulting material as characterized in the terms of crystallinity, functional groups absorption bands shift, thermal stability, morphology and surface area. Based on the catalytic test the compound is effective as the catalyst for esterification reaction of palmitic acid and ethanol with catalyst loading of 0.6 % to give maximum conversion at 69.20 %, as shown by GC-MS analysis.

5. References

[1] Dincer I 2000 *Renew. Sustain. Ener. Rev* **4**(2) 157–175
[2] Bu Q, Lei H, Zacher A H, Wang L, Ren S, Liang J, Wei Y, Liu Y, Tang J, Zhang Q, and Ruan R 2012 *Bioresour. Technol* **124** 470–477
[3] Hassan M H and M A Kalam 2013 *Procedia Eng* **56** 39–53
[4] Ejikeme P M, Anyaogu I D, Ejikeme C L, Nwafor N P, Egbuonu C A C, Ukogu K, and Ibemesi J A 2010 *E-Journal Chem* **7**(4) 1120–1132
[5] Basha S A, Gopal K R, and Jebaraj S 2009 *Renew. Sustain. Ener. Rev* **13**(6–7) 1628–1634
[6] Marchetti J M, Miguel V U, and Errazu A F 2007 *Renew. Sustain. Ener. Rev* **11**(6) 1300–1311
[7] Lee A F, Bennett J, Manayil J C, and Wilson K 2014 *Chem. Soc. Rev* **43** 7887–7916
[8] Thanh L T, Okitsu K, Van L B, and Maeda Y 2012 *Catalysts* **2**(1) 191–222
[9] Zhou H C, Long J R, and Yaghi O M 2012 *Chem. Rev* **112** 673–674
[10] Čejka J 2012 *Angew. Chem. Int. Ed* **51**(20) 4782–4783
[11] Cirujano F G, Corma A, and Xamena F X L 2015 *Chem. Eng. Sci* **124** 52–60
[12] Lee J, Farha O K, Roberts J, Scheidt K A, Nguyen S T, and Hupp J T 2009 *Chem. Soc. Rev* **38**(5) 1450–1459
[13] Kim M and Cohen S M 2012 *Cryst. Eng. Comm* **14**(12) 4096–4104
[14] Bai Y, Dou Y, Xie L H, Rutledge W, Li J R, and Zhou H C 2016 *Chem. Soc. Rev* **45**(8) 2327–2367
[15] Cavka J H, Jakobsen S, Olsbye U, Guillou N, Lamberti C, Bordiga S, and Lillerud K P 2008 *J. Am. Chem. Soc* **6** 13850–13851
[16] Kandiah M, Nilsen M H, Usseglio S, Jakobsen S, Olsbye U, Tilset M, Larabi C, Quadrelli E A, Bonino F, and Lillerud K P 2010 *Chem. Mater* **22**(24) 6632–6640
[17] Feng D, Gu Z Y, Li J R, Jiang H L, Wei Z, and Zhou H C 2012 *Angew. Chemie - Int. Ed.* **51**(41) 10307–10310
[18] Liang W, Chevreau H, Ragon F, Southon P D, Peterson V K, and D'Alessandro D M 2014 *Cryst. Eng. Comm.* **16**(29) 6530
[19] Ren J, Segakweng T, Langmi H W, Musyoka N M, North B C, Mathe M, and Bessarabov D 2014 *Int. J. Mater. Res.* **105**(5) 516–519

[20] Stassen I, Styles M, Assche T V, Campagnol N, Fransaer J, Denayer J, Tan J C, Falcaro P, Vos D D, and Ameloot R 2015 *Chem. Mater* **27**(5) 1801–1807
[21] Užarević K, Wang T C, Moon S Y, Fidelli A M, Hupp J T, Farha O K, and Friščić T 2016 *Chem. Commun* **52** (10) 2133–2136
[22] Reinsch H, Bueken B, Vermoortele F, Stassen I, Lieb A, Lillerud K P, and Vos D D 2015 *Cryst. Eng. Comm.* **17** (22) 4070–4074
[23] Morris W, Volosskiy B, Demir S, Gándara F, McGrier P L, Furukawa H, Cascio D, Stoddart J F, and Yaghi O M 2012 *Inorg. Chem* **51**(12) 6443–6445
[24] Xi F G, Yang Y, Liu H, Yao H F, and Gao E Q 2015 *RSC Adv.* **5**(97) 79216–79223
[25] Yang Y, Yao H F, Xi F G, and Gao E Q 2014 *J. Mol. Catal.* A Chem **390** 198–205
[26] Vermoortele F, Bueken B, Voorde B V D, Vandichel M, Houthoofd K, Vimont A, Daturi M, Waroquier M, Speybroeck V V, Kirschhock C E, and Vos D E D 2013 *J. Am. Chem. Soc* **66** 0–3
[27] Carmo A C, Souza L K C., Costa C E F, Longo E, Zamian J R, and Filho G N R 2009 *Fuel* **88** (3) 461–468

Acknowledgements

The authors thank the Ministry of Research, Technology and Higher Education of the Republic of Indonesia (Student Creativity Program/ PKM Grant 2016 and KI 2017) for the financial support.

IOP Conf. Series: Materials Science and Engineering **214** (2017) 012007 doi:10.1088/1757-899X/214/1/012007

From Cell to Module: Fabrication and Long-term Stability of Dye-sensitized Solar Cells

N M Nursam[1,*], J Hidayat[1], L Muliani[1], P N Anggraeni[1], L Retnaningsih[1], and N Idayanti[1]

[1]Research Center for Electronics and Telecommunications, Indonesian Institute of Science, Kompleks LIPI Building 20 Level 4, Cisitu, Bandung – Indonesia 40135

Email: nata002@lipi.go.id

Abstract. Dye-sensitized solar cell (DSSC), which has been firstly developed by Graetzel et al back in 1991, has attracted a considerable interest since its discovery. However, two of the main challenges that the DSSC technology will have to overcome towards commercialization involve device scale-up and long-term stability. In our group, the fabrication technology of DSSC has been developed from laboratory to module scale over the past few years, nevertheless, the long-term stability has still became a major concern. In this contribution, the long-term DSSC performance in relation to their scale-up from cell to module is investigated. The photoelectrode of the DSSCs were fabricated using nanocrystalline titanium dioxide materials that were subsequently sensitized using ruthenium-based dye. Additionally, $TiCl_4$ pre- and post-treatment were carried out to enhance the overall device efficiency. When fabricated as cells, the DSSC prototypes showed relatively stable performance during repeated tests over three months. In order to increase the output power of the solar cells, the DSSCs were then connected in a Z-type series connection to obtain sub-module panels. The DSSC sub-modules exhibit poor stability, particularly as indicated by the significant decrease in the short circuit current (I_{SC}). Herein, the effect of photoelectrode and sealant materials as well as module design are investigated, highlighting their profound influence upon the DSSC efficiency and long-term stability.

1. Introduction

Among various types of solar cells, DSSC combines both the promise of high efficiency and low fabrication cost. Additionally, they can be fabricated over flexible substrate and can offer transparency and workability to be integrated into buildings as smart window, of which those could not be accommodated by the first two generations of photovoltaic technology, that is, the silicon and thin-film based solar cells. Since its breakthrough in 1991 [1], the performance of DSSC has progressed considerably. DSSC with an efficiency of ~13 % has recently been reported by Graetzel and co-workers [2], which constitutes the highest efficiency reported to date. Due to their attractive appearance and affordability reasons, considerable effort has therefore been devoted to enable a commercial upscaling of this type of solar cell.

DSSC panel has been reported to deliver more electricity than the silicon and thin-film solar cells when exposed to low light condition. The upscaling of DSSC therefore holds a promising prospect for commercialization. However, the reports in this area [3-6] have not been as extensive as those reported in the area of cells optimization, with the total number of works almost 100 times lower than the latter [7]. One of the early work on DSSC panel was reported by Kay and Gratzel in 1996 [8], demonstrating the fabrication of six DSSC cells connected to each other on a single FTO substrate with a series connection, showing an overall module efficiency of ~5.3%. Large area DSSC modules or panels often suffer from long-term stability issue and also lack in performance compared to their lab-scale devices. In term of commercial applications, these two major issues are crucial in addition to the power conversion efficiency. As a comparison, most silicon solar cell manufacturers would give a warranty of 80-90% of the initial performance after 20 years of operation. Therefore, as the DSSC technology progress from laboratory bench to practical applications, scalability and durability are the two paramount aspects that need to be tackled.

In this contribution, we present a comparative and stability study of DSSC sub-modules with various sealing materials. To date, little effort has been undertaken to investigate the stability of DSSC modules in relation to the selection of sealant materials. For instance, Sommeling *et al.* [5] and Kato *et al.*[4] reported the effects of electrolyte compositions and types upon the long-term performance of DSSC cells, respectively, whereas Sastrawan and co-workers [9] reported the long-term stability of glass-frit sealed DSSC modules without comparing with other types of sealant. The main functions of the sealing materials in DSSC are to protect the conductor grid (typically, in the form of silver) from the highly reactive electrolyte and to prevent any leaking of electrolytes between the neighboring cells or toward out of the module. Ideally, the sealing material must be chemically inert against the I^-/I_3^- redox couple, while maintain its mechanical and chemical stability under various testing conditions.

2. Experimental Section

2.1. Photoelectrode and Counter-electrode Preparation

The photoelectrode and counter-electrode of the DSSC sub-modules were fabricated on fluorine-doped tin oxide (FTO) coated glass with a conductivity of 15 Ω/sq. First, a diamond cutter was used to remove the FTO layer on the photoelectrode side following the module design as shown in Figure 1. All of the FTO glasses were cleaned by ultrasonicating in soap water, isopropanol and ethanol, consecutively, for 10 min. The TiO_2 paste (Dyesol 18NR-AO) was deposited on the photoelectrode two times using screen printing method (Nylon #325 mesh) to obtain TiO_2 films with a thickness of 8-10 μm. The active area for each sub-module was 2×5 cm^2. Each sub-module consists of two individual cells. All samples were dried in an oven at 100 °C for 10 min between each deposition and then annealed in conveyer belt furnace (Radiant Technology Corporation) at 500 °C for 15 min. All samples then received $TiCl_4$ post-treatment by heating in 40 mM $TiCl_4$ solution at 70 °C for 30 min, followed by another calcination at 500 °C for 15 min. After cooling, the TiO_2 film coated FTO were soaked in Z907 dye solution ($C_{42}H_{52}N_6O_4RuS_2$, Dyesol) in ethanol (20 mg/100 mL) overnight. Before cell construction, the sensitized samples were rinsed thoroughly in ethanol to remove any excess dyes and then dried naturally. All samples turned to maroon color after the immersion.

For the preparation of the counter electrode, the pre-drilled FTO substrates were sputtered with platina target using DC-sputtering process. The coating process was carried out for 20 min with a power of 5 W, initial pressure of 6.6×10^{-3} Pa, argon gas pressure of 5.3×10^{-1} Pa, and rotation speed of 5 rpm.

2.2. Module Assembly

Figure 1 shows the schematic design of the sub-module configuration used in this work. All of the sub-modules were assembled using Z-type series connection, wherein each sub-module comprised of two cells with an active area of 1×5 cm^2. There were three types of module assembled in this work that were categorized according to the type of sealant material. The sealing compounds were printed in both electrodes through a stainless steel screen with #200 mesh. In the first sub-module, namely Module A, the sub-module was sealed using hermetic sealing compound or also known as glass frit (Dyesol), which was made from a mixture of part A and part B with weight compositions of 4:3, 4:2 and 4:1 (see Table 1 for details). These sub-modules were labeled as Module A1, A2 and A3, respectively. Meanwhile, the sealing compound in Module B consisted of Amosil 4R and Amosil 4H

(Solaronix) with weight ratios of 1:1, 4:3 and 4:2 (Table 1). These sub-modules were labeled as Module B1, B2 and B3, respectively. In Module C, a commercial silicone rubber sealant was printed without any additives. Then, silver paste (Hunan LEED Electronics, Inc. Co. Ltd.) was coated into each electrode. Both sealant and silver paste were printed with a line-width of 1 mm, while the gaps between the neighboring lines were maintained at approximately 0.3 mm.

Table 1. Mass compositions used for the sealing materials used for DSSC sub-modules

No.	Glass frit		Amosil		Silicon rubber
	Part A (g)	Part B (gr)	4R (g)	4H (g)	
1.	0.500	0.375	0.500	0.500	used as is
2.	0.500	0.250	0.500	0.375	
3.	0.500	0.125	0.500	0.250	

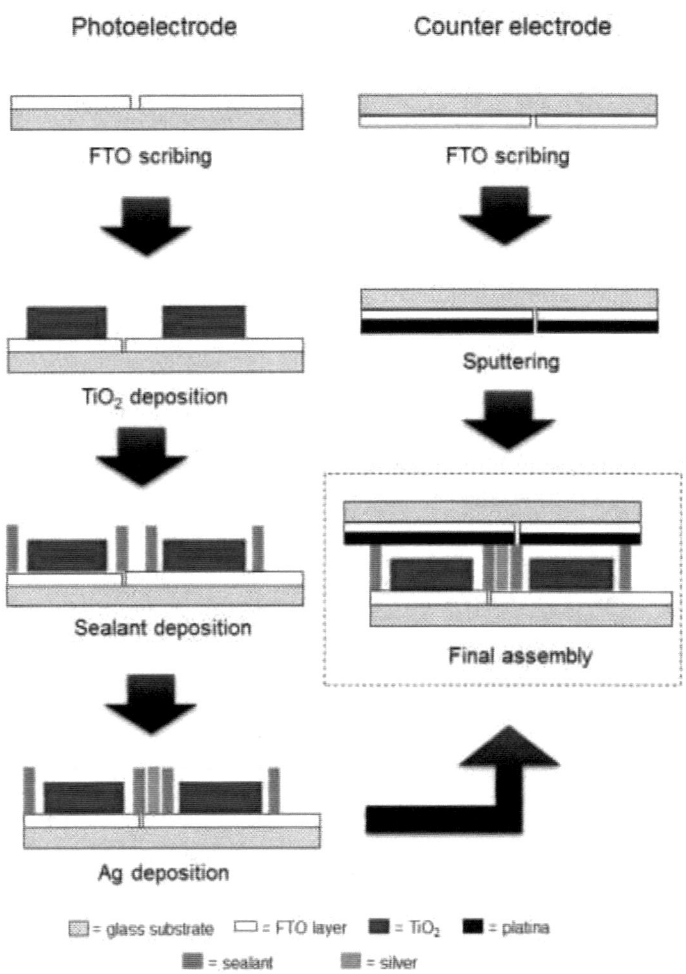

Figure 1. Schematic drawing of the assembly process of DSSC sub-modules.

Both electrodes were subsequently pressed together using paper clip to strengthen the bonding overnight. Once the electrodes have been strongly attached to each other, liquid electrolyte EL-HPE (Dyesol) was injected through the pre-drilled holes. Finally, all of the pre-drilled holes were sealed with glass frit and left to solidify in air. In order to compare the long-term performance of DSSC cell and sub-module, a single cell was prepared following the same procedures as above. The fabrication of this cell can be found elsewhere [10]. This cell was subsequently assembled and sealed using hot melt foil with a thickness of 50 μm (Surlyn, Dyesol). The sealing process was followed by heating the sample at 120 °C for 35 min to melt the foil and to make both electrodes adhered to each other.

2.3. Characterization

The current-voltage (*I-V*) characteristics of sub-modules were measured using a National Instrument source meter under AM1.5G solar simulator (Oriel, Newport USA) and analyzed using Lab-View software. The irradiation intensity was 500 W/m^2 at 25 °C. The efficiency of the cell and sub-modules was determined using the following equation:

$$\eta = \frac{V_{OC} I_{SC} FF}{P_{in}} \times 100\%$$

where V_{OC} and I_{SC} are the open circuit voltage and the short circuit current, respectively, FF is the fill factor, and P_{in} is the power input received during the illumination.

3. Results and Discussion

In this work, hermetic sealing compound (also known as glass frit), Amosil, and silicon rubber was chosen instead of widely used Surlyn hot melt foil due to two reasons. First, the application of glass frit in epoxy form upon large area module is more feasible as it can be deposited using screen printing method. Contrarily, the use of Surlyn would require high precision during the cutting and attachment following the desired pattern. It should be noted, however, that the printing of glass frit needs to be carried out as quickly as possible to avoid fast solidification of the compound. The second reason is related to thermal stability. As Surlyn has low melting point, it will not withstand any high temperature test above 60 °C, thus, glass frit offers better temperature resistibility. Likewise, Amosil and silicone rubber also could stand high temperature (i.e. up to 300 °C), thus rendering them as better sealant than Surlyn. In addition, glass frit, Amosil and silicon rubber can be cured at room temperature, thus avoiding the need for a heating that may degrade the organic components existing in the dye materials. The glass frit and Amosil were opaque-grey in color, while the silicon rubber sealant was transparent.

The photograph of the sub-modules fabricated herein is displayed in Figure 2. All of the sub-modules were opaque and have similar appearance apart from the sealant materials. Figure 3 shows the *I-V* curves of all samples measured on the first day, that is, one day after the assembly has been completed. The corresponding *I-V* parameters are summarized in Table 2. It can be seen that the sub-module fabricated using Amosil demonstrated superior performances compared to the rest of the sub-modules. This was indicated by their average photovoltaic efficiency that was higher than 3 %, regardless of the different element compositions. Overall, the sub-module assembled using silicon rubber and glass frit with 4:1 ratio showed the lowest efficiency. It was suspected that the incomplete curing process was likely to be the reason for such poor performances, which was specifically attributed by the low current generation. The incomplete sealant hardening process may have caused the electrolyte to react with the sealant materials, thus causing insufficient regeneration of the redox components. Moreover, there were higher chance of electrolyte leaking and reacting with silver, thus hampering charge collection and rendering low output current.

IOP Conf. Series: Materials Science and Engineering **214** (2017) 012007 doi:10.1088/1757-899X/214/1/012007

Figure 2. Photograph of the sub-modules sealed with glass frit and Amosil.

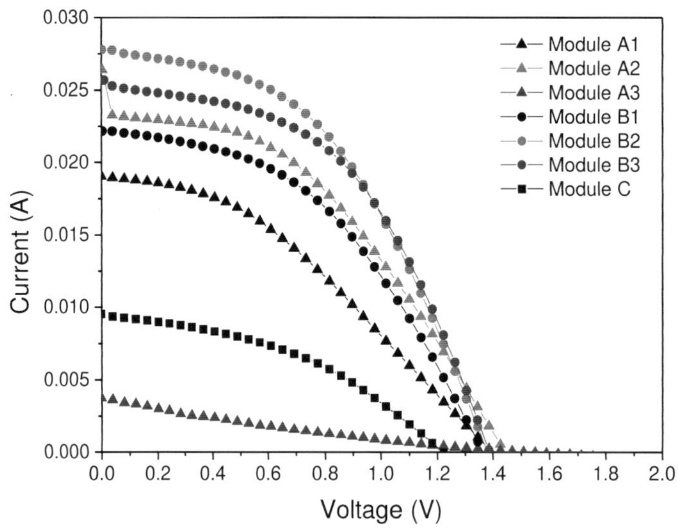

Figure 3. *I-V* curves of the sub-modules assembled using various sealant materials.

Table 2. Electrical characteristics of the sub-modules as measured using *I-V* measurement

Parameter	Glass Frit			Amosil			Silicon rubber
	A1	A2	A3	B1	B2	B3	
P_{max} (watt)	0.009	0.014	0.001	0.014	0.018	0.017	0.005
V_{mp} (volt)	0.733	0.815	0.651	0.815	0.855	0.896	0.733
I_{mp} (ampere)	0.013	0.018	0.002	0.017	0.021	0.019	0.007
FF	0.373	0.384	0.202	0.408	0.379	0.401	0.429
η (%)	1.962	2.918	0.221	2.716	3.579	3.478	0.939

For the long-term stability test, the I-V curves were measured after storing the samples in petri dishes for one week at room temperature without controlling the humidity. It can be seen in Figure 4 that most of the sub-modules were substantially deteriorated, especially when compared to the single cell sample that only suffered from less than 10% reduction in efficiency after one week and was relatively stable after more than 60 days of observation (see inset in Figure 4). Compared to a cell, there are more complicated factors that could attribute to the very low device stability in a module. A single DSC cell does not contain any current collecting component, i.e. silver, that is required to connect the adjacent cells. Thus, there is lower possibility of corrosion. Contrarily, the selection of sealant material in a module is very crucial as the sealant should be able to protect the silver from the electrolyte, in addition to prevent the electrolyte from leaking out. The trend that shows significant decrease in the efficiency (as shown in Figure 4) indicates that both Amosil and silicon rubber could not provide a good sealing ability. The poor stability shown by the sub-modules sealed either with Amosil or silicon rubber was attributed to the electrolyte leaking. This was apparent from the photograph in Figure 5. A visual inspection revealed that there were distinct discolorations around the edges of the sealing, indicating electrolyte depletion in those areas.

From the comparison of the relative η (that is η from day 7 divided by the η_0), the sub-module that was sealed using glass frit with 4:3 ratio demonstrated the best stability out of all samples with an efficiency decrease of approximately 26 % after one week of storage. There is also an indication that the decrease in glass frit part B tends to decrease the long-term stability of the module, and contrarily, the decrease in Amosil 4H tends to increase the long-term stability of the module. However, further tests with more variations in the glass frit and Amosil compositions will be required to confirm these trends. There were also no specific trends showing that the best performing sub-module in the initial measurement consistently showed the best performance after one week. This suggests that the photovoltaic output does not necessarily correlate with the long-term stability. Overall, the sub-modules that were prepared using Amosil have shown the most superior performance, but the best stability was achieved by the sub-modules containing glass frit with 4:3 ratio.

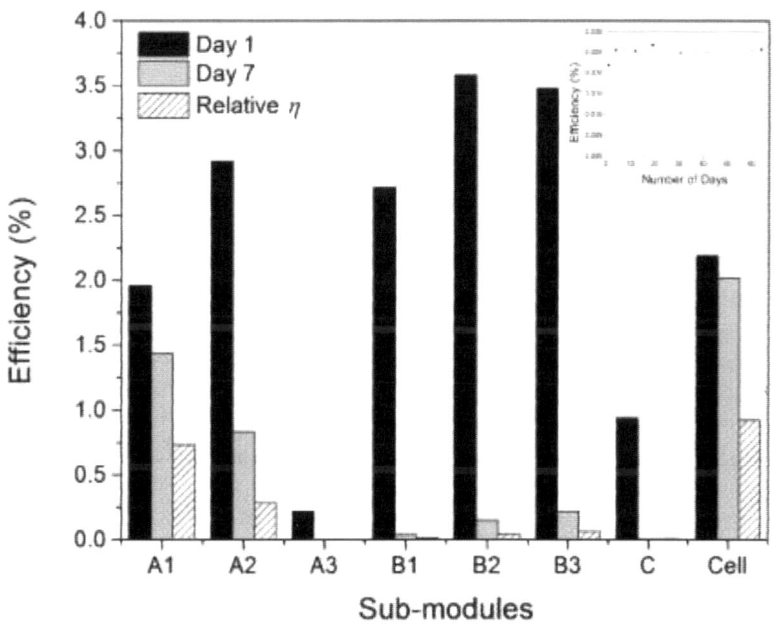

Figure 4. Comparison between the efficiency of sub-modules with various sealant and single cell with Surlyn as the sealant material. The efficiency was also compared on the first day of the assembly (black bars) and after 1 week (patterned bars). The relative η was obtained as a ratio between the efficiency from day 7 and day 1. The time-dependant efficiency of the single cell is shown in the inset.

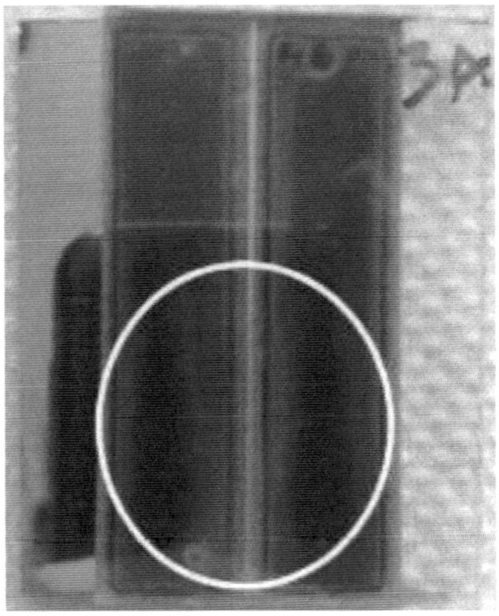

Figure 5. Photograph showing electrolyte leaking observed in the sub-module sealed with Amosil with 1:1 ratio.

4. Conclusions

Dye-sensitized solar cell (DSSC) sub-modules were fabricated using various sealant materials and tested after one-week storage to monitor their stability. A single cell sealed with hot melt foil was also fabricated as an additional comparison. It was shown that the use of glass frit, Amosil and silicon rubber sealant is applicable for the upscaling of DSSC from cell size to sub-modules with a total active area of 2×5 cm^2. The main technology used for the deposition of the sealant was screen printing, proving that this technology is a low-cost yet facile technique that fully supports the upscaling of DSSC.

During the first *I-V* measurement, the sub-modules sealed with Amosil showed superior performances compared to that of sub-modules sealed with glass frit and silicon rubber. Second measurements carried out after one week of storage, however, revealed that the sub-modules sealed with glass frit demonstrated better stability with the least degradation in the sub-module's efficiency. Electrolyte leaking was suspected as the main culprit for the deterioration. In contrary, the single cell sealed with hot melt foil showed a relatively stable performance. Our study showed that the encapsulation quality of glass frit, Amosil and silicon rubber still are still incomparable with the hot melt foil, although the latter has been proven to be difficult for DSSC upscaling. Thus, further study to optimise the glass frit composition as well as finding the appropriate treatment for glass frit is required to further improve the long-term stability of DSSC modules.

References

[1] O'Regan B and M Graetzel 1991 *Nature* **353** 737

[2] Mathew S, A Yella, P Gao, R Humphrey-Baker, B F E Curchod, and N Ashari-Astani 2014 *Nat. Chem* **6** 242

[3] Hinsch A, J M Kroon, R Kern, I Uhlendorf, J Holzbock, A Meyer, *et al.* 2001 *Prog. Photovoltaics Res. Appl* **9** 425

[4] Kato N, K Higuchi, H Tanaka, J Nakajima, T Sano, and T Toyoda 2011 *Sol. Energy Mater. Sol. Cells* **95** 301

[5] Sommeling P M, M Spath, H J P Smit, N J Bakker, and J M Kroon 2004 *J. Photochem. Photobiol. A* **164** 137

[6] Xue G, Y Guo, T Yu, J Guan, X Yu, J Zhang, *et al.* 2012 *Int. J. Electrochem. Sci* **7** 1496

[7] Fakharuddin A, R Jose, T M Brown, F Fabregat-Santiago, and J Bisquert 2014 *Energy Environ. Sci* **7** 3952
[8] Kay A and M Graetzel 1996 *Sol. Energy Mater. Sol. Cells* **44** 99
[9] Sastrawan R, J Beier, U Belledin, S Hemming, A Hinsch, R Kern, *et al.* 2006 *Sol. Energy Mater. Sol. Cells* **90** 1680
[10] Nursam N M, L Muliani, and J Hidayat 2011 *Jurnal Teknologi Indonesia* **34** 85

The Effect of Nd Content Variation on Crystal Structure and Microstructure to Improve Magnetic Properties Performance

Nenen Rusnaeni[1*], Priyo Sarjono[1], Taufik[2], Hanifah[2], Muljadi[1]
[1] Research Center for Physics, Indonesian Institute of Sciences, Kawasan Puspiptek Serpong,Tangerang Selatan, Indonesia, Tel. 021-7560570
[2] Department of Physics, University of Sumatera Utara, Kampus USU, Jalan Bioteknologi Medan 20155 Indonesia

*Email: nenenrusnaeni@yahoo.com

Abstract. $Nd_2Fe_{14}B$ magnet powder has been fabricated by mixing Fe, B, and Nd powder based on their stoichiometric proportion at 650°C for 2 hours. Heat treatment series were applied at 720°C for one hour and continued with annealing at 100°C for 4 hours on different Nd stoichiometric variations; 26.6 %wt, 32.6 %wt, and 40.6 %wt. The results obtained from the X-ray Diffraction (XRD) phase analysis of Nd-rich Nd-Fe-B alloy suggested that the alloy has monophasic composition with hard magnetic $Nd_2Fe_{14}B$ phase as a composition. The mean of the crystal grain size was determined using XRD analysis, while the microstructure and composition of the alloys were analyzed using the Scanning Electron Microscope – Energy Dispersive X-ray (SEM-EDX). The magnetic properties characterization were determined using the Vibrating Sample Magnetometer (VSM), which indicates that the sample with 40.6 %wt Nd was able to achieve the highest remanence of 446.50 G and BHmax of 17.83 kGOe. Despite the high remanence result, the coercivity and BHmax value of the third sample was still lower than the commercial's but has adequate potential value.
Keywords : Nd-Fe-B alloys; Nd-rich; phase composition; crystallite size

1. Introduction

Neodymium Iron Boron (NdFeB) is well known as rare earth alloy magnets made from neodymium, iron and boron elements and formed the $Nd_2Fe_{14}B$ tetragonal crystalline structure which has high coercivity and large energy product [1]. The permanent magnet Nd-Fe-B has become an important part in the everyday life as its being used in electric motors, generators, and electronics equipment that has been developed by Japan Company since 1982. The Industries demand of permanent magnets in Indonesia is very high while 100% permanent magnet products in Indonesia are imported with 80% of them came from China since they produced low cost global [2].

Therefore, research in permanent magnet material field to increase magnetic performance such as magnetic induction remanence (Br), coercivity (Hc), and maximum energy product (BHmax) is done in order to produce local permanent magnet product [3,4,5]. The stoichiometric alloys are investigated by some intergranular exchange between the grains of the same hard magnetic phase. One of these studies are the enhancement of non-ferromagnetic phase densification through enrichment Nd (Nd-rich) in order to distinct the separation between $Nd_2Fe_{14}B$ magnetic grains. Phase composition and grain size which has direct influence on magnetic properties were observed on different Nd contents to investigate the change in structure while fabricating the $Nd_2Fe_{14}B$ alloy powder for the permanent magnet.

2. Experimental

The $Nd_2Fe_{14}B$ powder were prepared in series of heat treatment at 300°C (1 hour), 650°C (2 hours), and 720 °C (1 hour), before being dropped to100 °C. Powder Nd, Fe, and B with a purity of 99.99% were used as starting materials, and the composition of each powder was calculated using the stoichiometric ratio with 26.6 %wt Nd; 72.32%mol Fe; and 1%mol B (sample 1). The sample 2 and sample 3 have Nd-rich with 32.6 %wt and 40.6 %wt respectively. Phase composition was determined by using X-ray diffraction (XRD) whereas mean crystal grain size was calculated from the XRD data. The microstructure and composition analysis were characterized by using SEM-EDX and the magnetic properties were determined by using VSM.

3. Results and Discussion

The XRD patterns result of the $Nd_2Fe_{14}B$ powder (Sample 1), and the Nd-rich $Nd_2Fe_{14}B$ alloys powder (sample 2 and sample 3) are presented on figure 1.

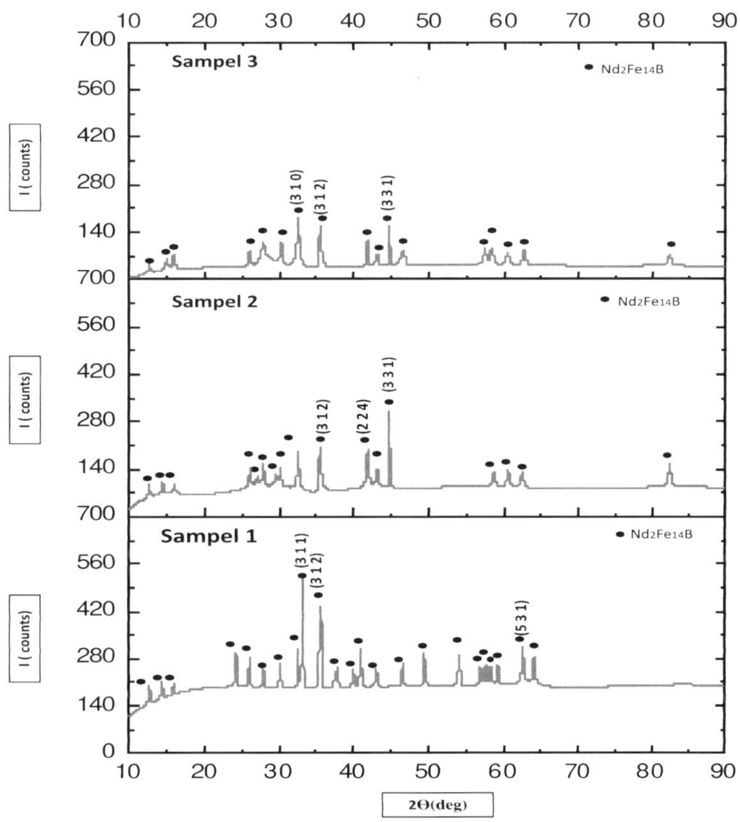

Figure 1. XRD patterns of the $Nd_2Fe_{14}B$ powder (sample 1) and $Nd_2Fe_{14}B$ powder Nd-rich (sample 2 and 3).

$Nd_2Fe_{14}B$ crystal structure analysis using XRD aims to observe the phases that are formed on the sample powder. The peaks matched with NdFeB cards numbered 04-004-9492 were marked with black round which show that the third sample in the figure 1 above are $Nd_2Fe_{14}B$ crystalline alloy powder. At the same time, the phase composition and crystalline structure produced shows that the alloy has almost monophasic composition with $Nd_2Fe_{14}B$ dominating the magnetic phase both in stoichiometric and Nd-rich sample. The mean crystallite grain size was calculated using the Scherrer formula from the highest peak of the XRD patterns (the mean grain sizes were calculated only for $Nd_2Fe_{14}B$ phase).

Table 1. "Crystallite size" of the $Nd_2Fe_{14}B$ powder (sample 1) and $Nd_2Fe_{14}B$ powder Nd-rich (sample 2 and 3)

Sample	"Crystallite Size" (μm)
1	0,272
2	0,310
3	0,148

The crystallite sizing results in sample 1, sample 2, sample 3 are respectively around to 0.272 μm, 0.310 μm and 0.148 μm. In sample 2, the mass of Nd element is raised by 23%, crystal planes diffracted are reduced and the crystalline grain size increased by 1.5 times. In the third sample, the mass of Nd element is raised by 53%, crystal planes diffracted are the same as in sample 2 but the crystalline grain size reduced by half.

Figure 2. SEM micrographs of $Nd_2Fe_{14}B$ powder Nd-rich (sample 2).

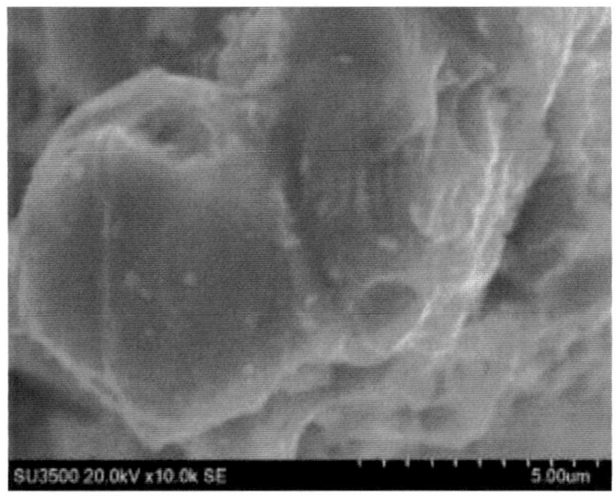

Figure 3. SEM micrographs of $Nd_2Fe_{14}B$ powder (sample 1).

Figure 2 shows white color which show Nd element on the surface of the $Nd_2Fe_{14}B$ alloy grains. The whiter the color show greater concentration of Nd (see Table 2).

Table 2. The Nd content by EDX Result

% atom Element	Sample 1	Sample 2	Sample 3
Nd	0.41	24.44	34.40

Meanwhile, figure 3 shows the $Nd_2Fe_{14}B$ crystalline surface morphology by SEM. Dark gray color indicates the $Nd_2Fe_{14}B$ tetragonal structure formed in stoichiometric state. The magnetic property of the powder was tested by using vibrating sample magnetometer (VSM) and is presented on figure 4. The result shows that the NdFeB magnets is categorized as hard magnet material powder by having 301 G in sample 1 and 413 G in sample 2 and 447 G in sample 3 of remanence (Br) value. Saturated remanence is enough well around 4 kG as it can be seen in Table 3. The enhancement of Nd content such as coercivity (Hc) and BHmax value is expected to increase after the compaction treatment.

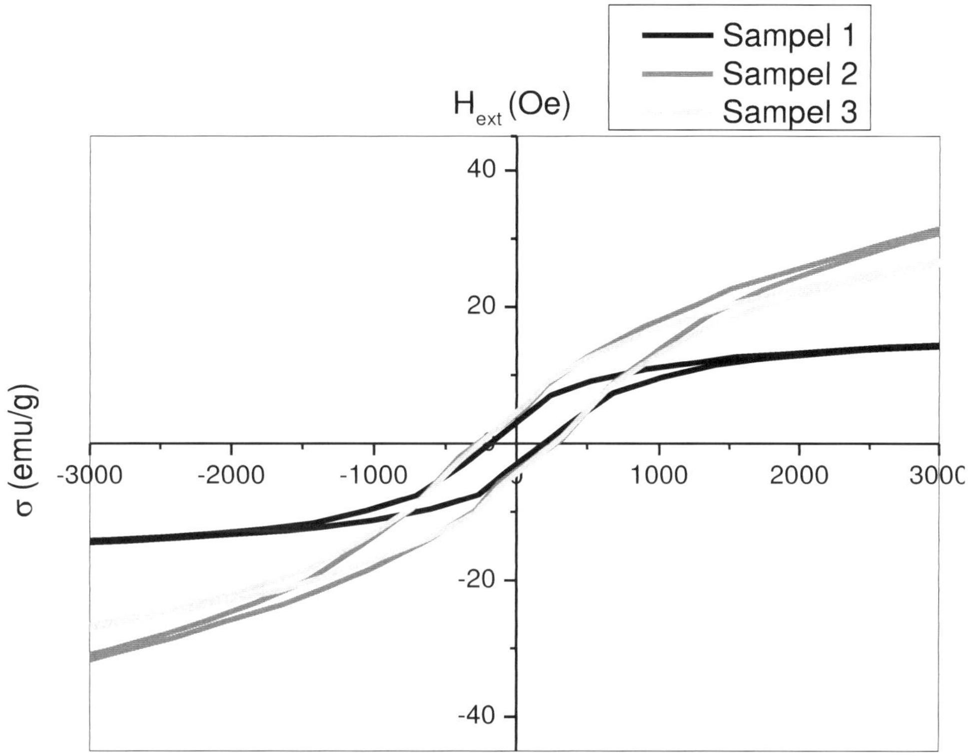

Figure 4. Hysteresis loops of of the $Nd_2Fe_{14}B$ powder (sample 1 black) and $Nd_2Fe_{14}B$ powder Nd-rich (sample 2 and 3).

Table 3. Magnetic Properties of samples 1, 2 and 3 from hysteresis loops

Sampel	Br (G)	Bs (G)	Hc (Oe)	BH_{max} (KGOe)
Sampel 1	301,24	1584,22	188,89	8,56
Sampel 2	413,36	4431,92	274,60	16,59
Sampel 3	446,50	3496,01	261,10	17,83

The heat treatment mechanism has direct influence on the formation of $Nd_2Fe_{14}B$ alloys crystallized grains [5, 6]. The remanence-enhancement effect is not yet increase because the ideal conditions to achieve effective ferromagnetic exchange between the grains of magnetic phases are not fulfilled and the grain size must be below 100 nm [7,8]. Although for sample 3 appear that crystallite size become smaller but the optimal magnetic state as the distinct separation between $Nd_2Fe_{14}B$ magnetic grains and the identification phase composition must be detailed again.

4. Conclusion

$Nd_2Fe_{14}B$ magnetic alloys powder has been fabricated by mixing the Fe, B, and Nd powder with heat treatment series on different Nd stoichiometric variations; 26.6 %wt, 32.6 %wt, and 40.6 %wt. The results obtained from the XRD phase analysis of Nd-rich Nd-Fe-B alloy suggested that the alloy has monophasic composition with hard magnetic $Nd_2Fe_{14}B$ phase as a composition. The mean crystal grain size was determined by using XRD analysis tend to approach nanocrystalline alloy powder, while the microstructure and composition of the alloys were analyzed using the SEM-EDX. The magnetic properties was determined using the Vibrating Sample Magnetometer (VSM) which indicates that the sample with 40.6 %wt Nd was able to achieve the optimal magnetic state with highest remanence of 446.50 G and BHmax of 17.83 kGOe. Despite the high remanence result, the coercivity and BHmax value of the third sample was still lower than the commercial's but has adequate potential value.

References

[1] Jan F. Herbst, John J. Croat, and Frederick E. Pinkerton 1984 *Physical Review B* **29** (7)
[2] Walter T. Benecki 2013 Magnetics Conference February 7-8 Orlando, Florida
[3] Masato Sagawa, Setsuo Fujimura, Hitoshi Yamamoto, and Yutaka Matsuura 1984 *IEEE Transactions on Magnetics* **20** (5) 1584-1589
[4] Ishfaq Ahmad Shah, Tahir Abbas, Zaka Ullah, Najam ul Hassan, Abdur Rauf, Kaleem Ullah, Shahzad Naseem 2015 *American Journal of Physics* **8** (4) 185-190
[5] M. Kazmerczak, K. Pawlik, P. Pawlik, P. Gebara, A. Przybyl, J.J. Wyslocki 2015 *Acta Physica Polonica* A **128**
[6] J. M. D. Coey 2011 *IEEE Transactions on Magnetics* **47** (12) 4671
[7] M. Kamierczak, K. Pawlik, P. Pawlik, P. Gbara, A. Przyby, J.J. Wysocki 2015 *Acta Physica Polonica A* **128** (1)
[8] T. ˇ Z´ak 2007 Proceedings of the CSMAG'07 Conference, Koˇsice, July 9–12
[9] A. Grujić, T. Žák, N. Talijan, J. Stajić-Trošić, V. Ćosović, 2009, *Science of Sintering* **41** 337-345

Acknowledgments

This work was supported and as part of Magnetic Research Group of PPF-LIPI. The author is grateful to Utami Sastramihardja B.Eng (Hons) MAusIMM for the editing and helpful comments.

IOP Conf. Series: Materials Science and Engineering **214** (2017) 012009 doi:10.1088/1757-899X/214/1/012009

Effect of Sintering Temperature to Physical, Magnetic Properties and Crystal Structure on Permanent Magnet BaFe12O19 Prepared From Mill Scale

Ramlan[1], Muljadi[2*], Priyo Sardjono[2], Fakhili Gulo[3], Dedi Setiabudidaya[1]

[1]Physics Department FMIPA Sriwijaya University,Palembang Indonesia
[2]Research Center for Physics, Indonesian Institute of Science
[3]S3 MIPA Sriwijaya University, Palembang Indonesia

[*]Email: muljadi2002@yahoo.com

Abstract. Permanent magnet of Barium hexa Ferrite with formula $BaFe_{12}O_{19}$ has been made by metallurgy powder method from raw materials : Barium carbonate ($BaCO_3$ E-merck) and Iron Oxide (Fe_2O_3 from mill scale). Both of raw materials have been mixed with stoichiometry composition by using a ball mill for 24 hours. The fine powder obtained from milling process was formed by using a hydraulic press at pressure 50 MPa and continued with sintering process. The sintering temperature was varied : 1150°C, 1200°C, 1250°C and 1300°C with holding time for 1 hour. The sintered samples were characterized such as : physical properties (bulk density , porosity and shrinkage), magnetic properties (flux density, remanence, coercivity and magnetic saturation) by using VSM and crystal structure by using XRD. According characterization results show that the crystal structure of $BaFe_{12}O_{19}$ does not change after sintering process, but the grain size tends to increase. The optimum condition is achieved at temperature 1250°C, and at this condition, the sample has characterization such as : bulk density =4.35 g/cm^3, porosity= 1.03% and firing shrinkage= 11.63%, flux density = 681.1 Gauss, remanence (σr) = 20.78 emu/g, coercivity (Hc) =2058 Oe and magnetic saturation (σs) 45.16 emu/g.

1. Introduction

Barium hexaferrites with formula $BaFe_{12}O_{19}$ are well known as the hexagonal M-type hard ferrites, They are also called as ferrite magnets and could not be easily replaced by any other magnets. Barium ferrite has relatively a high curie temperature, a high coercive force and a high magnetic anisotropy field [1,2]. The hexagonal M-type hard ferrites have attracted much attention as the most widely used permanent magnets, which account for about 90% of the annual production of the permanent magnets due to the good combination of high magnetic properties, chemical stability and low cost [3]. The properties of extrinsic material Barium hexa Ferrite depend on kinds of raw materials or impurity, composition, grain size and kinds of processing route [4]. Barium hexaferrites has a hexagonal crystal structure with lattice parameters a = 5.888 Å and c = 23.228 Å[5]. This material has many applications in magnetic recording media, ferrofluids, sensors, microwave absorbing materials, motors and generators [6]. Many methods of synthesis Barium hexa Ferrite have been developed with a low production cost, the most popular method is the technique of powder metallurgy or conventional method [7]. Raw materials in the form of oxides such as iron oxide and barium carbonate used to manufacture Barium Hexaferrite through conventional method [5]. This method consists of heating the mixture of barium carbonate and Fe_2O_3 precursors at temperature as high as 1100°C, in this condition, $BaCO_3$ will be changed to BaO, and BaO reacts with Fe_2O_3 to form $BaO.6Fe_2O_3$ or $BaFe_{12}O_{19}$, in accordance with the following equation [8]:

$$BaCO_3 + 6Fe_2O_3 \rightarrow 6Fe_2O_3 + BaO + CO_2 \quad (1)$$
$$6Fe_2O_3 + BaO \rightarrow BaO.6Fe_2O_3 \quad (2)$$

To obtain a fine powder BaO. $6Fe_2O_3$ with sizes approaching single domain particles are about < 1 μm [7], the powder should be milled in certain time. The reported theoretical calculated coercive force, saturation magnetization and curie temperature values for pure and single domain barium hexaferrite was 6700 Oe, 72 emu/g and 450°C, respectively [3, 9]. It is very difficult to obtain ultrafine and monodispersed particles that theoretical magnetic properties can be achieved. Janasi reported that the voercivity value of Ba-Ferrite prepared by conventional method was achieved about 162 kA/m (2,048 kOe), while Ba-Ferrite that prepared through cooprecipitation process has coercivity value about 444 kA/m [9].

The raw materials sources are widely found in nature such as hematite for Fe_2O_3 and barite resources as a source of Barium Carbonate[10]. The chemical routes for the production of permanent magnet Ba-Ferrite is needed more expensive raw materials compared to conventional methods. Many sources Fe_2O_3 apart from nature that they can be used as raw material for producing of permanent magnets Ba - Ferrite, one of which is the mill scale. A mill scale is a very attractive industrial waste due to its richness in iron oxide (about = 80 % Fe) [11]. The total content of iron oxide in the mill scale is quite high, so the mill scale can be used as an alternative raw material to produce barium hexa-ferrite. The need of iron oxide for the manufacture of barium hexa ferrite with stoichiometry composition is approximately 85.71% mole [12]. The purpose of this research is to determine the effect of the sintering temperature on physical, magnetic properties and microstructure of the permanent magnet $BaFe_{12}O_{19}$ prepared by using raw materials mill scale.

2. Experiments

Mixture of iron oxide from mill scale and barium carbonate $BaCO_3$ (99% purity) powders were used for synthesis of $BaFe_{12}O_{19}$ with stoichiometry composition (mole ratio BaO : Fe_2O_3 = 1 : 6). Both of raw materials have been mixed by using a ball mill for 24 hours. The mixed fine powder was measured by using TG/DTA simultans analyzer to know a calcination temperature (Tc). After the milling process, the powder was heated at a temperature Tc of about 1000°C for 2 hours with a heating rate of 10 °C/min using a Thermolyne furnace. The calcined powder was milled for 10 hours by using ball mill, then the sample powder was formed by using a hydraulic press at pressure 50 Mpa and continued with sintering process. The sintering temperature (Ts) was varied : 1150°C, 1200°C, 1250°C and 1300°C with heating rate of 5 °C/minute and holding time for 2 hour at Ts. The sintered samples were characterized such as : physical properties (bulk density , porosity and shrinkage), magnetic properties (flux density, remanence, coercivity and magnetic saturation) by using VSM and crystal structure by using XRD.

3. Results and Discussion

Table 1 shows chemical compositions of mill scale from Medan Nort Sumatera as obtained from XRF analysis and figure 1 shows the XRD patterns of mill scale.

Table 1. Chemical compositions of mill scale

Compunds	Fe (total)	FeO	CaO	SiO_2	MnO
% weight	82.17	66.79	0.36	0.16	0.17

The iron content in mill scale is enough high about 82.17 % wt, while the amounts of other elements are less than 0.4 % . The XRD pattern (figure 1) of the mill scale as raw material shows the presence of hematite (Fe_2O_3), magnetite (Fe_3O_4) and wustite (FeO) phases.

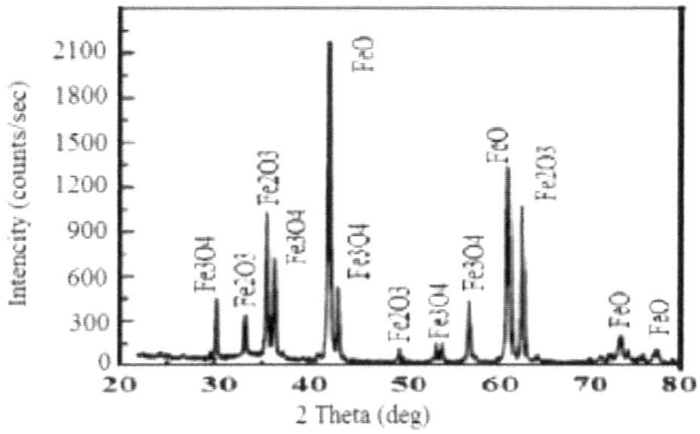

Figure 1. XRD patterns of mill scale.

Based on the results of XRD analysis of mill scale that raw materials can be used as raw material for the manufacture of barium hexa-ferrite material as a source of Fe_2O_3. FeO and Fe_3O_4 phase exist in the mill scale during the combustion process (calcination) will be oxidized to Fe_2O_3 and reacts with BaO to form $BaO.6Fe_2O_3$. The mixed raw materials ($BaCO_3$ and mill scale) were analyzed by using TG/DTA from room temperature to 1000°C under O_2 gas flow, and the result is shown at Figure 2. Analysis of TG / DTA of a mixture of $BaCO_3$ and mill scale is done only to a temperature of 1000°C, to determine the temperature phase formation $BaO.6Fe_2O_3$. The temperature that will be used as the basis for the calcination process. The result of DTA curve shows that there are 3 peaks endothermic, the first peak at 720°C indicates oxidation of mill sacle from Fe_3O_4 to Fe_2O_3. The second peak at 820°C is a decomposition process of $BaCO_3$ to BaO and CO_2. This can be proved on the TG curve that it has occurred fairly large mass reduction from temperature 700°C to 880°C. This mass reduction due to the decomposition of $BaCO_3$ to BaO and CO_2. The third DTA peak at 980°C indicates a solid reaction between BaO and Fe_2O_3 to form BaO.6 Fe_2O_3. The result of DTA curve shows that there are 3 peaks endothermic, the first peak at 720°C indicates oxidation of mill scale from Fe_3O_4 to Fe_2O_3. The second peak at 820°C is a decomposition process of $BaCO_3$ to BaO and CO_2, The third DTA peak at 980°C indicates a solid reaction between BaO and Fe_2O_3 to form BaO.6 Fe_2O_3.

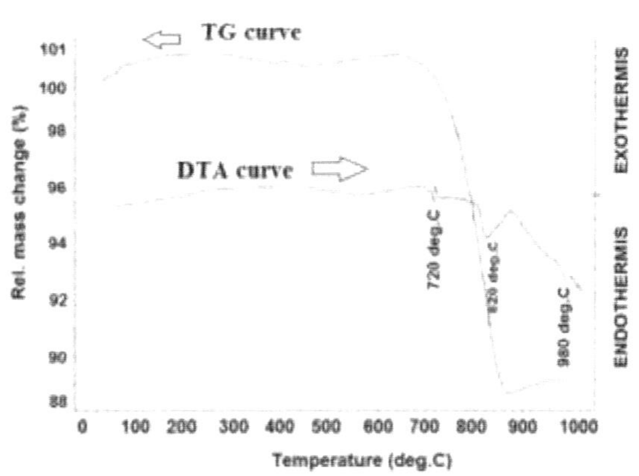

Figure 2. TG/DTA curve of mixed raw materials ($BaCO_3$ and mill scale).

According TG / DTA result, it can be determined the calcination temperature of mixed raw materials ($BaCO_3$ and mill scale) is 1000°C. The XRD results from a mixture of mixed raw materials ($BaCO_3$ and mill scale) which it has been calcined 1000°C are shown in figure 3. Based on matching of the XRD patterns (experiment) with the JCPDS card of $BaO.6Fe_2O_3$ (00039-1433) and JCPDS card Fe_2O_3 (87-1164) can be identified of two phases, namely Fe_2O_3 phase and $BaO.6Fe_2O_3$ phase. Based on matching of the XRD patterns (experiment) with the JCPDS card of $BaO.6Fe_2O_3$ (00039-1433) and

JCPDS card Fe_2O_3 (87-1164) can be identified of two phases, namely Fe_2O_3 phase and $BaO6Fe_2O_3$ phase.

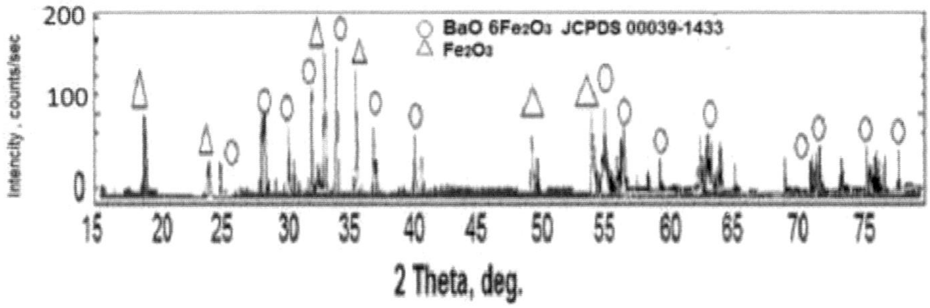

Figure 3. XRD patterns of sample powder after calcination at 1000°C.

Based on matching of the XRD patterns (experiment) with the JCPDS card of $BaO.6Fe_2O_3$ (00039-1433) and JCPDS card Fe_2O_3 (87-1164) can be identified of two phases, namely Fe_2O_3 phase and $BaO.6Fe_2O_3$ phase. Stages after calcination is the process of forming and sintering process, which sintering is a step compaction at high temperatures. Temperature sintering is done at a temperature above the calcination temperature and still below the melting temperature of the material. Determination of the optimal sintering temperature based on the analysis of measurement of the physical properties (bulk density and shrinkage) as well as magnetic properties. The sample powder has been pressed at 50MPa to form pellets and have been continued with the sintering process at temperature : 1150, 1200, 1250 1nd 1300 °C. The sintered samples were measured bulk density, porosity and firing shrinkage to evaluation densification results as function sintering temperature. The results of measurement of bulk density, porosity and firing shrinkage are shown at Figure 4 and 5.

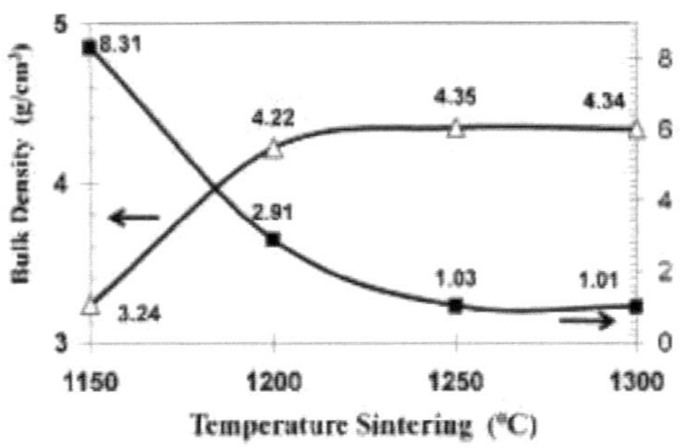

Figure 4. Bulk density and porosity curves as function sintering temperature.

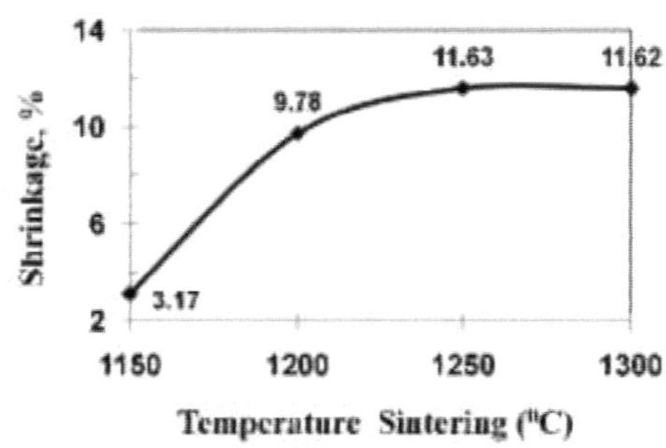

Figure 5. Firing shrinkage curves as function sintering temperature.

Figure 4 shows that bulk density values tend to increase sharply from the temperature 1150°C to 1200 °C. This is a beginning of the sintering process, where the grains closer together and begin a reduction of pores and also it has seen a sharp decline for porosity values. The changes in bulk density values from the temperature of 1200 °C to 1300 °C is not too big and tend to constant, it is due to end the sintering process. Also, porosity value from 1250 °C to 1300 °C tend to constantly. The evaluation of firing shrinkage value is important because the data of firing shrinkage can be used for manufacturing of real product. The firing shrinkage value as seen at figure 5 tends to increasing sharfly from temperature 1150 °C to 1250 °C , and became constant from temperature 1250°C to 1300°C. This is because the process densification during the sintering process has been achieved with a value of firing shrinkage about 11.63% . The XRD resutls of sintered samples are shown at figure 6.

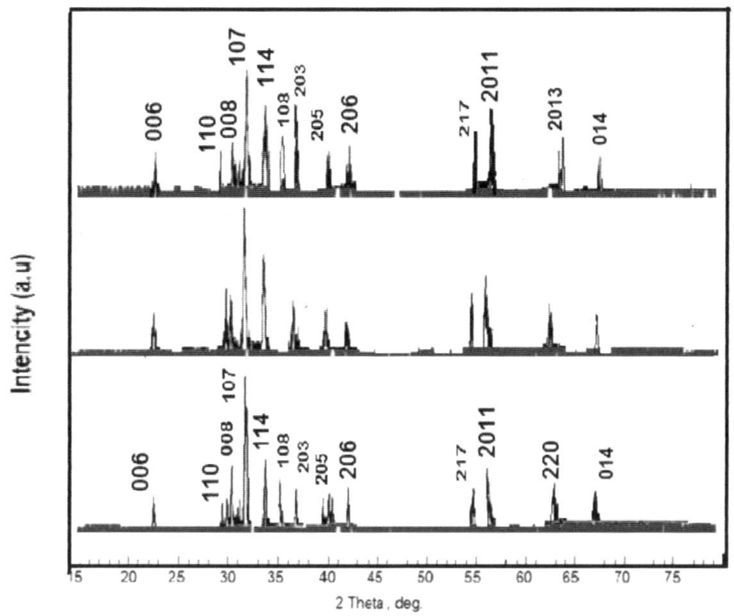

Figure 6. XRD patterns of sample after sintering at temperature 1200 °C, 1250°C and 1300°C.

The XRD patterns from expeimence for sample after sintering 1200 °C, 1250°C and 1300°C are matched with JCPDS card of $BaO.6Fe_2O_3$ (00039-1433) and it is found single phase of $BaO.6Fe_2O_3$ with hexagonal structure. In this case, the crystal structure with variations of sintering temperature was unchanged and still hexagonal $BaO.6Fe_2O_3$. Based on XRD patterns at figure 6, it can be calculated crystallite size of $BaO.6Fe_2O_3$ by using Sherrer formula [10]. Based on XRD patterns at figure 6, it can be calculated crystallite size of $BaO.6Fe_2O_3$ by using Sherrer formula [10]. Based on XRD patterns at figure 6, it can be calculated crystallite size of $BaO.6Fe_2O_3$ by using Sherrer formula [13]. Here is the Scherrer equation to calculate the crystallite size :

$$L = \frac{k\,\lambda}{B\,cos\theta} \tag{3}$$

The crystallite size (L in nm) of $BaO.6Fe_2O_3$ phase were calculated from most intense peaks in XRD patterns. Here, "k" is Scherrer constant and its value is '0.94' for our material, 'B' is full width at half maximum (FWHM) of respective peaks, 'θ' is Bragg's angle and 'λ' is wavelength of Cu-k α radiation (1.5406 Å). Calculation of the crystallite size of $BaO6Fe_2O_3$ phase was done at $2\,\theta = 32.20^o$. The results of calculationn of the crystallite size of $BaO6Fe_2O_3$ phase at sample after sintering process are shown at table 2.

Table 2. Value of crystallite size of $BaO.6Fe_2O_3$ phase

Sample	crystallite size of $BaO.6Fe_2O_3$ (nm)
After sintering 1200 °C	36.46
After sintering 1250 °C	41.94
After sintering 1300 °C	75.80

According the result of calculated crystallite size (Table 2), that crystal size is quite small in range 36.46 to 75.80 nm. Table 2 shows that the crystallite size of $BaO.6Fe_2O_3$ phase increases with increasing of sintering temperature, this means that there is a grain growth during the sintering

IOP Conf. Series: Materials Science and Engineering **214** (2017) 012009 doi:10.1088/1757-899X/214/1/012009

process, in which the crystal size after sintering at 1300°C be nearly doubled, this is caused during the sintering process of grain growth occurs. The large grain growth is not as expected, generally can degrade the magnetic properties of the material. The measurement of magnetic properties was done by using Gaussmeter to know the flux magnetic value and by using VSM to know the value of Remanence (Mr), Coecivity force (Hc), Magnetic saturation (Ms) and Energy product (BHmax) and the results are shown in figure 7 and 8. According to figure 7, The flux magnetic value tends to increase from sintering temperature 1150 °C to 1250°C, and the flux magnetic value tends to decrease from 1250 °C to 1300 °C . The highest value of flux magnetic is achieved about 681.10 Gauss at samples with sintering temperature 1250°C. The decreasing of the flux magnetic value of the sample after sintering 1300°C is due to the growth of crystallite value.

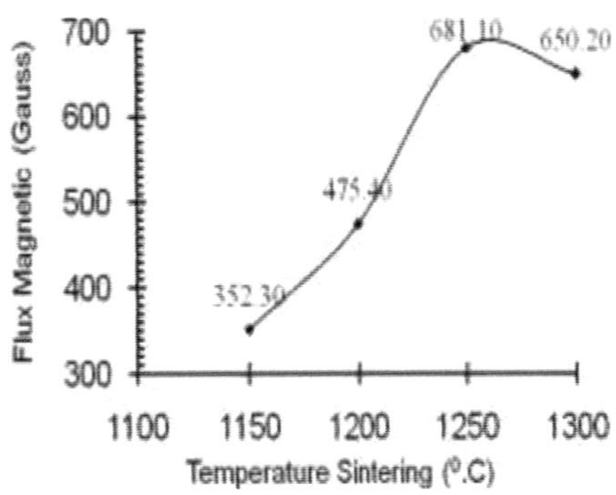

Figure 7. Flux Magnetic curve as function sintering temperature.

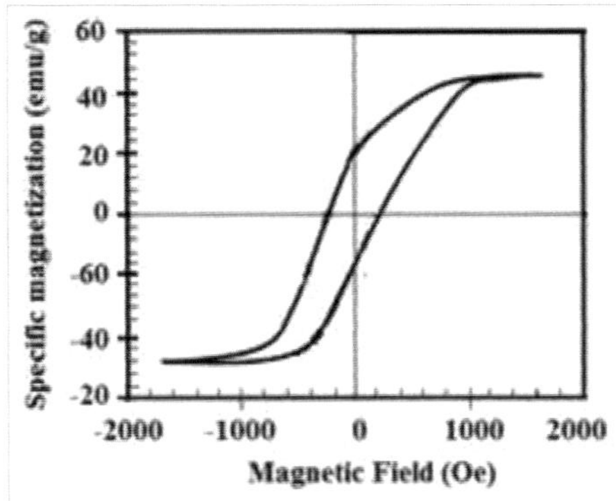

Figure 8. The VSM curve for sample afters sintering at 1250°C.

This is because the size of the crystals produced after sintering 130 exceeds the size of single domain. If the crystal size is larger than the size of single domain Ba hexa ferrite, it will affect the magnetic properties. The VSM results of the sample after sintering 1250°C is shown in Figure 8, it can be found value of magnetic properties such as : remanence (Mr) = 20.78 emu/g, coercivity (Hc) =2058 Oe and magnetic saturation (Ms) 45.16 emu/g. The results of this experiment shows that the magnetic properties (flux magnetic value or remanence) of permanent magnet $BaO6Fe_2O_3$ made from millscale is 681.10 Gauss and has density value about 4.35 g/cm^3. This research achieve a higher yield for the value of Flux magnetic or remanence and bulk density when compared with studies of other authors [14] that use the same method.

4. Conclusion

A permanent magnet $BaO.6Fe_2O_3$ with crystal structure hexagonal has successfully made from raw materials of mill scale, with the optimal sintering temperature 1250°C and its properties as follows: bulk density = 4.35 g/cm^3, porosity= 1.03% and firing shrinkage= 11.63%, flux density = 681.1 Gauss, remanence (σr) = 20.78 emu/g, coercivity (Hc) =2058 Oe and magnetic saturation (σs) 45.16 emu/g. Sintering temperature does not influence of crystal structure, but it can give effect on growth of crystallite size.

5. References

[1] Verma, OP. Pandey, Puneet Sharma 2000 *Indian Journal of Engineering and materials Sciences* **7** 364-369

[2] R. Nowosielski, R. Babilas, G. Dercz , L. Pajk, J. Wrona, 2007, *Archives of Materials Science and Engineering* **28** (12) 735-742.
[3] A. Mahgoob and A. Y. Hudeish 2012 *American Academic & Scholarly Research Journal* **4** (5)
[4] Mokhtar, Mustaffa HJ. Abdullah & Sahrim HJ. Ahmad, 2012, *Sains Malaysiana* **41**(9) 1125 131
[5] Li Jie, Zhang Huai-Wu, Li Yuan-Xun,Liu Ying-Li and Ma Yan-Bing 2012 *Chin. Phys. B* **21** (1) 017501-1 - 017501-4
[6] M.J. Molaei, A. Ataie, S. Raygan, S.J. Picken, E. Mendes, F.D. Tichelaar 2012 *Powder Technology* **221** 292–295
[7] Rahmat Doni W, Azwar Manaf, P. Sardjono 2013 *International Journal of Basic & Applied Sciences IJBAS-IJENS* **13** (04) 65-68
[8] M.N. Shipko, V.G. Kostishyn1, D.N. Chitanov, 2015, *Journal of Nano - and Electronic Physics* **7** (4)04046-1 04046-4
[9] S.R. Janasi, M. Emura, F.J.G. Landgraf, D. Rodrigues 2002 *Journal of Magnetism and Magnetic Materials* **238** 168–172
[10] M.J. Molaei, A. Ataie, S. Raygan, S.J. Pickenb, E. Mendes, F.D. Tichelaar 2012 *Powder Technology* **221** 292–295
[11] N. M. Gaballah, A. F. Zikry, M. G. Khalifa, A. B. Farag, N. A. El-Hussiny, M. E. H. Shalabi 2013 Journal of Inorganic Non-Metallic Materials **3** 23-28
[12] R. Nowosielski, R. Babilas, J. Wrona 2007 Journal of Achievements in Materials and Manufacturing Engineering **20** 307-310
[13] Ishfaq Ahmad Shah, Tahir Abbas, Zaka Ullah1, Najam ul Hassan, Abdur Rauf, Kaleem Ullah,
[14] Shahzad Naseem 2015 *Armenian Journal of Physics* **8** (4) 185-190
[15] Arie Fiandimas, Azwar Manaf 2003 *Jurnal Sains Materi Indonesia* **5** (1) 45-50

IOP Conf. Series: Materials Science and Engineering **214** (2017) 012010 doi:10.1088/1757-899X/214/1/012010

Novel Solar Photocatalytic Reactor for Wastewater Treatment

Sutisna[a,b], M Rokhmat[a], E Wibowo[a], R Murniati[a], Khairurrijal[a], and M Abdullah[a*]

[a]Department of Physics, Faculty of Mathematical and Natural Sciences, Bandung Institute of Technology, Jalan Ganeca 10 Bandung 40132 Indonesia.
[b]Department of Physics, Faculty of Mathematical and Natural Sciences, Jember University, Jalan Kalimantan 37 Jember Indonesia.

* E-mail: din@fi.itb.ac.id

Abstract. A new solar photocatalytic reactor (photoreactor) using TiO_2 nanoparticles coated onto plastic granules has been designed. Catalyst granules are placed into the cavity of a reactor panel made of glass. A pump is used to circulate wastewater in the photoreactor. Methylene blue (MB) dissolved in water was chosen as the wastewater model. The performance of the photoreactor was evaluated based on changes in MB concentration with respect to time. The photoreactor showed a good performance by degrading 10 L of MB solution up to 96.54% after 48 h of solar irradiation. The photoreactor was scaled up by enlarging the panel area to twice its original size. The increase in the surface area of the reactor panel and therefore of the mass of catalyst granules and reactor volume led to a three-fold increase of the photodegradation rate. In addition, the MB degradation kinetics were also studied. Data analysis confirmed the applicability of the pseudo-first-order Langmuir-Hinshelwood model. The proposed photoreactor has great potential for use in large-scale wastewater treatment.

1. Introduction

Currently, advanced oxidation processing is probably the most widely used method in wastewater treatment to degrade organic pollutants that are difficult to degrade with conventional methods due to their high chemical stability [1]. A number of oxidation processes, using oxidants such as TiO_2/UV, H_2O_2/UV, photo-Fenton and ozone, can be used for this purpose. The essence of this approach is to produce hydroxyl radicals ($^{*}OH$) that destroy pollutants in the wastewater. The attack of the pollutants by hydroxyl radicals is not very selective, which is very useful in wastewater treatment. The major drawback of advanced oxidation processes (AOPs) is their high cost when using UV radiation from lamps or ozone production [1]. Therefore, researches now are more focused on AOPs that can be driven by solar irradiation, i.e. light with a wavelength longer than 300 nm and heterogeneous catalyst with TiO_2/UV [2-4].

Designs of solar photoreactors using TiO_2 as a photocatalyst, whether directly dispersed (slurry photoreactor) or immobilized on a solid support (immobilized photoreactor), are still being developed [2,5]. In a slurry photoreactor, the catalyst particles are freely dispersed in the fluid phase. This is a simple technique resulting in a large surface area for adsorption and reaction [6]. However, the main drawback of using nano- or micro-particles in suspension is the requirement of post-treatment for separation and recycling of the catalyst, potentially making the treatment more complex and expensive [7]. An immobilized photoreactor is a good choice for practical application to overcome the deficiencies of a slurry photoreactor. Other than that, the immobilized TiO_2 can be used repeatedly and is easy to reproduce [8].

Lately, various designs of immobilized photoreactors using sunlight have been reported [2,7]. The photoreactors, although quite promising for large-scale application, use an immobilization technique and design that are not simple, causing high-cost operation. In addition, most of the reported photoreactor designs use TiO_2 catalyst coated onto the reactor body, so it is impractical to reproduce the catalyst.

This paper proposes a solar photoreactor design that is simpler than those previously reported by other researchers [2,7], both with respect to the installation and the catalyst immobilization technique used. The TiO_2 nanoparticles are deposited on the plastic granules using a combination of electrostatic and thermal treatment. The proposed solar photoreactor itself consists of a reactor panel made of glass. Catalyst granules are loaded into the cavity of the reactor panel where the wastewater is circulated. The reactor panel is mounted obliquely and the wastewater is circulated using a pump. The immobilized catalyst can be separated from the reactor body and facilitate catalyst reproduction. An increase in photoreactor efficiency (scale-up) by expanding the size of the reactor panel was observed.

2. Materials and method

2.1. Materials

In this study, MB purchased from Sakura, Bandung, Indonesia was used as the pollutant model. Technical grade anatase TiO_2 (Bratachem, Indonesia) was used as the photocatalyst (average particle size: 166 nm, density: 4.32 g cm^{-3}, BET surface area: 7.005 m^2 g^{-1}, average pore diameter: 55.31 nm, pore volume: $9.68x10^{-2}$ cc/g). Compared with pure TiO_2, this nanoparticle is much less expensive and has a fairly wide absorption band (from about 200 to 800 nm) [9]. Low-density transparent granules (density: 0.817 g cm^{-3}, average diameter: 3.5 mm, melting point ranging from 130-170 °C and heat deflection temperature/HDT: 12 °C (at 0.45 MPa)) made of Cosmoplene (from Polyolefin Company Pte. Ltd., Singapore), were used as supporting material for the TiO_2 nanoparticles.

2.2. Catalyst preparation

A very simple coating method was used to coat the surface of the plastic granules with TiO_2 nanoparticles. This coating method is a combination of electrostatic and thermal treatment mechanisms. The coating process was initiated by generating electric charges on the plastic granules by stirring them in a glass container. The friction among the plastic granules and the glass wall produces electron transfer from the glass to the plastic. TiO_2 powder (with a mass ratio of 1: 100) was sown onto the charged plastic granules and then stirred. The electric induction causes the TiO_2 particles to attach themselves to the surface of the granules. In the next step, the TiO_2-covered granules were heated in a stainless still cylinder that was rotated inside an electric oven at a temperature of 100 °C for 75 minutes. This process was done to stick TiO_2 particles strongly to the surface of the granules. The coating process was terminated by washing the granules using distilled water and drying.

2.3. Photoreactor design

Two parameters must be considered in designing a photoreactor: the ratio of reactor volume to total volume and amount of illuminated catalyst [10]. Based on these parameters, the proposed photoreactor is expected to improve the efficiency of the photocatalytic reaction. The hollow rectangular reactor panel is made of transparent glass at the top and a mirror at the bottom. The use of the mirror is meant to improve the efficiency of photon absorption by the catalyst. Photons reflected by the mirror illuminate the bottom of the catalyst granules so that the illuminated surface area of the catalyst granules is improved. The catalyst granules (a) were loaded into the cavity of the reactor panel and kept inside by installing a filter at the bottom edge of the reactor panel. A control channel (b) was mounted at the center of the reactor panel to keep the waste solution from overflowing, which would cause flow resistance by the catalyst granules. Wastewater was supplied to the reactor panel through a pipe (c) with tiny channels to emit water. A test solution was flowed between the catalyst granules, going out of the reactor panel through a pipe at the bottom's edge (d). The reactor panel was made in two sizes: 48 x 56 x 1 cm^3 (small) and 68 x 76 x 1 cm^3 (large). An illustration of the reactor panel is shown in Figure 1.

Figure 1. Installation of the reactor panel in the solar photoreactor.

A solar photoreactor is a system that consists of a reactor panel and a water circulator. Wastewater is circulated from a container (e) through the reactor panel using an electric pump. In the reactor panel, the wastewater will come in contact with the catalyst granules creating a photocatalytic reaction.

2.4. Performance test of the proposed photoreactor
The performance of the photoreactor was tested in a photodegradation experiment using 10 L MB compound. In this experiment, the decomposition rate of the polluted water was compared between a control reactor (without catalyst) and the photoreactor (with catalyst). All of the experiments were carried out in Bandung, Indonesia at latitude = -6.9° and longitude = 107.6°. Different geographical locations have different solar intensity. The experiments were run in May-June 2015 between 08:00 o'clock and 16:00 o'clock. During the photocatalytic process, the solar illumination changed with time at an average intensity of about 622 W/m^2. An attempt was made to improve the performance of the photoreactor by enlarging the size of the reactor panel. An increase in the area of reactor panel increases the reactor's volume and the number of catalyst granules it can contain, which was expected to result in an increase in the efficiency of the photoreactor.

2.5. Characterization methods
The surface area and the pore size distribution of the TiO_2 were analyzed with N_2 physisorption at 77.35 K using a Nova 1200e surface area analyzer (Quantachrome Instruments). During the degradation process, the absorbance of the test solution was measured to determine the reduction of the MB compound. Absorbance measurement was done using a UV-Vis spectrophotometer (Mikropack Brands, NanoCalc 2000, Florida, USA). The distribution of sunlight intensity was measured using a TM-206 solar power meter (TENMARS, Taiwan).

3. Results and discussion

3.1. Optimization of catalyst mass in the reactor panel
Before performing the photocatalytic experiment, optimization of the catalyst mass in the reactor panel was performed. As previously explained, the presence of the catalyst granules will

impede the flow of wastewater in the reactor panel. In certain circumstances, the retarding force will cause the reactor volume to exceed the capacity of the reactor panel so that the water will overflow through the back of the reactor panel. For the small reactor panel, the masses of the catalyst were varied at 300, 350, 400, 450, 500, 550, and 600 g. Figure 2 represents the optimization of the catalyst mass for the small reactor panel. It can be seen that a catalyst mass of 550 g was the optimum mass of catalyst in the reactor panel. When the catalyst mass exceeded 550 g, the water overflowed out of the reactor panel.

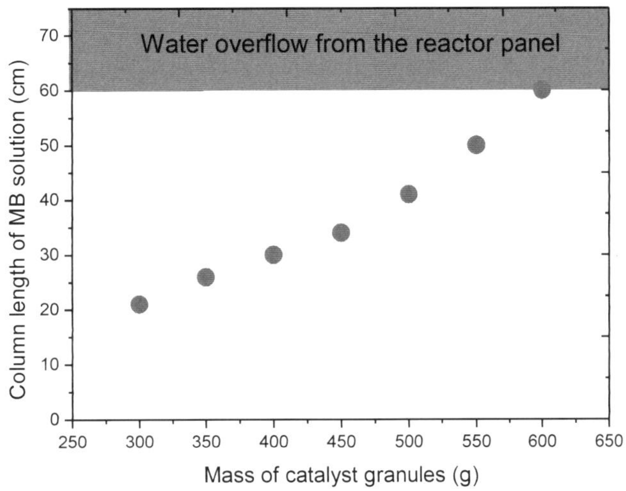

Figure 2. Effect of catalyst granule mass on column length of MB solution in the reactor panel.

3.2. Decolorization of MB solution

The experiments were aimed at determining the effect of the photocatalyst on the decomposition rate of the MB. The degradation rate of the MB was compared between the photoreactor and a control reactor. In the control reactor, the MB solution was circulated through the reactor panel that was filled with pure plastic granules (without TiO_2 coating). The experiments were carried out from 08:00 o'clock to 16:00 o'clock each day. During the experiment, every 2 hours about 5 ml of test solution was sampled for absorbance measurement. The experimental parameters are listed in Table 1.

Table 1. Experiment parameters for MB degradation

Experiment parameters	Small reactor panel	Large reactor panel
Size of panel reactor	48x56x1 cm^3	68x76x1 cm^3
Initial concentration	25 mg L^{-1}	25 mg L^{-1}
Optimum mass of catalyst	550 g	900 g
Total volume	10 L	10 L
Reactor volume	1.25 L	3.25 L
Circulation rate	57.45 ml/s	57.45 ml/s
The tilt angle of the reactor panel	20°	20°

Figure 3 shows the change in color of the test solution before and after 48 h of treatment by the small photoreactor. In this experiment, the effect of the catalyst was evaluated by comparing the MB degradation in the photoreactor with the control reactor. In the control reactor, the reactor panel was filled with pristine plastic granules (uncoated TiO_2). Based on the UV-Vis absorption spectra of the test solution in the figure, it appears that there was a significant difference in the peak decline of the absorption spectra between the MB solution treated in the photoreactor (Figure 4b) and in the control reactor (Figure 4a). During 48 h of irradiation, the MB solution in the photoreactor was almost entirely reduced, indicated by the flat absorbance curve. Meanwhile, in the same time interval, the control reactor showed a much lower performance in degrading the MB compound via photolytic reaction. In addition, as shown in Figure 4b, the peak of the absorption spectra of the MB at 663 nm decreased considerably and a narrow blue-shift (from 663 to 651 nm) occurred with increasing irradiation time (from 0 to 48 h). Figure 5 shows the shift of the MB absorption spectra as a function of irradiation time. The shift of the absorption peaks toward greater energy often refers to an hypsochromic effect. This phenomenon is caused by a demethylation process on the N-methyl group of MB leading to a change of the MB structure [11].

Figure 3. MB photodegradation experiment using the proposed photoreactor: (a) initial condition, (b) after 48 h of irradiation.

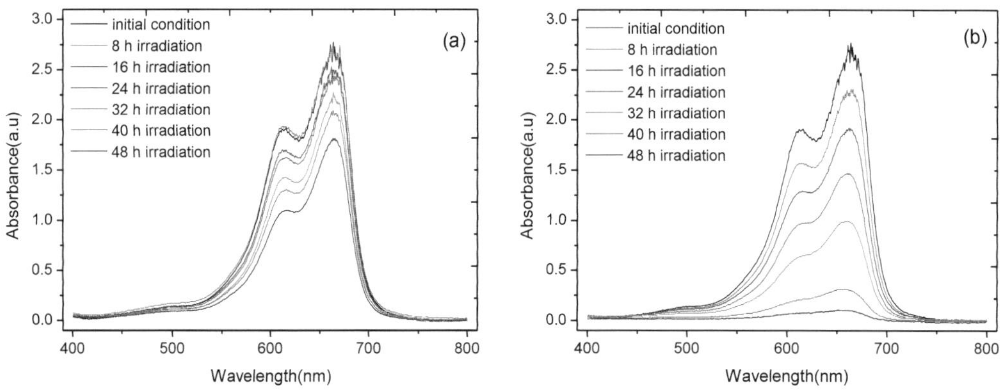

Figure 4. UV-Vis absorption spectra of MB during the degradation process (a) in the control reactor, and (b) in the small photoreactor.

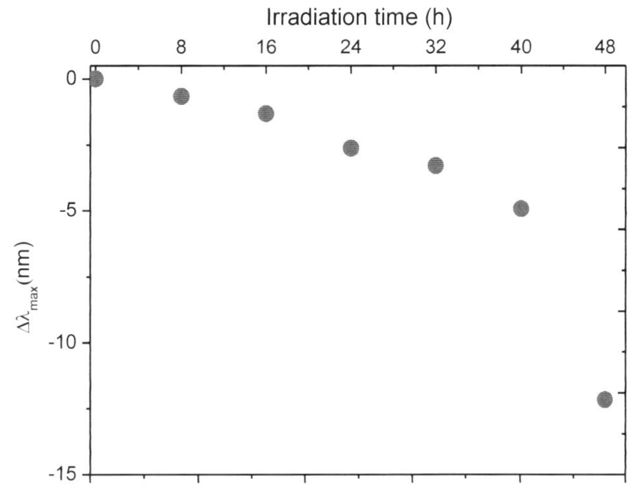

Figure 5. Blue shift (Δλ in nm) in the MB absorption band wavelength (initially centered at 663 nm) as a function of irradiation time.

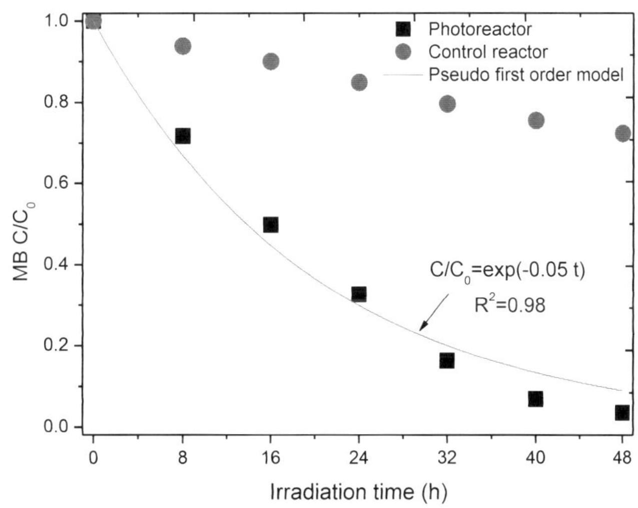

Figure 6. MB degradation in the small photoreactor. Experimental conditions: C_0 = 25 mg L^{-1}, total volume = 10 L, pH = 7.03.

The reduction of the MB concentration in the presence of sunlight, with and without catalysts, is shown in figure 6. The correlation of the kinetic data with the pseudo-first-order kinetic model was studied. As shown in Figure 6, the kinetics were well fitted with the pseudo-first-order Langmuir-Hinshelwood (L-H) model given by the following equation:

$$C/C_0 = \exp(-k_{app}t) \quad (1)$$

where C_0 is the initial concentration of MB, C is the concentration at time t and k_{app}= 0.05 h^{-1} is the apparent first-order constant [12]. There are two mechanisms of MB degradation, direct photolysis and photocatalysis. Based on Figure 6, the proposed photoreactor showed an excellent performance with the ability to degrade 96.54% of MB for 48 h irradiation. MB degradation in the control reactor (photolytic reaction) took place much more slowly than that in the photoreactor. In the same time interval, it only decomposed MB at 27.71%. From this result it can be concluded that the presence of the catalyst played a very significant role in degrading the MB compound.

3.3. Scaled-up photoreactor

In this study, an increase in the performance of the proposed photoreactor was attempted by enlarging the size of the reactor panel. A larger size of the reactor panel will have an impact on the reactor's volume and its capacity to accommodate the catalyst granules. With the addition of 20 cm to both the length and the width of the panel reactor, the optimum mass of catalyst granules that could be accommodated in the reactor panel increased to 900 g, while the reactor volume increased to 3.25 L. Optimization of the catalyst mass in the large reactor panel was performed in the same manner as for the small reactor panel, where the catalyst mass was varied from 500 to 1000 g. The increase in size of the reactor panel, and therefore the reactor volume and the mass of catalyst granules, successfully increased the photodegradation rate, which was more than tripled compared to the previous rate. As shown in Figure 7, the larger photoreactor took 16 h to degrade 97.46% of MB compound. Meanwhile, in order to degrade 96.54% of MB, the smaller photoreactor took 48 h, which is three times longer. By comparing the rate constants of both photoreactors, it was found that the rate constant of the larger reactor (0.117 h^{-1}) was more than twice that of the small reactor (0.05 h^{-1}). This is a very promising result towards implementing this type of photoreactor as an industry-scale wastewater treatment alternative. Besides expanding the size of the reactor panel, scaling up the photoreactor can also be done by increasing the number of reactor panels. This will be part of our future work.

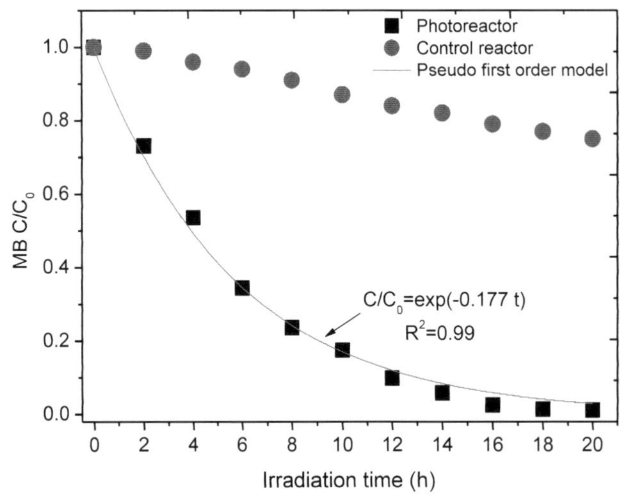

Figure 7. MB degradation in the large photoreactor. Experimental conditions: C_0 = 25 mg L^{-1}, total volume = 10 L, pH = 7.03.

4. Conclusion

The present research demonstrated that the proposed photoreactor with TiO_2 nanoparticles coated onto plastic granules as catalyst, can effectively degrade MB under sunlight irradiation. The performance of the photoreactor was increased by enlarging the size of the reactor panel. A three-fold increase in performance of the photoreactor was achieved by doubling the reactor panel area. The scaling-up of the photoreactor can still be improved by arranging several reactor panels in series. The kinetics of MB degradation were also investigated. Data analysis showed a match with the pseudo-first-order

Langmuir-Hinshelwood model. This study showed that the proposed photoreactor has great potential to be implemented in large-scale wastewater treatment.

5. References

[1] Blanco-Galvez J, Fernandez-Ibanez P and Malato-Rodriguez S 2007 *ASME J. Sol. Energy Eng*
[2] Khalilian H, Behpour M, Atouf V and Hosseini S N 2015 *Solar Energy* **112** 239
[3] Zahedi F, Behpour M, Ghoreishi S M and Khalilian H 2015 *Solar Energy* **120** 287
[4] Santiago D E, Espino-Estévez M R, González G V, Arana J, González-Díaz O and Dona-Rodríguez J M 2015 *Applied Catalysis A: General* **498** 1
[5] Spasiano D, Marotta R, Malato S, Fernandez-Ibanez P and Di Somma I 2015 *Applied Catalysis B: Environmental* **170-171** 90
[6] Du P, Carneiro J T, Moulijn J A and Mul G 2008 *Appl. Catal. A* **334** 119
[7] Byrne J A, Fernandez-Ibanez P A, Dunlop P S M, Alrousan D M A and Hamilton J W J 2011 *International Journal of Photoenergy* **2011** Article ID 798051 12 pages
[8] Sutisna, Rokhmat M, Wibowo E, Murniati R, Khairurrijal and Abdullah M 2015 *Advanced Material Researches* **1112** 149
[9] Abdullah M, Nurmawati I, Subianto H, Khairurrijal and Mahfudz H 2010 *Jurnal Nanosains & Nanoteknologi* **3** 10
[10] Keane D A, McGuigan K G, Ibáñez P F, Polo-López M I, Byrne J A, Dunlop P S M, O'Shea K, Dionysiou D D and Pillai S C 2014 *Catal. Sci. Technol.* **4** 1211
[11] Zuo R, Du G, Zhang W, Liu L, Liu Y, Mei L and Li Z 2014 *Advances in Materials Science and Engineering* **2014** Article ID 170148 7
[12] Souzanchi S, Vahabzadeh F, Fazel S and Hosseini S N 2013 *Chemical Engineering Journal* **223** 268

Acknowledgments
This work was supported by the Ministry of Research and Higher Education, Republic of Indonesia through Hibah Unggulan Perguruan Tinggi No. 310y/I1.C01/PL/2015 and Hibah Disertasi Doktor No. 187Z/UN25.3.1/LT/2016.

Effect of The Addition of PEG and PVA Polymer for Gel Electrolytes in Dye-Sensitized Solar Cell (DSSC) with Chlorophyll as Dye Sensitizer

Seni Ramadhanti S[1], Nurrisma Puspitasari[1] and Endarko[1,*]

[1]Physics Department, Institut Teknologi Sepuluh Nopember Surabaya
Jl. Arif Rachman Hakim, Sukolilo Surabaya (60111), Indonesia

*Email: endarko@physics.its.ac.id

Abstract. Dye-sensitized Solar Cell (DSSC) is a third-generation solar cell that consists of a working electrode, electrolyte and counter electrode. One of the most important parts of DSSC is an electrolyte that roles as a medium and regenerates the electron transport of electrons in the dye. However, the liquid electrolyte has a lack of stability in long-term use and easily evaporate or leak in DSSC. Therefore, this study aims to investigate an effect of the addition of polymer material such as PEG 1000, 4000 and PVA 60000 for fabricating a gel electrolyte to solve the problems of liquid electrolyte. The synthesized TiO_2 nanoparticles used in this study was prepared using co-precipitation (CPT) method which produces TiO_2 anatase phase with a crystal size of 11.1 nm. DSSC has been successfully conducted and analyzed to evaluate its performance. The results showed that the efficiency of DSSC cells using gel electrolyte prepared with PVA 60000 was better than a liquid electrolyte, PEG 1000, 4000, with the efficiency could be obtained at 0.083, 0.018, 0.033, and 0.054%, respectively. The results demonstrated that the addition PEG and/or PVA could be enhanced the performance of DSSC due to gel electrolyte produced current and voltage more stable compared to the liquid electrolyte.

1. Introduction

The fabrication of conventional silicon solar cell has developed and used rapidly, but it still has several problems due to its expensive production and procurement of main materials costs. On the other side, Dye-sensitized solar cell (DSSC), known as Grätzel cells, has become a promising photovoltaic device to generate energy [1-3]. The previous study has reported that DSSC was considered as a prominent alternative to cover up the problems of the conventional silicon solar cells due to its simple fabrication process, relatively high energy-conversion efficiency, low cost, and flexibility [1-9].

To achieve a higher energy conversion efficiency, various studies have been conducted by modifying the material of photoanode, dye sensitizer and electrolyte [10]. Study of [11] had used nanostructured TiO_2 film combining commercial titania powder with a titanium alkoxide sol as semiconductor on photoanode which was achieved high efficiency of 1.9%. Some studies had used natural dye sensitizer which contains anthocyanins and chlorophyll, extracted from the plant sources include leaves, fruits, flowers, etc. [2]. [12] successfully extracted 20 natural dyes from the plant sources including leaves, flowers, fruits, traditional Chinese medicines, beverages and used them as sensitizers to make DSSC. The photoelectrochemical performance of the DSSC based on these dyes showed that the open circuit voltages (*Voc*) varied from 0.337 to 0.689 V, and the short circuit photocurrent densities (*Jsc*) ranged from 0.14 to 2.69 mA cm^{-2}. Specifically, a high *Voc* of 0.689 V was obtained from the dye extracted from mangosteen pericarp sensitizer. The photo-to-electric conversion efficiency of the DSSC sensitized by the ethanol extract of mangosteen pericarp without purification reached 1.17%. [13] reported that DSSCs were assembled by using the bougainvillea flowers, red turnip and the purple wild Silician prickly pear fruit juice extract as natural sensitizers of TiO_2 films. The best overall solar energy conversion efficiency of 1.7% was obtained, under AM 1.5 irradiation, with the red turnip extract, that showed a remarkable Jsc of 9.5 mA/cm^2 and a high Incident Photon to Charge Carrier Efficiency (IPCE) value (65% at λ = 470 nm). Also the purple extract of the wild Silician prickly pear fruit showed interesting performances, with a Jsc of 9.4 mA/cm^2, corresponding to a solar to electrical power conversion of 1.26%. Rosella, the kind of flowers, was employed as photosensitizer in DSSC by [14] and energy conversion efficiency of 0.70% was repoted when the aqueous dye was extracted at 50 ^oC instead of 100 ^oC and pH of the dye was adjusted from 3.2 to 1.0. Moreover, DSSC stability was also improved by the changes in condition. [15] reported that DSSC made of Fe^{2+}/Fe^{3+}-based electrolyte and *Reseda luteola* extract as the natural dye has high efficiency of 2.17%. In the current study, a series of natural dyes extracted from the plant sources of Alfalfa's leaf as a chlorophyll was used as sensitizer in manufacturing DSSC.

The conventional DSSC have limited long-term stability due to the unsatisfied liquid electrolytes performance. It is easy to evaporate and leak after long-term use [16]. It needs the improvement in electrolyte making by adding polymer material like a gel maker which has low volatility, high ionic conductivity at constant temperature and good stability for long-term use [17]. This has much-attracted attention recently to find out the best polymer materials. DSSC using electrolyte gel have been investigated by several researchers. [16] used polymer gel electrolytes (PGE) of PVDF-HFP/PVA blends and SiO_2 and TiO_2 nanoparticles as photoanode. The concentration of polymer blend was fixed at 10 wt% with ratios of 8:2. The TiO_2 and SiO_2 nanoparticles was varied at 0.25 - 2.0 wt%. It was found that the best efficiencies of the DSSCs using PGE containing 0.25 wt% TiO_2 and 0.5 wt% SiO_2 were found to be 2.71±0.10% and 3.03±0.09%, respectively. The natural polymer solutions were used in other studies that prepared with electrospun fibers, such as poly(athylene oxide) (PEO) in distilled water [18], polyurethane in N,N-dimethylformamide (DMF) [19], poly(έ-caprolactone) (PCL) in acetone [20], PVDF in acetone/ N,N-dimethylformamide (DMAc) (7:3 by weight) [21] and regenerated cellulose in 2:1(w/w) acetone/DMAc [22].

This research had used several types of polymers namely poly(ethylene glycol) (PEG) and poly(vinyl alcohol) (PVA). PEG is a synthetic polymer that has stable properties and can bind other materials such as pigments. PEG has an excellent solubility in water and having hydroxyl groups in its chemical structure. Then, PVA is a soluble polymer in water that is very attractive material for easy preparation of gel based electrolyte [23, 24]. PVA also has some good advantages such as high mechanical strength, biodegradable, biocompatible, non-toxic, and excellent ionic conductivity [25, 26]. PVA has been used to fabricate lithium battery [27], electrochemical capacitor [28] and supercapacitor [29] that are environmental friendly electrochemical devices. Therefore, this paper had investigated the effect of the addition of PEG, and PVA polymer to the performance of DSSC.

2. Materials and Method

2.1. Materials

Chlorophyll dye used in this study was extracted from *Medicago sativa* (Alfalfa's leaf) and was purchased from K-Link Norway AS (Norway). Hereinafter, the main materials for synthesizing and fabricating of TiO_2 in this study namely $TiCl_3$ solution (15%), NH_4OH (25%), and aquadest. Meanwhile, PEG (Mw approx. 1000 and 4000), PVA fully hydrolyzed (Mw approx. 60000), potassium iodide (99%), acetonitrile, iodine (>97%), ethanol (96%), Triton X-100 (purchased from Merck (Germany)), acetic acid, and chloroform were used for fabrication of electrolytes. Black Carbon (200 mesh) asthe counter electrode, and Indium Tin Oxide (ITO) glass (42 Ω sq^{-1} dimension of 2×2 cm^2) as transparent conducting oxide substrates have been used in this study.

2.2. Synthesis of TiO_2 Nanoparticles

TiO_2 nanoparticles with anatase phase were synthesized by using co-precipitation (CPT) method. TiO_2 powder was made by mixing 20 mL of $TiCl_3$ in 100 mL of aquadest and then stirred for 1 hour. Subsequent, the NH_4OH solution was pipetted to that solution until pH 9. The solution was settled for 24 hours, and the sediment was calcined at 400°C for 3 hours to increase the crystallinity and to remove the carbonaceous materials present in TiO_2 [30].

2.3. Preparation Liquid and Gel Electrolyte

Three vessels of liquid electrolytes were prepared by mixing 3 g of Potassium Iodide, 3 mL of Iodine and 10 mL of Acetonitrile [31]. Then, 7 g of each polymer materials namely PEG 1000, 4000 and PVA 60000 were added into each vessel, respectively. Next, 25 mL of Chloroform was added and homogeneously stirred for 20 minutes at 80°C.

2.4. Fabrication of the DSSC

DSSC cells consist of a working electrode, gel electrolyte, and counter electrode. The part of working electrode including 1×1 cm^2 active area of ITO glass coated with a TiO_2 nanoparticles paste that has been soaked in the dye sensitizer. The ITO glass had a positive potential (anode). The counter electrode consists of ITO glass coated black carbon (cathode). The distance of 2 cm was applied between two electrodes and an apparent area of the film was 1.5×1.5 cm^2. The gel electrolyte was spilled on the both of electrode layer. All components were assembled by sandwiching that illustrated in Figure 1.

2.5. Measurement and characterization

The current density-voltage (I-V) characterization of DSSC performed using Solar Simulator on input voltages up to 10 volts with a scale of 0.5 Volt. A xenon lamp as a source of light with the power density of 100 mW/cm^2 was used to evaluate the performance of DSSC. The performance of the DSSC cell was performed by analyzing the parameters of solar cells namely voltage open-circuit (V_{oc}), the current short-circuit (I_{sc}), Maximum Power Point (*MPP*), the voltage and current at MPP (V_{MPP} and I_{MPP}), fill factor (*FF*) and efficiency (η). Crystal phase and particle size of TiO_2 were analyzed using Match!® and MAUD (*Material Analysis Using Diffraction*), the software for crystal phase and size identification, respectively. The powder diffraction datas from X-Ray Diffraction (XRD) can be used in that software package [32]. The absorbance spectrum of chlorophyll dye was characterized by Spectrophotometer UV-Vis with a wavelength range from 300 to 700 nm.

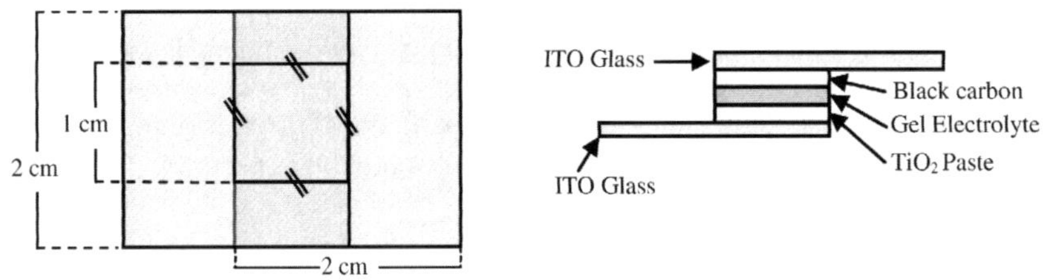

Figure 1. Schematic of cell assembly with gel electrode.

3. Results and Discussion

3.1 Analysis of Phase and Particle Size of TiO_2

The semiconductor of TiO_2 as photoanode has a very important role in improving the efficiency of DSSC. Characterization of TiO_2 was performed by X-Ray Diffraction (XRD) to determine the crystal structure of TiO_2 which has three phases namely rutile, anatase, and brookite phase. The result of XRD test was shown in Figure 2.

XRD pattern in Figure 2 indicates that the nano-TiO_2 has mainly anatase phase. The diffraction peak at $2\theta = 25.3, 37.7, 48.04, 53.89, 55.09, 62.63,$ and 68.78^o are identified in the spectrum of dense TiO_2 anatase and attributed to diffraction faces of (011), (004), (020), (015), (121), (024), and (116) respectively. In the present study, anatase phase was required because it has better photoactive and photocatalytic properties than other phases for achieving high energy efficiency [33]. In addition, the particle size characteristic was also needed in the DSSC applications. That had been analyzed using MAUD software. The synthesized TiO_2 was known as nanoparticle size of 11.15 nm of this research. In the previous study [34] was stated that the suitable particle size for applications DSSC was less than 30 nm.

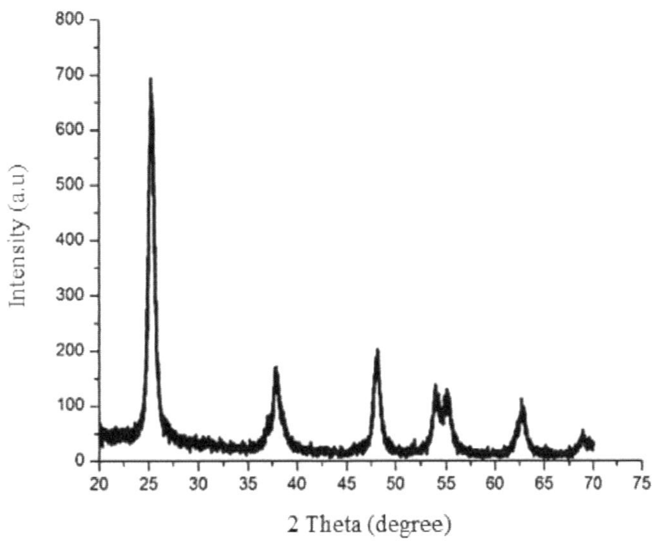

Figure 2. Crystal diffraction pattern of TiO_2 anatase phase.

3.2 Absorbance Analysis of Chlorophyll Dye

Dye plays a role as an absorber of photons energy that carried by light. Dye sensitizers can optimally absorb energy in a wide wavelength range, especially in a visible region which has the highest intensity of electromagnetic waves.

From Figure 3, it can be seen that chlorophyll of *Medicago sativa* leaves has wide absorption in the UV region (about 200-399 nm) of the light with optimum absorbance value at 290 nm (3.03 a.u). It can be also able to absorb visible light of solar radiation with optimum absorbance value at 400 nm (2.2 a.u) although in smaller absorbance region than UV light. Due to that ability, chlorophyll dye is expected to have higher photon-to-current energy conversion efficiency.

3.3 The Effect of Gel Electrolytes on DSSC Performance

DSSC electrolyte has a function as electron transport medium and regenerating the excited electrons. The liquid electrolyte has many disadvantages (like have been written before) that were caused by the type of solvent (organic solvent) that could evaporate during the irradiation process. If it occurs, the energy generation will be interrupted.

Therefore, this study used a gel electrolyte that has a capability to keep the DSSC performance for long periods without re-fabrication, beside to prevent the electrolyte leakage. The electrolyte gel was made by adding a polymer material namely PEG 1000, 4000 and PVA 60000 on the liquid electrolyte. The DSSC performance had tested and compared based on its current density-voltage and efficiency; liquid electrolyte was used as a control. The comparison DSSC performance using those electrolytes was summarized in Table 1.

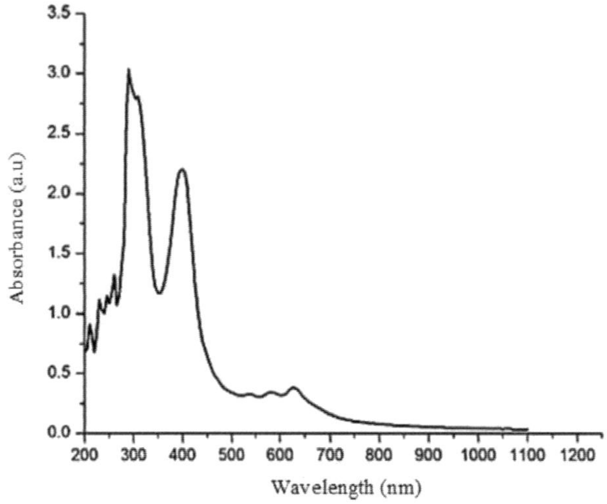

Figure 3. Absorption spectra of chlorophyll dye using UV-Vis Spectrophotometer.

Table 1. Comparison of 1 cm^2 active area of DSSC performance

Cell Names	V_{OC} (Volt)	J_{SC} (mA/cm^2)	*FF* (%)	η (%)
Liquid Electrolyte	0.421	0.088	49.21	0.018
PEG-1000	0.521	0.133	48.47	0.034
PEG-4000	0.701	0.13	59.27	0.054
PVA-60000	1.41	0.098	59.84	0.083

From Table 1, it can be concluded that the four DSSC cells have been giving good results. The highest efficiency value was at 0.083% for PVA 60000 gel electrolyte with a fill factor of 59.84%. Fill factor is a dimensionless quantity that states the maximum power ratio of the generated solar cells to the multiplication of the V_{oc} and J_{sc} [35]. Furthermore, the current density greatly affected on the DSSC efficiency than fill factor and voltage. The greater value of current density will generate the greater efficiency because it can increase the specific surface area that led to sufficient absorption of dye molecules [36]. The generated current and voltage value was plotted as shown in Figure 4.

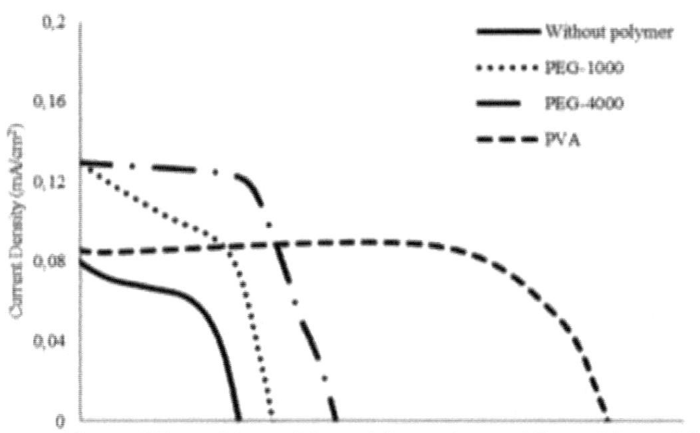

Figure 4. Characteristic curve current density-voltage variations on DSSC with the addition of PEG and PVA polymer on the electrolyte.

Figure 4 shows the characteristic of current density-voltage of DSSC with under intensity light of 100 mW/cm^2. Based on Figure 4, DSSC using a liquid electrolyte has a lower current density than that of addition of PEG and PVA polymer. That addition of PEG or PVA could improve the performance and efficiency of DSSC because gel electrolyte could produce more stable current and voltage than the liquid electrolyte.

The average fill factor value of those DSSCs was about 50%. It was still too small because of its large series resistance that was affected by ITO glass resistance, electrolyte, and the counter electrode. ITO glass resistance and interface between the electrolyte and the black carbon (catalyst) influence the electrons transfer at short circuit conditions and facilitate the process of recombination. When the series resistance has greater value, the electron transfer process would be longer and the value of the fill factor became smaller [37].

4. Conclusion

The present work has investigated the performance of DSSC prepared with gel electrolytes formed from the addition of PEG 1000, 4000, or PVA 60000 into the liquid electrolyte. The results showed that the addition of PEG or PVA to liquid electrolyte could be enhancing the performance of DSCC.The best efficiency result of DSSC's performance can be achieved from DSSC prepared with a mixture of liquid electrolyte and PVA 60000, with the value obtained of 0.083%.

Acknowledgments
The author would like to thank the Physics Department, Institut Teknologi Sepuluh Nopember Surabaya for supporting the research. Special thanks to my parent, Endarko, Ph.D and Nurrisma Puspitasari, M.Si as my supervisor.

References

[1] O'Regan & Gratzel M 1991 *Nature* **353** 737–740
[2] M Hamadanian, J Safaei-Ghomi, M Hosseinpour, R Masoomi and V Jabbari 2014 *Mater. Sci. Semicon. Process* **27** 733–739
[3] M Gratzel 2006 *Research and Appl.* **14** 429–442
[4] J Chen, C Chen, S Wu, J Li, C Wu and K Ho 2008 *Sol. Energy Mat. Sol Cells* **92** 1723–1727
[5] C Tu, K Liu, A Chien, C Lee, K Ho and K Lin 2008 *Eur. Polym. J* **44** 608–614
[6] K Lee, C Hu, H Chen, and K Ho 2008 *Sol. Energy Mater. Sol. Cells* **92** 1628–1633
[7] Y Chiba, A Islam, Y Watanabe, R Komiya, N Koide and L Han 2006 *Japan. J. Appl. Phys* **45** L638–L640
[8] S Zhang, X Yang, Y Numata and L Han 2013 *Energy Environ. Sci* **6** 1443–1464
[9] T Y Ma, Y S Wei, T Z Ren, L Liu, Q Guo and Z Y Yuan 2010 *ACS Appl. Mater. Interfaces* **2** 3563–3571
[10] Lee B K and Kim J J 2009 *Current Appl. Phys* **9** 404–408
[11] P Faralas, A I Kontos, A G Kontos, D S Tsoukleris, M C Bernard and N Spyrellis 2008 *J. Mater. Process. Technol* **196** 243–248
[12] H Zhou, L Wu, Y Gao and T Ma 2011 *J. Photochem. Photobiol. A: Chem* **219** 188–194
[13] G Calogero and G D Marco 2008 *Sol. Energy Mater. Sol. Cells* **92** 1341–1346
[14] K Wongcharee, V Meeyoo and S Chavadej 2007 *Sol. Energy Mater. Sol. Cells* **91** 566–571
[15] S Sönmezoğlu, C Akyurek and S Akin 2012 *J. Phys. D: Appl. Phys* **45** 425101
[16] S Saengsuwan, N Tiautit, C Puratane and S Panpinit 2014 *Energy Procedia* **56** 378–385
[17] C Zhang, S Gamble, D Ainsworth, A M Z Slawin, Y G Andreev, and P G Bruce 2009 *Nature Mat* **8** 580-584
[18] H Q Hou, Jun Z, Reuning A, Schaper A, Wendorff J H & Greiner A 2002 *Macromolecules* **35** 2429-2431
[19] Hu Y J, Chen B, and Yuan Y 2007 *J. Cent. South Univ.Technol* **14** 47-49
[20] Huang H T & Wunder S L 2001 *Journal of the Electrochemical Society* **148** A279-A283

[21] Huang Z M, Zhang Y Z, Kotaki M & Ramakrishna S 2003 *Composites Science and Technology* **63** 2223-2253
[22] Huynh W U, Dittmer J J & Alivisatos A P 2002 *Science* **295** 2425-2427
[23] Datta A 2007 *Characterization of Polyethylene Glycol Hydrogels for Biomedical Application* (India: B. E. University of Pune)
[24] M A Careem, M F Aziz, M H Buraidah and A K Arof 2015 *Electrochimica Acta* **182** 217–223
[25] Y T Jia, J Gong, X H Gu, H Y Kim, J Dong and X Y Shen 2007 *Carbohydrate Polymers* **67** 403–409
[26] S Rajendran, M Sivakumar and R Subadevi 2004 *Solid State Ionics* **167** 335–339
[27] J Chatterjee, T Liu, B Wang and J P Zheng 2010 *Solid State Ionics* **181** 531–535
[28] C Yun, T Qiang-qiang and X Yu-xing 2012 *Chinese J. Pro. Eng* **12** 665–671
[29] H Yu, J Wu, L Fan, K Xu, X Zhong, Y Lin and J Lin 2011 *Electrochimica Acta* **56** 6881–6886
[30] K Mi-Ra, L J Kook, J Bo-Hwa, J Sung-il, K Young-Guen, J Yong-Wook and L Su-Bin 2009 *J. Ind. Eng. Chem.* **15** 724-729
[31] Halme J 2002 *Dye-Sensitized Nanostructured and Organic Photovoltaic Cells: Technical review and Preliminary Test* (Espoo: Helsinki University Of Technology)
[32] L Lutterotti, S Matthies, and H R Wenk 1999 *Proceedings of the 12th International Conference on Textures of Materials (ICOTOM '99)* **1** 1599
[33] Rajashekar R K 2011 *Preparation of Nanocrystalline Titanium Dioxide Particles From New-Zealand Ilmenite* (New Zealand: Massey University)
[34] C Guozhong, T P Chou, Q Zhang, B Russo and G E Fryxell 2007 *J. Phys. Chem. C* **111** 6296–6302
[35] Liu W, Hua L, Dai S, Guo L, Jiang N, and Kou D 2010 *Electrochimica acta* **55** 2338–2343
[36] H D Jang, E H Jo, H Chang, S K Kim, R Ki-Min and J Kim 2014 *Materials Letters* **131** 244–247
[37] Nogueira AF, Paoli M-AD, Montanari I, Monkhouse R, Nelson J, and Durrant JR 2001 *J. Phys. Chem. B* **105** 7517-7524

Progress on Zeolite-membrane-aided Organic Acid Esterification

I G B N Makertiharta and P T Dharmawijaya

Chemical Engineering Department, Institut Teknologi Bandung, Jl. Ganesha 10, Bandung, Jawa Barat, 40132, Indonesia

E-mail: makertia@che.itb.ac.id

Abstract. Esterification is a common route to produce carboxylic acid esters as important intermediates in chemical and pharmaceutical industries. However, the reaction is equilibrium limited and needs to be driven forward by selective removal one of the products. There have been some efforts to selectively remove water from reaction mixture via several separation processes (such as pervaporation and reactive distillation). Integrated pervaporation and esterification has gained increasing attention towards. Inorganic zeolite is the most popular material for pervaporation due to its high chemical resistant and separation performance towards water. Zeolite also has proven to be an effective material in removing water from organic compound. Zeolite can act not only as selective layer but also simultaneously act as a catalyst on promoting the reaction. Hence, there are many configurations in integrating zeolite membrane for esterification reaction. As a selective layer to remove water from reaction mixture, high Si/Al zeolite is preferred to enhance its hydrophilicity. However, low Si/Al zeolite is unstable in acid condition due to dealumination thus eliminate its advantages. As a catalyst, acid zeolites (e.g. H-ZSM-5) provide protons for autoprotolysis of the carboxylic acid similar to other catalyst for esterification (e.g. inorganic acid, and ion exchange resins). There are many studies related to zeolite membrane aided esterification. This paper will give brief information related to zeolite membrane role in esterification and also research trend towards it.

1. Introduction

Esters have a lot of important applications in industry mainly for fragrances and flavors. For example, combination of esters with other organic compounds will resemble fruit flavors [1]. Esters are also used in the synthesis of polymers (such as the Dacron® polymer), medicines [2] and biodiesels [3]. Generally, esters are produced via esterification reaction [4].

Esterification of organic acid with alcohol is an example of equilibrium-limited reaction. To drive the reaction to favor the product formation, an excess of alcohol can be used. However, the usage of excess alcohol requires further separation process and increase the operation cost. Another approach to enhance the reaction is via water removal from reaction mixture using reactive distillation or membrane processes such as pervaporation (PV) and vapor permeation (VP). PV and VP offers more efficient energy usage and environmentally friendly.

Nevertheless, both cases of membrane processes use a stirred tank type reactor that gives less capacity than a flow type reactor. Hence, PV system is difficult to be used in the reaction due to its direct contact with strong acid that damage the membrane layer unless solid catalysts such as zeolites and ion-exchange resins are employed. On the other hand, VP suffer from limited water fraction in the vapor phase is due to the gas-liquid phase equilibrium. After reaction, a post treatment to remove acid from reaction mixture is required. However, there are no attempt to present vapor-phase esterification and in-situ water vapor separation in the flow-type membrane reactor. Because of the harsh condition, no zeolite membrane can withstand it [5].

During PV or VP, the permselective membranes are used to separate water from reaction mixture[6, 7]. The driving force of water permeance through membrane is the chemical potential gradient across the membrane. In the past decade, PV and VP have been proven effective for solvent dehydration and organic separation [8-10]. VP has the advantage of indirect contact between acid and membrane material plus usually VP has better flux than PV due to higher temperature employed. However, VP have higher energy requirement compared with PV in order to vaporize the reaction mixture or reactant prior to contact with membrane.

Membrane reactors employed to carry out equilibrium-limited esterification are an attractive alternative due to several reasons such as [11]:

(1) Separation efficiency is not limited by relative volatility as in distillation
(2) Only permeate change phase into vapor so that the energy consumption is lower than distillation
(3) Pervaporation could be carried out at reaction temperature making it simpler in operation.

In the literatures, esterification reaction coupled with pervaporation usually was carried out in a batch reactor using homogeneous or heterogeneous catalyst [11, 12]. A part of reaction mixture is then fed into polymeric pervaporation membrane module, and refluxing the retentate to the reactor. This configuration has been used for many pairs of organic acid and alcohol such as acetic acid with methanol [13, 14] or borneol [15], oleic acid with various alcohols [16], and propanoic acid with propanol [17]. Some modifications have also been done to this configuration such as placing the membrane inside the reactor [18, 19], or enhancing the separation of water by the employment of adsorbent in the permeate side [20].

Among membrane materials, zeolite membranes have high permeability and chemical resistant suitable for esterification. Not only as selective layer, zeolites can also have a role as a catalyst for esterification. Esterification reaction mixture consists of water, alcohol, carboxylic acid, and catalyst. This reaction can gain benefit over selective removal of water from reaction mixture by pushing equilibrium towards product formation. Zeolite membrane's first commercial application was solvent dehydration especially ethanol dehydration. Water from Ethanol-water mixture can be effectively separated via pervaporation using zeolite membrane. Because of said reasons, zeolite membrane possesses a big potential to assist esterification reaction.

There are a lot of works related to organic acid esterification using zeolite membrane reactor. However to the best of our knowledge, there are currently no review paper that summarize research efforts in this field although the significance of the process. This paper will discuss general research trend related to zeolite membrane reactor concept for esterification of organic acid. Special attention will be given to the progress of zeolite selective layer

2. Membrane Reactor Concept for Organic Acid Esterification

Membrane technology has rapid development recently primarily due to its advantages and potential to various applications in various sectors [21-25]. Besides providing selective layer to one of reaction component, membrane can also play a role as catalyst support even active catalytically itself. Recently, integration of membrane and reaction vessel into membrane reactor (MR) has gained increasing attention. Some advantages in using membrane reactors are removal of an intermediate (or final) product, control of reactant addition, control gas and catalyst contact, and combine multiple reaction at the same system [26, 27].

A respective ester can be formed directly from an alcohol and a carboxyl acid by the addition of heat. This catalyst-free reaction is desirable because there is no catalyst to be separated from the reaction medium and products. Catalyst, mainly acid or base compounds, is usually employed to this reaction to enhance the reaction and eliminate the use of thermal energy,. The general esterification reaction is:

$$R_1COOH + R_2OH \longleftrightarrow R_1COOR_2 + H_2O \quad (1)$$

Because esterification is a reversible reaction, selective removal of reaction product(s) from reaction mixture can be beneficial to shift the reaction towards product formation. Pervaporation is a membrane based separation where a permselective membrane is in contact with a liquid mixture (feed) on one side and vacuum condition is applied to the other side to promote permeation of feed components. After permeating across the membrane, permeating compounds will be condensed in a cold trap. By selection of membrane type, it is possible to control which compound will permeate through the membrane and which compound that will remained as retentate stream [15].

Membrane reactors can be categorized by the catalytic activity of the membrane [28-33]. Catalytically active membranes work simultaneously as catalysts and separators while inert membrane reactors help the equilibrium of reaction only by removing the products or byproducts [34]. Hydrogen selective membrane for hydrogen production are more widely tested application of membrane reactor [35, 36], meanwhile water-permeable membranes are tested to study esterification reactions [37, 38]. Several membrane reactor configurations to carry away esterification reaction are depicted in Figure 1.

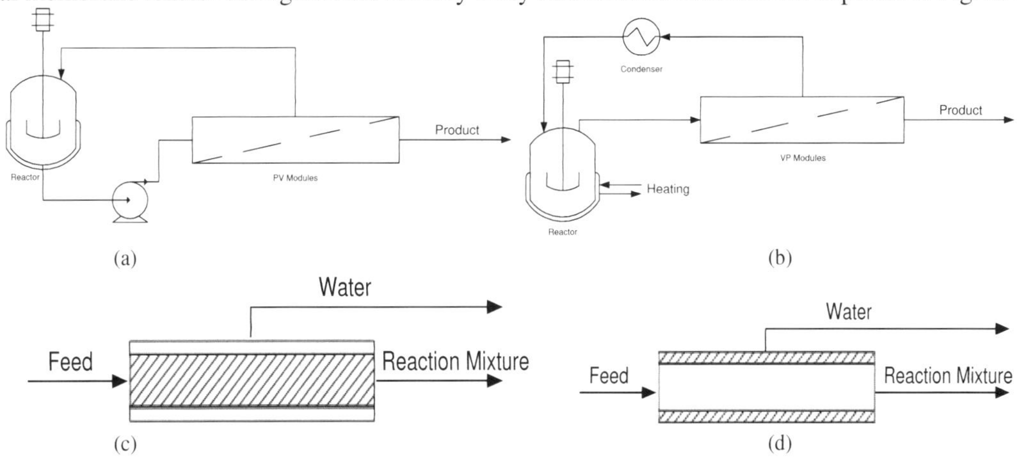

Figure 1. Different configuration of membrane reactor aided esterification reaction: (a) semi-batch pervaporation; (b) semi-batch vapor permeation; (c) Catalytic membrane reactor (d) Catalytically active membrane.

Zeolite membranes are suitable materials for membrane assisted esterification [38, 39] due to their potential in high permeability and high acid stability. For example, permeability of A-type zeolite membrane was found to be 0.14 mol m^{-2} s^{-1} for 78 mol% ethanol solution at 348 K [40]. However, the membrane has very low acid stability [41]. On the other hand, T-, MOR-, and SOD-type zeolite membranes have high acid stability [42-44]. Nevertheless, their permeation flux was much lower than A-type zeolite membrane. Membranes with high permeation flux and acid stability are necessary for membrane-assisted esterification [34].

3. Performance of Zeolite as Selective Layer

The main objective of zeolite selective layer in Esterification membrane reactor is to remove water from reaction mixture. Thus, hydrophilic zeolite membrane is desirable for this task. On the other hand, zeolite must sustain highly acidic reaction mixture condition due to organic acid and the presence of acidic catalyst. Said properties are related to Al content in the zeolite shown by its Si/Al ratio. While lower Si/Al ratio tend to produce more hydrophilic zeolite, serious dealumination by acid condition occur to zeolite that contain more Al [11]. This problem usually coped by using moderate Si/Al ratio zeolite or preventing contact between zeolite and reaction solution by vapor permeation.

During the last 20 years, zeolite membranes have gained increased attention to various applications. Simultaneously, their performances have been much improved in terms of permeance and selectivity. Popular zeolite framework used as composite membranes are LTA, FAU, MFI, etc. specifically, LTA have already been in practical use for solvent dehydration [45].

IOP Conf. Series: Materials Science and Engineering **214** (2017) 012012 doi:10.1088/1757-899X/214/1/012012

Table 1. Esterification performance carried out using zeolite membrane

Membrane	Catalyst	Temperature (°C)	Reactant	Organic Acid Conversion (%)	Water/reactant separation factor		Reference
					Alcohol	Acid	
MOR	H-ZSM-5	65	EtOH/H-Ac	22.8	2.9	2.6	[46]
LTA	PTSA	70	EtOH/H-Lac	>90	1,000	N/A	[39]
Zeolite T	Amberlyst 15	70	EtOH/H-Ac	>90	260	182	[47]
MOR	Amberlyst 15	85	EtOH/H-Ac	89.4	192	98	[11]
LTA	Amberlyst 15	85	EtOH/H-Ac	53.1	32	18	[11]
LTA	PTSA	81-92	IPA/H-Propionic	>90	200-10,000	N/A	[48]
CHA	H_2SO_4	92	IPA/H-Adipic	88-89	N/A	N/A	[34]
SOD	Amberlyst 15	90	EtOH/H-Ac	>90	~10,000	N/A	[38]
Na-CHA	H-MOR	130	EtOH/H-Ac	76	2850	∞	[5]
MOR	H_2SO_4	85	EtOH/H-Ac	76.76	N/A	852	[49]

Zeolite membranes have pores in the range of molecular sizes, which implies that they are capable of very specific interactions (e.g., selective adsorption, molecular sieving) with the target molecules. Moreover, their characteristics make them ideal candidates to integrate reaction and separation [28]. The available works that employ zeolite membrane reactors in esterification reaction mostly used a discontinuous configuration.

Tanaka et al. [47] used zeolite T membranes and Amberlyst-15 catalyst for esterification of acetic acid and ethanol. The result shows an acetic acid conversion of almost 100% after 8 h. Iglesia et al. compared mordenite and LTA membranes for esterification of acetic acid with ethanol using Amberlyst-15 catalyst in a continuous membrane reactor [11]. Mordenite and LTA was able to cross the equilibrium limitation in less than one day of experiment. Mordenite membranes showed better resistance to acid even after five days of experiment with high separation factors of H_2O/EtOH and H_2O/HAc. On the other hand, LTA conversion dropped dramatically due to the instability of the zeolite to acid condition.

As seen on Table 1, LTA, MOR, and CHA are common framework used as separative layer while Amberlyst or PTSA are common catalyst. The usage of MOR and CHA can be directly contacted with acidic reaction mixture due to its higher acid stability. However, their water/alcohol separation factor is relatively low compared with LTA membrane which is popular due to its high hydrophilicity and separation factor towards water. Nevertheless, the acid conversion of most work tend to approach completion in terms of high conversion at about >90% while at similar temperature the equilibrium conversion is about ~70%. On the other hand, the choice of catalyst favor ion exchange resin Amberlyst-15 or homogeneous catalyst. These catalysts show almost complete reaction with the assistance of zeolite membrane.

Continuous esterification reaction has been done by several authors mostly by employing membrane reactor. Overall, continuous reactor showed higher activity to batch or semi-batch process in steady state condition. An exception to this trend is when high Si/Al zeolite such as LTA was used. Some authors reported a huge decline in reactor performance after some time due to dealumination experienced by said framework [11]. Itoh et al. [5] studied continuous vapor permeation membrane reactor for esterification of acetic acid and ethanol. The reaction was carried out using various catalyst and CHA membrane at atmospheric temperature. The result suggest that H-MOR is the most active catalyst and both CHA and H-MOR have good durability over long term esterification reaction with no decline observed for 8 hour operation.

4. Zeolite as a Catalyst for Esterification

As for Esterification reaction catalyst, acid or ion exchange resins are usually used. These catalysts donate proton for the autoprotolysis of carboxylic acid. Previously, inorganic acids such as H_2SO_4, HCl, HI or $ClSO_3OH$ [50, 51] or organic acids such as p-toluene sulphonic acid (PTSA) [39, 52] are used to perform the reaction. Base catalyst may also be used with additional formation of long chain soaps [53]. However, the use of homogeneous catalysts in chemical industry are no longer interesting since they are miscible with reaction medium even though they have lower cost and higher performance. Their miscibility with reaction medium causes more problems such as additional needs of downstream separation process and also corrosion of devices in contact with the acid. Therefore, heterogeneous catalysts such as ion exchange resins [54-58], and zeolites [59-61] have gained increasing attention to carry away esterification reaction. Zeolites for esterification catalyst are popular to upgrade bio-oil into biodiesel or for hydrotreating feed [62-64]. Among zeolites, MFI and FAU, especially zeolite Y, are the most popular framework for esterification reactions. The performance of zeolite depends mainly on their acidity and pore size related with mass transfer of reactants. Peters et al. [61] compares catalytic activity of various catalyst for the esterification of acetic acid with butanol. The best result in terms of activity was gained for ion exchange resins, followed by sulphated zirconia and zeolites. The results also show that zeolite H-USY (with Si/Al=20) has the highest activity among zeolites [11].

Kirumakkin et al. [65] compared the performance of Hβ, HY, and HZSM5 zeolite for esterification of benzyl alcohol with acetic acid. The result showed that the conversion follows the order of Hβ>HY>HZSM5. The conversion of Hβ reached 75% at 130°C while the other zeolite performance capped around 60% conversion. However, ester selectivity over ether formation is better using HZSM5 zeolite where the pores are smaller than Hβ and HY. Hβ and HZSM5 have stronger acidity compared with HY that they possess higher potential in catalyzing esterification reaction [66]. It is widely accepted that esterification activity using zeolite catalyst is affected by surface acidity and pore structure related to reactant mass transfer. Different reactants and/or product may have different optimum framework to carry away the reaction. In addition, for catalytically active membrane, different framework might have to be prepared as distinguish layers because separation performance of highly active framework might not be sufficient.

5. Summary and Future Outlook

General aspects of zeolite membrane aided organic acid esterification have been reviewed. Zeolite not only has role on removing water from reaction mixture but also catalytically boost the reaction kinetics. A lot of works have shown the effectiveness of zeolite based membrane reactor to substitute conventional processes. Hopefully, the aid of zeolite membrane can be implemented to the synthesis of various esters.

In the future, this technology needs some improvement in the subject of membrane (materials, methods, and reproducibility), and cost optimization (mainly for industrial implementation). The numbers of researches that study various aspects of this technology suggest that it is ready for further industrial implementation. Nevertheless, the implementation of this technology depends mainly to energy source cost because its main advantage lies on lower energy consumption compared with conventional separation systems [67].

Further study on membrane material, configuration, and catalyst selection for various type of target ester should be done to quantitatively and economically reflect the progress of zeolite membrane aided organic acid esterification to be implemented in industry.

References

[1] Greever J C 1995 *J. Chem. Educ* **72** A151
[2] Uragami T, Kishimoto J and Miyata T 2012 *Catal. Today* **193** 57
[3] Mazubert A, Poux M and Aubin J 2013 *Chem. Eng. J* **233** 201
[4] Zhang W, Na S, Li W and Xing W 2015 *Ind. Eng. Chem. Res* **54** 4940
[5] Itoh N, Ishida J, Sato T and Hasegawa Y 2016 *Catal. Today* **268** 79

[6] Krasiński A, Wierzba P, Grudzień A, Hajmowicz H, Zawada K and Synoradzki L 2016 *Chem. Process Eng.* **37** 121
[7] Truong H, Rode S, Roizard D, Mouzon-Pelletier S and Tretjak S 2013 *Sep. Purif. Technol.* **120** 24
[8] Li W, Liu W, Xing W and Xu N 2013 *Ind. Eng. Chem. Res.* **52** 6336
[9] Peng X, Lei Q, Lv G and Zhang X 2012 *Sep. Purif. Technol.* **89** 84
[10] Ameri E, Moheb A and Roodpeyma S 2011 *Korean J. Chem. Eng.* **28** 1593
[11] de la Iglesia Ó, Mallada R, Menéndez M and Coronas J 2007 *Chem. Eng. J.* **131** 35
[12] Tanaka K, Yoshikawa R, Ying C, Kita H and Okamoto K-i 2002 *Chem. Eng. Sci.* **57** 1577
[13] Bagnell L, Cavell K, Hodges A M, Mau A W and Seen A J 1993 *J. Membr. Sci.* **85** 291
[14] Assabumrungrat S, Phongpatthanapanich J, Praserthdam P, Tagawa T and Goto S 2003 *Chem. Eng. J.* **95** 57
[15] Izák P, Mateus N M, Afonso C A and Crespo J G 2005 *Sep. Purif. Technol.* **41** 141
[16] Dörmő N, Bélafi-Bakó K, Bartha L, Ehrenstein U and Gubicza L 2004 *Biochem. Eng. J.* **21** 229
[17] David M-O, Gref R, Nguyen T and Neel J 1991 *Chem. Eng. Res. Des.* **69** 335
[18] Liu Q, Zhang Z and Chen H 2001 *J. Membr. Sci.* **182** 173
[19] Xuehui L and Lefu W 2001 *J. Membr. Sci.* **186** 19
[20] Park B-G and Tsotsis T T 2004 *Chem. Eng. Process.* **43** 1171
[21] Drioli E, Stankiewicz A I and Macedonio F 2011 *J. Membr. Sci.* **380** 1
[22] Strathmann H, Grabowski A and Eigenberger G 2013 *Ind. Eng. Chem. Res.* **52** 10364
[23] Wenten I 2016 *Desalination* **391** 112
[24] Himma N F, Anisah S, Prasetya N and Wenten I G 2016 *J. Polym. Eng.* **36** 329
[25] Khoiruddin K, Hakim A and Wenten I 2014 *Membr. Water Treat.* **5** 87
[26] Criscuoli A, Basile A, Drioli E and Loiacono O 2001 *J. Membr. Sci.* **181** 21
[27] Li K 2007 *Ceramic membranes for separation and reaction*: John Wiley & Sons)
[28] Coronas J and Santamarıa J 1999 *Catal. Today* **51** 377
[29] Julbe A, Farrusseng D and Guizard C 2001 *J. Membr. Sci.* **181** 3
[30] Wenten I and Widiasa I 2002 *Desalination* **149** 279
[31] Wenten I G, Hakim A and Aryanti P 2014 *Teknik Kimia Institut Teknologi Bandung* 11
[32] Wenten I G, Julian H and Panjaitan N T 2012 *Desalination* **306** 29
[33] Subagjo S, Prasetya N and Wenten I 2015 *J. Membr. Sci. Res.* **1** 79
[34] Hasegawa Y, Abe C, Mizukami F, Kowata Y and Hanaoka T 2012 *J. Membr. Sci.* **415** 368
[35] Itoh N 1987 *AlChE J.* **33** 1576
[36] Bracht M, Alderliesten P, Kloster R, Pruschek R, Haupt G, Xue E, Ross J, Koukou M and Papayannakos N 1997 *Energy Convers. Manage.* **38** S159
[37] Kita H, Sasaki S, Tanaka K, Okamoto K-i and Yamamoto M 1988 *Chem. Lett.* 2025
[38] Khajavi S, Jansen J C and Kapteijn F 2010 *Catal. Today* **156** 132
[39] Jafar J J, Budd P M and Hughes R 2002 *J. Membr. Sci.* **199** 117
[40] Sato K, Sugimoto K and Nakane T 2008 *J. Membr. Sci.* **307** 181
[41] Hasegawa Y, Nagase T, Kiyozumi Y, Hanaoka T and Mizukami F 2010 *J. Membr. Sci.* **349** 189
[42] Cui Y, Kita H and Okamoto K-I 2004 *J. Membr. Sci.* **236** 17
[43] Khajavi S, Jansen J C and Kapteijn F 2009 *J. Membr. Sci.* **326** 153
[44] Li G, Kikuchi E and Matsukata M 2003 *Sep. Purif. Technol.* **32** 199
[45] Morigami Y, Kondo M, Abe J, Kita H and Okamoto K 2001 *Sep. Purif. Technol.* **25** 251
[46] de la Iglesia Ó, Irusta S, Mallada R, Menéndez M, Coronas J and Santamaría J 2006 *Microporous Mesoporous Mater.* **93** 318
[47] Tanaka K, Yoshikawa R, Ying C, Kita H and Okamoto K-i 2001 *Catal. Today* **67** 121
[48] Ameri E, Moheb A and Roodpeyma S 2010 *Chem. Eng. J.* **162** 355
[49] Zhu M-H, Feng Z-J, Hua X-M, Hu H-l, Xia S-L, Hu N, Yang Z, Kumakiri I, Chen X S and Kita H 2016 *Microporous Mesoporous Mater.*
[50] Mäki-Arvela P, Salmi T, Sundell M, Ekman K, Peltonen R and Lehtonen J 1999 *Appl. Catal. A* **184** 25
[51] Mekala M and Goli V R 2015 *Chin. J. Chem. Eng.* **23** 100

[52] Xu Z, Afacan A and Chuang K T 1999 *Can. J. Chem. Eng.* **77** 676
[53] Han Y, Lv E, Ma L, Lu J, Chen K and Ding J 2015 *Energy Convers. Manage.* **106** 1379
[54] Mazzotti M, Neri B, Gelosa D and Morbidelli M 1997 *Ind. Eng. Chem. Res.* **36** 3163
[55] Okuhara T, Kimura M, Kawai T, Xu Z and Nakato T 1998 *Catal. Today* **45** 73
[56] Pipus G, Plazl I and Koloini T 2000 *Chem. Eng. J.* **76** 239
[57] Peters T A, Benes N E and Keurentjes J T 2007 *Appl. Catal. A* **317** 113
[58] Calvar N, Gonzalez B and Dominguez A 2007 *Chem. Eng. Process.* **46** 1317
[59] Wu K-C and Chen Y-W 2004 *Appl. Catal. A* **257** 33
[60] Kirumakki S R, Nagaraju N and Chary K V 2006 *Appl. Catal. A* **299** 185
[61] Peters T A, Benes N E, Holmen A and Keurentjes J T 2006 *Appl. Catal. A* **297** 182
[62] Dong T, Yu X, Miao C, Rasco B, Garcia-Pérez M, Sablani S S and Chen S 2015 *RSC Adv.* **5** 84894
[63] Milina M, Mitchell S and Pérez-Ramírez J 2014 *Catal. Today* **235** 176
[64] Narkhede N and Patel A 2013 *Ind. Eng. Chem. Res.* **52** 13637
[65] Kirumakki S R, Nagaraju N and Narayanan S 2004 *Appl. Catal. A* **273** 1
[66] Kirumakki S R, Nagaraju N, Chary K V and Narayanan S 2003 *Appl. Catal. A* **248** 161
[67] Genduso G, Luis P and Van der Bruggen B 2015 *Membrane Reactors for Energy Applications and Basic Chemical Production*, pp 565

IOP Conf. Series: Materials Science and Engineering **214** (2017) 012013 doi:10.1088/1757-899X/214/1/012013

Simulation of Water Gas Shift Zeolite Membrane Reactor

I G B N Makertiharta, Z Rizki, Megawati Zunita and P T Dharmawijaya

Chemical Engineering Department, Institut Teknologi Bandung, Jl. Ganesha 10, Bandung, West Java, 40132, Indonesia

E-mail: megazunita.105@gmail.com

Abstract. The search of alternative energy sources keeps growing from time to time. Various alternatives have been introduced to reduce the use of fossil fuel, including hydrogen. Many pathways can be used to produce hydrogen. Among all of those, the Water Gas Shift (WGS) reaction is the most common pathway to produce high purity hydrogen. The WGS technique faces a downstream processing challenge due to the removal hydrogen from the product stream itself since it contains a mixture of hydrogen, carbon dioxide and also the excess reactants. An integrated process using zeolite membrane reactor has been introduced to improve the performance of the process by selectively separate the hydrogen whilst boosting the conversion. Furthermore, the zeolite membrane reactor can be further improved via optimizing the process condition. This paper discusses the simulation of Zeolite Membrane Water Gas Shift Reactor (ZMWGSR) with variation of process condition to achieve an optimum performance. The simulation can be simulated into two consecutive mechanisms, the reaction prior to the permeation of gases through the zeolite membrane. This paper is focused on the optimization of the process parameters (e.g. temperature, initial concentration) and also membrane properties (e.g. pore size) to achieve an optimum product specification (concentration, purity).

1. Introduction

At the present, a lot of researchers focus on searching alternative renewable energy sources. Hydrogen becomes one of the potential subjects. Hydrogen as an energy source took the attention of many researchers due to its ease of obtaining yet simple production process, as well as zero pollution emissions [1]. This condition then challenges researchers to continuously improve the effectiveness and efficiency of the hydrogen production process.

The water gas shift (WGS) reaction is used widely in industry. Commonly, this process is performed in two consecutive stages to achieve both fast reaction kinetics and high conversion. High temperature (300 - 450°C) reactions is performed prior to low temperature (150 - 300°C) one [2, 3]. Due to the exothermic reaction, the low temperature condition may increase the conversion. This conventional operation has major drawbacks related to the two steps complex operation, large catalyst consumption and high steam recycling.

Improving the hydrogen production via WGS reaction can be performed by integrating the reactor with the membrane separation process resulting the membrane reactor [4, 5]. The research of integrative process which focused on reactor using hydrogen selective membrane has been considered as an alternative strategy for process optimization [6, 7].

The membrane that is used for WGS reaction requires some important properties including the thermal stability, good hydrogen permeability, stable against chemical and also highly selective for hydrogen. Researchers reported that those properties can be found in zeolite [8-10] which makes the Zeolite Membrane Water Gas Shift Reactor (ZMWGSR) an interesting topic to be investigated further. In that scheme, the hydrogen produced inside the reactor will be directly separated from other gases in the stream [11, 12], which can be seen in **Error! Reference source not found.**.

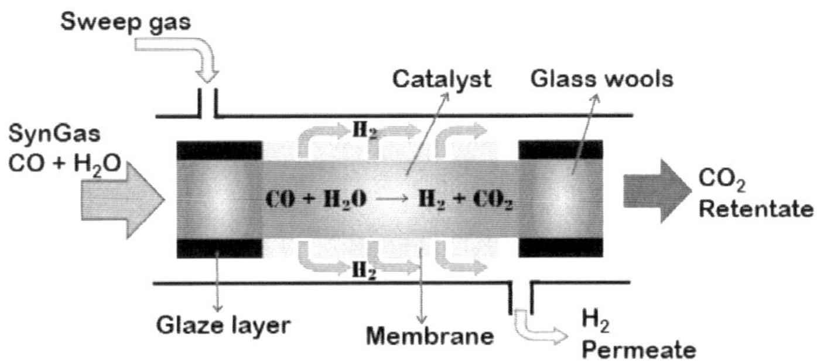

Figure 1. Scheme of Tubular Membrane Reactor for WGS Reaction.

As in many other reactors, a catalyst in WGS reactor is a key-point to achieve a good reactor performance [13]. Current development for research concerning WGS reaction is by directly combining both catalyst and H_2 separator. Conventionally, Pd, Fe-Cr oxides catalyst are used in WGS reaction which operates at temperature above 40°C [1, 3, 4].

This paper studies a simulation approach of ZMWGSR which more focus on the reaction conditions. The variables studied includes CO/H_2O ratio, pressure and temperature while specifically observed the effects on hydrogen recovery and separation factor.

2. Zeolite membrane reactor for water gas shift reaction

For years, researchers have been working on the development of ZMWGSR. The zeolite itself has specific characters regarding to its pore size and structure that can be utilized in certain purpose. Moreover, zeolite is also more stable in high temperature which makes it suitable for WGS reactor. Several researchers have performed modification method to increase the hydrogen production [12, 14]. Typical zeolite used in ZMWGSR are the MFI (ZSM-5 or silicate-1) (diameter 0,53 x 0,56 nm) and DDR (deca-dodecasil 3R) (diameter 0,46 x 0,44) [15, 16]. The MFI zeolite has given a new outlook in high temperature hydrogen production using WGS reaction. However, the MFI zeolite still has a very large pore size thus the permeance is high with low selectivity [17]. Therefore, Masuda et al [18] modified the zeolite membrane using silane to reduce the pore size.

Previous studies mentioned that the MFI type zeolite reactors could not be directly used for WGS performance prediction because the difference in permeation and separation mechanism [15, 17, 18]. Hence, this paper studies the influence Hence, this paper studies the influence of some parameters such as pressure, temperature effect, CO/H_2O ratio to the performance of zeolite membrane reactor in WGSR. Practically, WGS zeolite membrane reactor is operated at high pressure about 20 atm. However, the effect of operation pressure, temperature effect, CO/H_2O ratio on WGS reaction in a tubular zeolite membrane is still not clear.

3. Reaction and permeation model

WGS reaction is an exothermic reaction which conversion is limited by the equilibrium condition, see equation ((1)). In this simulation, some parameters were set in the range of commonly used value which have been reported. The operating condition used in this simulation is shown in Table 1.

$$CO + H_2O \leftrightarrow CO_2 + H_2, \Delta H_{298.15\,K} = -41.2 kJ/mol \qquad (1)$$

The kinetic of WGS reaction had been modelled using various equation. More specifically, the empirical power-law model is the most common one due to its simplicity yet effective. In this paper, a

IOP Conf. Series: Materials Science and Engineering **214** (2017) 012013 doi:10.1088/1757-899X/214/1/012013

kinetic model using Fe-Cr-Cu catalyst developed by Hla et al. [1] is used. The power law model in such condition is shown by equation (2).

$$r_A = 10^{0.659} exp\left(\frac{-8.8 \times 10^4}{RT}\right) P_{CO}^{0.9} P_{H_2O}^{0.31} P_{CO_2}^{-0.156} P_{H_2}^{-0.05} \left(1 - \frac{1}{K_{eq}} \frac{P_{CO_2} P_{H_2}}{P_{CO} P_{H_2O}}\right) \quad (2)$$

Where r_A is the reaction rate (mol $g_{cat}^{-1}s^{-1}$), R is the gas constant, T is the temperature (K), P_i is the partial pressure of component i in the reaction zone (Pa). K_{eq} is the temperature-dependent thermodynamic equilibrium constant for WGS, which can be defined as [19]:

$$\ln(K_{eq}) = \frac{5693.5}{T} + 1.077 \ln(T) + 5.44\, x\, 10^{-4}\, T - 1.125\, x\, 10^{-7} T^2 - \frac{49170}{T^2} - 13.148 \quad (3)$$

Besides the reaction model, a permeation model needed to be considered in this simulation. This paper used the gas translational diffusion model proposed to predict diffusivity in zeolite pores for small gases diffusion with weak adsorption affinity with zeolite [20]. The permeance of each gas component in reaction mixture (H_2O, CO, H_2, CO_2, and Ar) can be expressed as [21]:

Permeation model of zeolite membrane in this paper used the gas translational diffusion model proposed to predict diffusivity in zeolite pores for small gases diffusion with weak adsorption affinity with zeolite [20]. The permeance of each gas component in reaction mixture (H_2O, CO, H_2, CO_2, and Ar) can be expressed as [21]:

$$j_i = \frac{\varphi}{\delta} \frac{\alpha}{\lambda} \left(\frac{8}{\pi RTM_i}\right)^{1/2} exp\left(\frac{-E_{d,i}}{RT}\right) \quad (4)$$

Where j_i is the permeance of component i through the membrane (mol$m^{-2}s^{-1}Pa^{-1}$), φ is the membrane porosity (~0.25) to tortuosity ratio, which for MFI membrane has the value of 0.05 [22], δ is the thickness of the membrane (m). α is the diffusion length or distance between two adjacent sites (m), which is about 1 nm for diffusion in MFI zeolite [22], k is the diffusion coordination number (4 for MFI), M_i is the molecular weight of the diffusing gas (g mol^{-1}), $E_{d,I}$ is the activation energy for diffusion of the gas I in the micropores (kJ mol^{-1}), which is determined by the relative size of diffusing gas molecule to the zeolite pore size. Typical values of $E_{d,I}$ are 13.39, 13.14, 12.55, and 14.97 kJ mol^{-1} for H_2, CO, CO_2 and Ar, respectively [17].

Table 1. WGS Membrane Reaction Condition

Parameters	Value
Temperature Range (^oC)	400-600
Feed inlet/outlet pressure (x10^5 Pa)	5 – 25
Permeate pressure (x10^5 Pa)	1
Feed H_2O/CO ratio ($R_{H2O/CO}$) (mol/mol)	0.33 – 3
Volumetric flow rate	1.1 – 5.5 x 10^{-7}
Ar sweeping flow rate (m^3(STP)/s)	1.1 – 5.5 x 10^{-7}
Fe/Ce catalyst (mg)	200
Inner Radial (m)	3.5 x 10^{-3}
Membrane thickness (m)	3 x 10^{-6}
Membrane length (m)	1.5 x 10^{-2}

4. Membrane reactor model

The model for membrane reactor was constructed based on the mass balance equations assuming the ideal gas behavior. The zone inside the reactor was categorized into two, inside the tube where the reaction occurs and at the shell side. The partial pressure profile for each component in both zone is expressed in equation (5) (tube side) and (6) (shell side). The *i* index indicates the component inside the tube while the *j* indicates the component at the shell side. The axial and radial position inside the reactor are annotated with *z* and *r*. R, T, and Q indicate the gas constant, isothermal process temperature and volumetric flow rate respectively. ρ_B is the packing density in the reactor and v is the stoichiometric coefficient. It was reported that the selectivity of H_2/H_2O was very low [23] resulting a negligible value of permeation. Thus, the water permeance was excluded in this simulation.

$$\frac{dP_i}{dz} = \frac{R.T}{Q}\,\pi r^2 \rho_B v_i r_A - 2\pi r j_i \Delta P_i \tag{5}$$

$$\frac{dP_j}{dz} = \frac{R.T}{Q}\,2\pi r j_j \Delta P_j \tag{6}$$

Equation 5 and 6 expressed the performance of ZMWGSR. Those equations would show a conventional WGS reactor behavior when the permeance (j_i) was set to be zero. Those equations are the differential equation which need to be solved simultaneously. Considering the components at the tube and shell side, there are 9 differential equations that need to be solved simultaneously. Those equations are solved using numerical method via Matlab®.

The performance of the membrane reactor is indicated by the CO conversion and the Hydrogen recovery which calculated via these equations.

$$X_{CO} = \frac{F_{CO}^{tube\ in} - F_{CO}^{tube\ out} - F_{CO}^{shell\ out}}{F_{CO}^{feed\ in}} \tag{7}$$

$$R_{H_2} = \frac{F_{H_2}^{shell\ out}}{F_{H_2}^{tube\ out} + F_{H_2}^{shell\ out}} \tag{8}$$

5. Results and discussion

5.1. Partial pressure profile

Each component in the conventional WGS reactor gives a sudden change in partial pressure which indicates a rapid reaction. Compared to the mass transport of each gas involved, the kinetic reaction is much higher which is typical for WGS reaction. Afterward, the partial pressure remains constant along the reactor since the equilibrium has been reached instantly. In a ZMWGSR, permeation of gases is expected. In the simulation, a decline of gases' partial pressure is observed. However, this kind of behavior is only appeared at low volumetric rate. As the inlet volumetric rate increase from 3,3 x 10^{-8} to 3,3 x 10^{-5} the partial pressure profile of each component along the ZMWGSR becomes similar with the conventional WGS reactor. This indicates that the mass transfer of the gases, especially hydrogen, through the membrane occurs slowly. Furthermore, the high flow rate also gives a flushed effect resulting more hydrogen flushed away with the stream resulting low recovery (Fig3). In term of CO conversion, a ZMWGS reactor shows a higher value while still affected by the volumetric rate. A relative increase of 9.2% is observed using volumetric rate of 3.3 x 10^{-8} m^3/s while only 0,07% at 3.3 x 10^{-5} m^3/s.

The reduction of partial pressure along the reactor at the ZMWGSR inside the tube is also followed by the increase of pressure at the shell side which indicates the transport phenomena occur. Figure 3 showed the partial pressure of each component at the shell side. It can be seen that the pressure of hydrogen is significantly higher than CO and CO_2. Moreover, the back flow of Argon as the sweep gas from the shell side is low. These indicate a good selectivity of zeolite to separate hydrogen from other gases involves in ZMWGSR system.

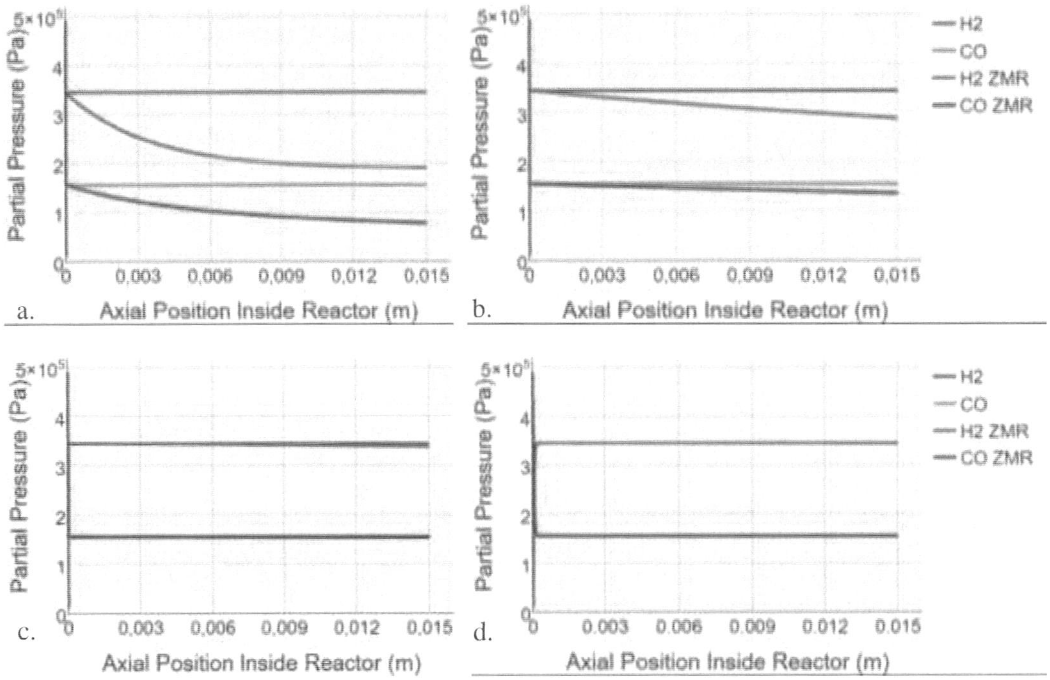

Figure 2. Partial Pressure of Hydrogen and CO inside the tube in Conventional WGS and ZMWGS Reactor with CO/H_2O ratio 1, Operating Pressure 10 bar, Operating Temperatur 500^{0}C, Tube and Sweep Gas Volumetric Flow of a. 3.3 x 10^{-8}, b. 3.3 x 10^{-7}, c. 3.3 x 10^{-6} and d. 3.3 x 10^{-5} m^3/s.

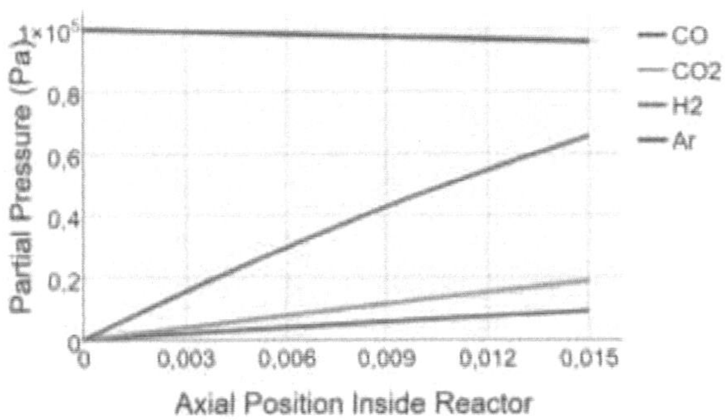

Figure 3. Partial Pressure Profile of Each Component at the Shell Side along the Reactor with CO/H_2O ratio 1, Operating Pressure 10 bar, Operating Temperature 500°C, Tube and Sweep Gas Volumetric Flow of 3.3 x 10^{-7} m^3/s.

5.2. Molar ratio effect

In term of partial pressure, the best profile is observed at an equimolar flow of CO to steam (Fig 4). The hydrogen pressure at the shell side decreases as the carbon to steal decrease. This reduction is also observed while the carbon to steam ratio increases.

On the other hand, the conversion in a ZMWGSR is highly dependent in the CO to steam ratio with a negative correlation. The highest conversion is observed at low CO to steam ratio which correlates to simulations performed by other authors [24, 25].

However, this variation does not show a big effect in H_2/CO_2 separation factor and the hydrogen recovery. The separation factor shows the lowest value at equimolar feed composition. Meanwhile the increase up to 1/3 or 3/1 ratio is less than 0,1%. Furthermore, the change of reactants ratio does not give an effect to the hydrogen recovery. Thus, the amount of hydrogen removed in the shell side is still proportional to which in the tube side.

IOP Conf. Series: Materials Science and Engineering **214** (2017) 012013 doi:10.1088/1757-899X/214/1/012013

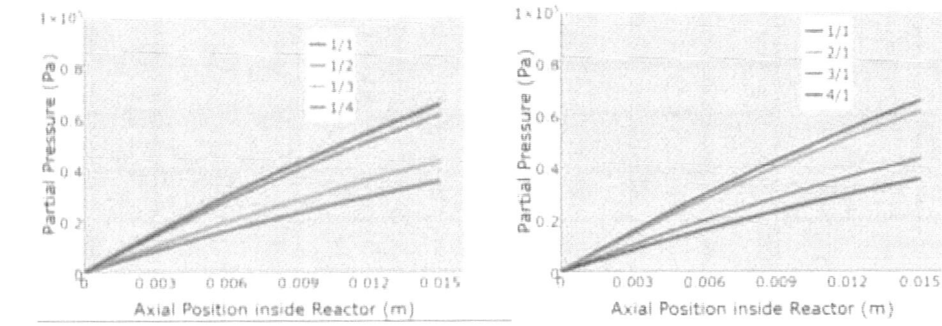

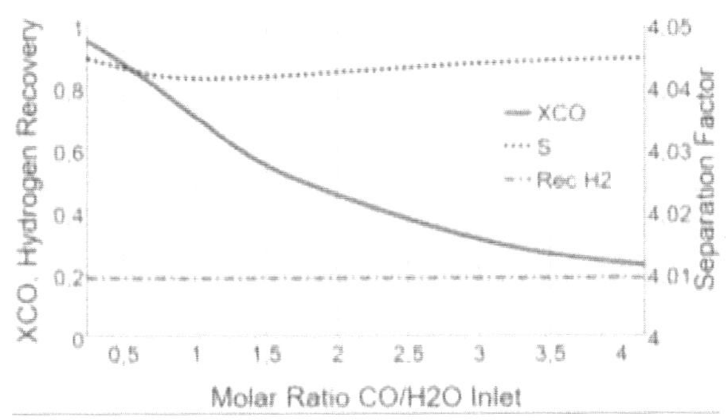

Figure 4. Molar Ratio effect to XCO, Hydrogen Recovery, and Separation Factor at ZMWGS Reactor with CO/H_2O ratio 1, Operating Pressure 10 bar, Tube and Sweep Gas Volumetric Flow of 3.3 x 10^{-7} m^3/s.

5.3. *Flow rate effect*

The system in ZMWGSR consists of two streams, the one inside the tube and another one at the shell. Both streams may affect the performance of the ZMWGSR itself. It is explained at the previous section that the permeation from the tube to the shell side is better at low volumetric rate. Figure 5 shows that the partial pressure of hydrogen at the shell side is higher when low flow rate is applied. At higher flow rate, more hydrogen will be flushed away before it could even travel through the membrane. Thus, the hydrogen recovery value becomes lower. At the opposite, the separation factor shows a positive correlation with the inner tube flow rate. (Figure 6.a)

The flush effect also occurs at the sweep gas. However, the flushed hydrogen at the shell side gives a good effect in the ZMWGSR system. Thus, increase the hydrogen recovery. Figure 6.b shows that the increase in the sweep gas flow rate is followed by the increase of hydrogen recovery. Meanwhile, the other parameters are seemed to be not affected. Furthermore, high sweep gas flow rate cuts off the chance of Argon to permeate to the tube side. This means the loss of sweep gas becomes lower.

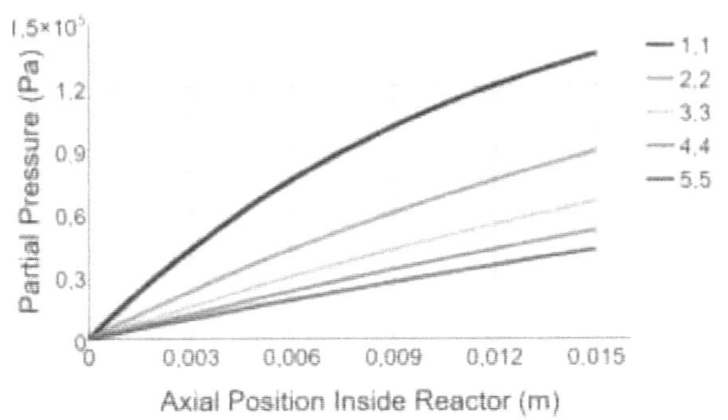

Figure 5. Hydrogen Partial Pressure Profile at the Shell Side along the Reactor with CO/H_2O ratio 1, Operating Pressure 10 bar, Operating Temperature 500^{0}C, Sweep Gas Volumetric Flow of 3.3 x 10^{-7} m^3/s, and the Inlet Tube Volumetric Flow ranges from 1.1 to 5.5 x 10^{-7} m^3/s.

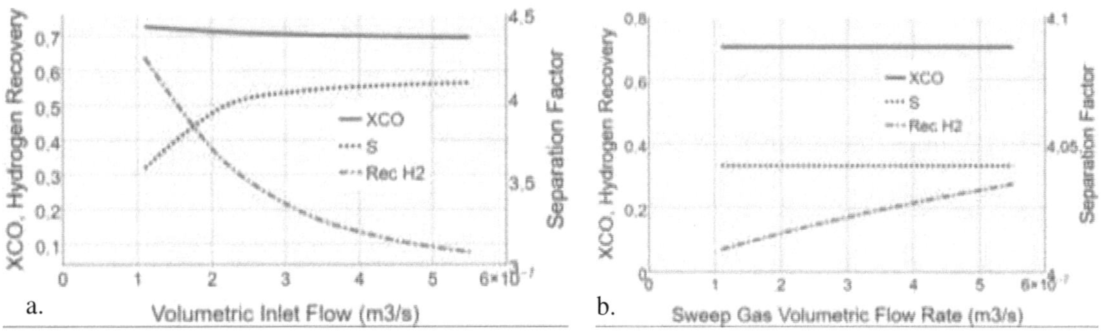

Figure 6. Effect of a. Inlet Tube and b. Sweep Gas Flow Rate to CO Conversion, Hydrogen Recovery and Separation Factor using CO/H_2O ratio 1, Operating Temperature 500^0C and Operating Pressure 10 bar.

5.4. Pressure and temperature effect

The inlet pressure at the tube side increases the partial pressure of hydrogen at the shell side. This corresponds with the flux equation which is pressure dependent. Increase in tube side pressure will increase the pressure difference between the tube and shell, thus enhance the transfer across the membrane. However, this increase seems to be proportional thus does not significantly affect the hydrogen recovery.

In this simulation, the CO conversion for both membrane reactor and conventional reactor does not differ so much. Hence, the conversion decreases as the temperature increase. Moreover, the hydrogen recovery and separation factor is also observed to be temperature-dependent. The separation factor has a maximum value at 550^0C, at a value of 4.045 while the hydrogen recovery has a positive correlation with temperature. A recovery of 0.23 is observed at 600^0C while using CO to steam ratio 1, at 10 bar.

5.5. Process optimization

It is observed in this simulation that the performance of a ZMWGSR is more sensitive to CO/H_2O ratio and the tube inlet flow rate. The tube inlet flow rate affects hydrogen recovery and separation factor in a different way, thus an optimum condition for both parameters may apply. Such condition is reached at flow rate of 2,2 x 10^{-7} m^3/s. The carbon to steam ratio gives the best performance at the lowest value the system can achieve.

The performance of a ZMWGSR may also affected by other parameters discussed in this paper. However, the effect is not as significant as the carbon to steam ratio and the inlet flow rate. The sweep gas flow rate improves the hydrogen recovery and potentially to be able for further increase outside the simulation range. However the cost of the sweep gas itself may be the constrain in increasing the sweep gas flow rate

6. Conclusion

A simulation of Zeolite Membrane Water Gas Shift (ZMWGS) reactor has been performed in this study. Integrating a zeolite membrane into a conventional WGS reactor increases the CO conversion. However the pressure profile across the ZMWGS reactor will be close to the conventional one when the flow rate is high.

This simulation also observes the effect of several operating parameters in the ZMWGSR performance. Parameters that affect the system significantly are found to be the carbon to steam ratio and the inlet flow rate. The optimum condition is achieved at flow rate of 2,2 x 10^{-7} m^3/s and low carbon to steam ratio. The pressure gives effect to hydrogen partial pressure at the shell side yet not affecting the hydrogen recovery. The optimum temperature for this process is observed at 550^0C.

7. References

[1] Lee J Y, Lee D-W, Lee K-Y and Wang Y 2009 Cr-free Fe-based metal oxide catalysts for high temperature water gas shift reaction of fuel processor using LPG *Catal. Today* **146** 260-4.

[2] Mendes D, Garcia H, Silva V, Mendes A and Madeira L M 2008 Comparison of nanosized gold-based and copper-based catalysts for the low-temperature water– gas shift reaction *Ind. Eng. Chem. Res.* **48** 430-9.

[3] Bi Y, Xu H, Li W and Goldbach A 2009 Water–gas shift reaction in a Pd membrane reactor over Pt/Ce 0.6 Zr 0.4 O 2 catalyst *Int. J. Hydrogen Energy* **34** 2965-71.

[4] Brunetti A, Caravella A, Fernandez E, Pacheco Tanaka D A, Gallucci F, Drioli E, Curcio E, Viviente J L and Barbieri G 2015 Syngas upgrading in a membrane reactor with thin Pd-alloy supported membrane *Int. J. Hydrogen Energy* **40** 10883-93.

[5] Baker R W 2004 *Membrane Technology and Applications, Second Edition,* pp 393-423.

[6] Ockwig N W and Nenoff T M 2007 Membranes for hydrogen separation *Chem. Rev.* **107** 4078-110.

[7] Gallucci F, Fernandez E, Corengia P and van Sint Annaland M 2013 Recent advances on membranes and membrane reactors for hydrogen production *Chem. Eng. Sci.* **92** 40-66

[8] Tavolaro A and Drioli E 1999 Zeolite membranes *Adv. Mater.* **11** 975-96.

[9] Kim S-J, Yang S, Reddy G K, Smirniotis P and Dong J 2013 Zeolite membrane reactor for high-temperature water-gas shift reaction: effects of membrane properties and operating conditions *Energy Fuels* **27** 4471-80.

[10] Basile A, Curcio S, Bagnato G, Liguori S, Jokar S M and Iulianelli A 2015 Water gas shift reaction in membrane reactors: Theoretical investigation by artificial neural networks model and experimental validation *Int. J. Hydrogen Energy* **40** 5897-906.

[11] Chein R Y, Chen Y C and Chung J N 2015 Sweep gas flow effect on membrane reactor performance for hydrogen production from high-temperature water-gas shift reaction *J. Membr. Sci.* **475** 193-203.

[12] Tang Z, Kim S-J, Reddy G K, Dong J and Smirniotis P 2010 Modified zeolite membrane reactor for high temperature water gas shift reaction *J. Membr. Sci.* **354** 114-22.

[13] Ratnasamy C and Wagner J P 2009 Water gas shift catalysis *Catal. Rev.* **51** 325-440.

[14] Hong M, Falconer J L and Noble R D 2005 Modification of zeolite membranes for H2 separation by catalytic cracking of methyldiethoxysilane *Ind. Eng. Chem. Res.* **44** 4035-41.

[15] Gu X, Tang Z and Dong J 2008 On-stream modification of MFI zeolite membranes for enhancing hydrogen separation at high temperature *Microporous Mesoporous Mater.* **111** 441-8.

[16] Wang H, Dong X and Lin Y S 2014 Highly stable bilayer MFI zeolite membranes for high temperature hydrogen separation *J. Membr. Sci.* **450** 425-32.

[17] Tang Z, Dong J and Nenoff T M 2009 Internal surface modification of MFI-type zeolite membranes for high selectivity and high flux for hydrogen *Langmuir* **25** 4848-52.

[18] Masuda T, Fukumoto N, Kitamura M, Mukai S R, Hashimoto K, Tanaka T and Funabiki T 2001 Modification of pore size of MFI-type zeolite by catalytic cracking of silane and application to preparation of H2-separating zeolite membrane *Microporous Mesoporous Mater.* **48** 239-45.

[19] Lima D F B, Zanella F A, Lenzi M K L and Ndiaye P M 2012 *Petrochemicals,* ed V Patel

[20] Xiao J and Wei J 1992 Diffusion mechanism of hydrocarbons in zeolites—I. Theory *Chem. Eng. Sci.* **47** 1123-41.

[21] Kanezashi M, O'Brien-Abraham J, Lin Y and Suzuki K 2008 Gas permeation through DDR-type zeolite membranes at high temperatures *AIChE J.* **54** 1478-86.

[22] Kanezashi M and Lin Y 2009 Gas permeation and diffusion characteristics of MFI-type zeolite membranes at high temperatures *J. Phys. Chem. C* **113** 3767-74.

[23] Dong T, Yu X, Miao C, Rasco B, Garcia-Pérez M, Sablani S S and Chen S 2015 Selective esterification to produce microalgal biodiesel and enrich polyunsaturated fatty acid using zeolite as a catalyst *RSC Adv.* **5** 84894-900.

[24] Hla S S, Morpeth L D and Dolan M D 2015 Modelling and experimental studies of a water – gas shift catalytic membrane reactor *Chem. Eng. J.* **276** 289-302.
[25] Dong X, Wang H, Rui Z and Lin Y S 2015 Tubular dual-layer MFI zeolite membrane reactor for hydrogen production via the WGS reaction : Experimental and modeling studies *Chem. Eng. J.* **268** 219-29.

IOP Conf. Series: Materials Science and Engineering **214** (2017) 012014 doi:10.1088/1757-899X/214/1/012014

Modification and Applications of Hydrophilic Polypropylene Membrane

Danu Ariono* and Anita Kusuma Wardani
Department of Chemical Engineering, ITB, Jl. Ganesha 10, Bandung 40132, Indonesia

* Email: danu@che.itb.ac.id

Abstract. Polypropylene (PP) is one of the most important polymers for microporous membrane due to its high void volume, well-controlled porosity, high thermal and chemical stability, and low cost. However, the hydrophobicity of PP becomes a limitation to broaden its applications. Furthermore, membrane fouling occurs more seriously on hydrophobic membranes than hydrophilic ones. To solve this problem, surface modifications have been developed to enhance PP membrane hydrophilicity without changing its bulk properties. Graft polymerization and plasma treatment are the most popular techniques for surface hydrophilization. Some studies showed that highly hydrophilic PP membranes with water contact angle less than 20° could be obtained by plasma treatment and graft polymerization. Furthermore, during plasma treatment, polar groups were formed on the PP membrane surface thus increased water uptake. To bring brief explanation on various research trends for PP modification, this paper provides a review of surface hydrophilization of microporous PP membrane, including plasma treatment and graft polymerization. The effects of surface modification on PP membrane performance such as porosity, water contact angle, and water flux are also discussed. In addition, the applications of modified PP membrane are presented as well.

1. Introduction

Membranes have been used for various separation processes, such as water and wastewater treatment, fuel cells, artificial organs, and industrial filtration. For each applications, the separation performances are determined by membrane characteristic (Ulbricht, 2006; Sykes *et al.*, 1982). For a polymer membrane, the performances depend much on the chemistry of the membrane material, the structures and properties of the porous layer, and the microstructures and surface properties of the top layer (Wan *et al.*, 2009).

Many of the polymers which are used for commercially available synthetic membranes are hydrophobic polymers. (Chung, 1998). Polypropylene (PP) is one of the polymer that has been used extensively as membrane material due to its chemical stability, low cost, good mechanical properties, and being readily available (Bhat *et al.*, 2003; Shahidi *et al.*, 2007; Yu *et al.*, 2007a; Yan *et al.*, 2008; Yang *et al.*, 2005; Meng *et al.*, 2014; Xu *et al.*, 2013; Feng *et al.*, 2011; Himma *et al.*, 2016b). PP membranes are used for a variety of industrial applications, such as sterilization of beverage and pharmaceutics, wastewater treatment, ultrapure water in the semiconductor industry, desalination of sea water, electrode separation, etc. (Bae *et al.*, 2001; Mulder, 1996). However, the application of PP membrane is restricted by its low energy surface and relatively high hydrophobicity that lead to membrane fouling. Therefore, surface modification is emerged to obtain hydrophilic PP membranes and increase antifouling characteristics.

IOP Conf. Series: Materials Science and Engineering **214** (2017) 012014 doi:10.1088/1757-899X/214/1/012014

Surface modification of polymer membranes has emerged to develop membranes separation performance and stability. Generally, the aim of membrane surface modification is to minimize fouling, modulate hydrophilicity/hydrophobicity, or enhance biocompatibility (Wan *et al.*, 2009; Bae *et al.*, 2001; Corn *et al.*, 1991; Guo and Ulbricht, 2010). Various techniques are available for surface modification of PP membranes, such as chemical, flame, corona, UV irradiation, and plasma treatment, etc.(Sykes *et al.*, 1982; Briggs *et al.*, 1979; Bhat and Upadhyay, 2002; Lanauze and Myers, 1990; Singh *et al.*, 2008; Lipnizki *et al.*, 1999). Among the various surface modification techniques, plasma treatment and graft polymerization are regarded as the most attractive methods for hydrophilization of PP membranes.

Hydrophilicity refers essentially to hydrogen bonding with water molecules. Therefore, having a high percentage of oxygen, either in the main chain or in pendant groups, contributes to the hydrophilicity of a material (Zou *et al.*, 2011). A thin layer of bound water exists at the surface of hydrophilic membranes. This layer prevent or reduce the undesirable adsorption or adhesion of foulants at the membrane surface (Kochkodan and Sharma, 2012). Thus, the anti-fouling property of membrane can be promoted by improving its hydrophilicity (Wenten, 1995).

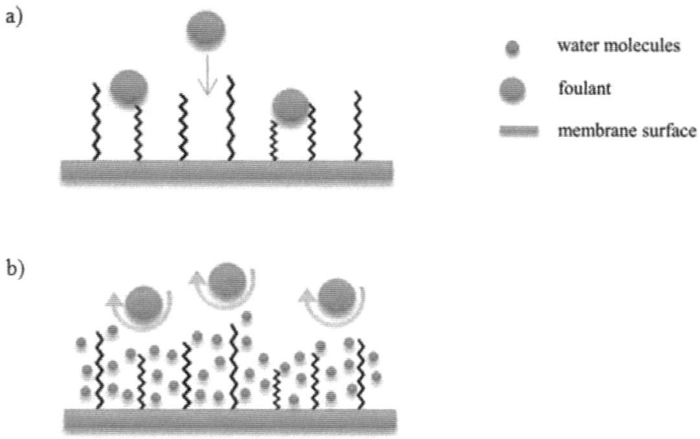

Figure 1. Interaction between foulant and membrane surface of: a) unmodified PP membrane and b) hydrophilic modified PP membrane, adapted from (Zhang *et al.*, 2016)

Some reviews relating to the modification strategies of PP membrane have been published. However, a review of surface hydrophilization of PP membrane has not been reviewed elsewhere. To provide an insight for the development of hydrophilic PP membrane in the future, this paper provides a review of surface hydrophilization of microporous PP membrane, including plasma treatment and graft polymerization. The effects of surface modification on PP membrane performance such as porosity, water contact angle, and water flux are also discussed. In addition, the applications of modified PP membrane are presented as well.

2. Plasma Treatment

Plasma is a quasi-neutral gas consists of electrons and ions as well as neutrals, atomic, and molecular species that exhibit a collective behavior in the presence of an electromagnetic field (Gomathi *et al.*, 2008; Ciszewski *et al.*, 2006). Plasma treatment processes have been extensively studied to modify the surface of PP membranes without affecting the bulk properties. The major effects from plasma treatment on the polymer surfaces are cleaning of organic contamination, micro-etching, cross-linking, inducing hydrophilic modification and improving adhesive properties of polymeric substrates (Bhat and Upadhyay, 2002; Liston, 1989; Hollahan *et al.*, 1969; Boyd *et al.*, 1997; Owens, 1975).

In general, the surface energy of untreated PP membranes is too small for good adhesion. For optimum adhesion, the surface energy of the membranes should be larger than that of the material to be bonded with. Similarly, for effective wetting by a liquid, the surface energy of the polymer membranes should exceed the surface tension of the liquid (Rajesh and Mark, 2003). By using plasma treatment, the overall surface energy of PP membrane can be increased.

Surface modification of PP membranes by plasma treatment can be achieved using gases such as air, oxygen, nitrogen, ammonia, argon, etc. (Yu *et al.*, 2007a; Liston, 1989; Boyd *et al.*, 1997; Rajesh and Mark, 2003; Yu *et al.*, 2008; Slepička *et al.*, 2010). Gas molecules in a plasma reactor are activated by the collision with electrons, positive ions, and metastable species that subsequently attack on PP membrane surface (Yasuda and Wang, 1985). Oxides and nitrides thus formed lead to accomplishment of surface hydrophilic modification. Therefore, the gases that used for the plasma treatment determine the hydrophilicity (Shahidi *et al.*, 2007; Bhat and Upadhyay, 2002; Chaivan *et al.*, 2005).

Table 1. Modified membrane characteristics after 20 minutes of plasma treatment

Plasma treatment	Average pore size (μm)	Porosity (%)	Water contact angle	Relative water flux, J/Jo	Ref.
N_2	-	-	86°	1.70	(Yu *et al.*, 2007a)
NH_3	0.10 ± 0.02	44.4 ± 2.7	58°	1.19	(Yan *et al.*, 2008)
Allylamine	-	-	70°	-	
Acrylic acid	-	-	42°	-	(Bae *et al.*, 2001)
H_2O	-	-	32°	-	
Air	0.09 ± 0.02	-	35°	1.30	(Yu *et al.*, 2008)
O_2	-	-	15°	-	(Yun *et al.*, 2004)
CO_2	0.10 ± 0.02	35.4 ± 1.8	76°	1.21	(Yu *et al.*, 2005)
CH_4-O_2	-	-	8°	-	(Tsai *et al.*, 2011)

The wetting properties are significantly affected according to the types of plasma treatment species. Water contact angle (WCA) measurements have been commonly used to characterize the hydrophilicity of the membrane surface. As shown in Table 1, plasma treatment can modified PP membranes to be hydrophilic ones. WCA hysteresis is happened due to the change of the surface roughness and hydrophilic characteristic. The WCA on a rough surface is smaller than that done on a smooth surface (Bae *et al.*, 2001). However, Yu et al. (Yu *et al.*, 2007a) and Yan et al. (Yan *et al.*, 2008) showed that the WCA increased with the increase of storage time. It indicates that hydrophilicity gained by plasma modification is not stable. The effect of plasma treatment can get lower and even disappear due to mobility of surface functionalities.

Beside decrease WCA, plasma treatment usually generates the presence of polar groups on the membrane surface. The presence of polar groups on a PP membrane surface also enhances the hydrophilic property of that PP membrane surface, thus easing water molecules to get through the membrane, which increases water uptake (Himma *et al.*, 2016a). Previous study (Hansen and Schonhorn, 1966) also showed that crosslinking initiated by activated species of inert gas as one of the reasons for improving surface cohesive bonding. Furthermore, the result from another study (Chung, 1991) indicated that bond-ability improves by intimate molecular contact and by inducing attractive force between the substrates. Molecular contact can be improved by inducing surface crosslinking, whereas attractive force between the substrates can be improved by surface hydrophilic modification.

Hydrophilic modification of the membrane surface by plasma polymerization also showed the effect of reducing organic absorption. Although the plasma polymerization treatment caused a reduction in initial membrane permeability, overall performance of the membranes showed almost constant fluxes for the whole experiment. This reduction of the initial flux of the treated membrane could respond to the small voids on the membrane surface that are partially filled with plasma polymer by the coating process (Zou *et al.*, 2011). Furthermore, the improvement of hydrophilicity varies according to plasma treatment time. Membranes with a shorter treatment time have better filtration results than those with a longer treatment time.

IOP Conf. Series: Materials Science and Engineering **214** (2017) 012014 doi:10.1088/1757-899X/214/1/012014

3. Graft Polymerization

Surface modification by graft polymerization is done by tethering polymer chains onto polymer membrane. It is a useful method for the development of membrane surface properties and bulky strength (Meng *et al.*, 2014; Kato *et al.*, 2003; Uyama *et al.*, 1998). Graft polymerization provides existing polymer membranes with new functionalities such as hydrophilicity, adhesion, biocompatibility, conductivity, anti-fogging, and anti-fouling. Monomer react with a membrane surface that has been activated, then another initiator is immobilized onto the membrane surface, where it will then be activated to initiate the polymerization process with monomer (Himma *et al.*, 2016a). Graft polymerization is very effective to modify PP membrane with stable hydrophilicity.

A variety of methods for graft polymerization onto PP membrane surfaces has been reported. The methods include chemical graft polymerization and grafting with the use of high energy radiation or oxidizing agents (Uyama *et al.*, 1998). Functional groups are first introduced to the polymeric substrate by UV irradiation, glow-discharge, or ozone and then are decomposed and initiated for polymerization. For materials with reactive groups on the surface, the desirable polymer can be easily tethered on through the free radical graft polymerization of a monomer or the chemical reaction of a polymer with functional end groups. Meanwhile, materials without reactive groups on the surface need plasma, gamma-ray irradiation and UV photo induced graft polymerizations (Liu *et al.*, 2004).

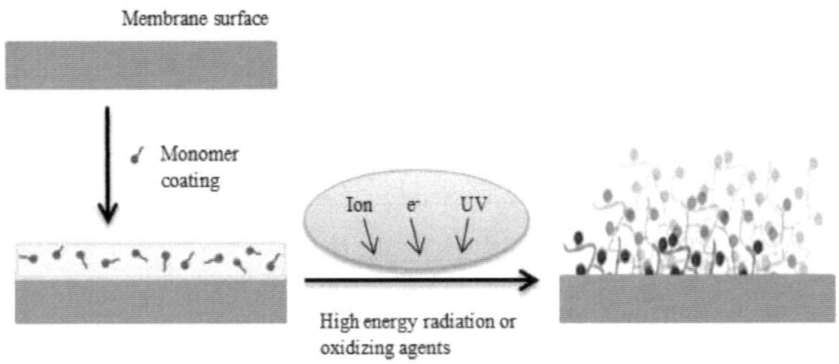

Figure 2. Illustration of PP membrane hydrophilization by graft polymerization, adapted from (Venault *et al.*, 2013)

Based on its grafting methods, graft polymerization can be divided into three types:

- Chemical compound induced. In chemical compound induced, a radical for the polymerization process is formed on the membrane surface through reaction with chemical compounds. One of the most attractive techniques for oxidizing a substrate surface is ozonation (Uyama *et al.*, 1998).
- Plasma induced. In plasma induced method, plasma is used to provide active sites either on the membrane surface or other initiators that can initiate the polymerization process. Plasma treatment is firstly performed to generate radicals on the membrane surface, which then the radicals initiates graft polymerization of monomers.
- Irradiation induced. Irradiation can be utilized to perform graft polymerization on a PP membrane surface by using ultra-violet (UV) and γ-ray as the initiators. PP membranes are first activated by thermal initiators or high energy irradiation, then followed by chemical grafting of hydrophilic polymer chains (Xu *et al.*, 2013). UV irradiation has been widely used for membrane preparation or modification due to its ability to induce cross-linking in the polymer matrix, thus the membrane selectivity can be improved.

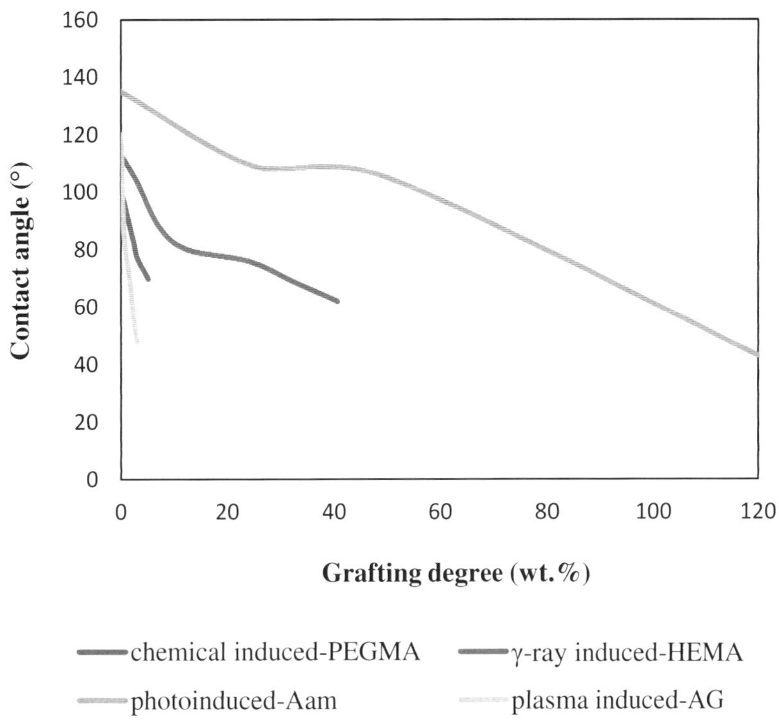

Figure 3. Effect of grafting degree on water contact angle for different methods and polymers (PEGMA: Poly-(ethylene glycol) methacrylate, HEMA: Hydroxyethyl methacrylate, Aam: Acrylamide, and AG: Allyl glucoside), adapted from (Jin *et al.*, 2012; Shim *et al.*, 2001; Yu *et al.*, 2007b; Nie *et al.*, 2003)

The existence of hydrophilic chains on the membrane surface facilitated the reduction of interfacial tension with water. As the result, WCA are drop after modified by graft polymerization. Some literature showed that WCA below 20° can be achieved by graft polymerization (Zou *et al.*, 2011; Zhao *et al.*, 2011; Yang *et al.*, 2010; Zhao *et al.*, 2010; Yang *et al.*, 2009). Furthermore, characteristics of modified PP membrane mainly depend on the grafting degree. The higher degree of graft polymerization provides a strong driving force to reduce the interfacial free energy by migrating and reorienting of the hydrophilic monomers (Jin *et al.*, 2012). As the result, higher grafting degree with a large coverage surface decreases the WCA (as shown in Figure 3).

The antifouling resistance of the modified membranes is also improved due to an increase in their hydrophilicity. Therefore, through surface grafting modification, improvement in the hydrophilicity of PP membrane may bring the increase of water permeation flux due to the balance between the pore-covering effect of the grafted chains and the increase of hydrophilicity. However, the water flux remain almost unchanged when the grafting degree is low, but the negative effect of pore-covering soon become the major factor with the increase of the grafting degree, causing the water flux dropping severely (Dai *et al.*, 2008). Thus, the precise design of a graft layer and well-controlled graft architecture is a very important issue to minimize the membrane fouling.

4. Applications

4.1 Water and wastewater treatment

With the increasing shortage of clean water sources worldwide, the development of advanced water treatment technologies of low-cost and high efficiency is urgent. There are many areas in which membrane water treatment can play an important role, such as drinking water treatment, brackish and seawater desalination and wastewater treatment and reuse (Yin and Deng, 2015). PP membranes have found potentially important applications in water treatment. However, its hydrophobicity and fouling is the major obstacle in the application. Surface modification of PP membrane is then undoubtedly necessary to avoid premature membrane deterioration due to fouling.

One of the most thorough applications studied is the membrane bioreactor (MBR), since fouling is the major concern of MBR (Le-Clech *et al.*, 2006). Previous study showed that the performance of modified hydrophilic PP membrane in MBR is enhanced as indicated by a decrease in fouling tendency (Chung and Lee, 1997), thus promoting it to address the fouling issues during MBR operation. When directly applied in MBR, this efficacy is proven by lower flux reduction and higher flux recovery.

4.2 Biotechnology

The choice of polymer membranes for various biotechnology applications depends on their surface properties. Specific surface properties like hydrophobicity, chemical structures, roughness, conductivity, etc. can be modified to meet the specific requirements (Gomathi *et al.*, 2008). At the most cases, PP membranes are not applicable for biotechnology applications due to the specificity of bio-fluids and biomolecules, which are highly environmentally sensitive. Biomolecules (e.g. enzyme) tend to irreversibly adsorb on the hydrophobic membrane surface. The adsorption causes membrane fouling and induces changes of the structures and functions of the biomolecules (Wan *et al.*, 2009). Therefore, surface modification of PP membrane is necessary to improve biocompatibility of membrane surfaces. Modification of PP membrane surfaces by plasma treatment can improve wettability, oxidize the surface, and enhance cell growth and adhesion (Gomathi *et al.*, 2008).

4.3 Medical

Most biomedical materials are used in constant contact with living systems, such as blood, cells, and tissues. Since the material surface can undergo unfavourable biological responses when in contact with a recipient, most of the conventional surfaces need to be modified so that the materials can function as designed (Uyama *et al.*, 1998). PP membranes, as one of the most popular membrane materials, have been used in many biomedical separation due to their random network of overlapped fibers, multiple connected pores, high thermal and chemical stability (Zhao *et al.*, 2011; Wang *et al.*, 2009; Zheng *et al.*, 2010). PP membranes have been applied for drug delivery, artificial organs (artificial kidney, liver, pancreas, lung, etc.), tissue regeneration, diagnostic devices, and other blood purification processes (e.g. plasma treatment and cell separation) (Wan *et al.*, 2009).

In a blood purification, hydrophilic property is required to some extent in order to prevent protein adsorption while allowing flux of aqueous solution (Gérard *et al.*, 2011). The same problems are also encountered in a blood oxygenator, where protein adsorption and platelet adhesion contribute in decreasing gas permeability. Bovine serum albumin (BSA) is one of the blood proteins that often used as the model to study the efficacy of a hydrophilized PP membrane in regard to its antifouling property. As PP membranes have been successfully hydrophilized, several promising results can be obtained as indicated in reduction of protein adsorption and increasing both permeate flux and membrane flux recovery (Himma *et al.*, 2016a).

4.4 Battery separator

Separator materials for alkaline batteries have received a great attention due to the desire to develop high energy per unit weight and volume battery packages. Battery separator is usually made from a porous membrane that placed between positive and negative electrodes in order to prevent them from contacting one another that can lead to short circuit. There are several important properties of membrane as a battery separator, such as good electric insulation, minimal electrolyte resistance, mechanically and chemically robust, and easily wetted by electrolyte solution (Arora and Zhang, 2004). Surface modified PP membranes can be used for this application since they fulfill basic requirements of a battery separator, such as presence of spontaneous, uniform and permanent wettable surface in order to accommodate and fully retain the aqueous electrolyte solution (Goel *et al.*, 2009).

5. Conclusion

The application of PP membrane is restricted by its hydrophobicity, thus surface modification have been developed to enhance PP membrane hydrophilicity without changing its bulk properties. Plasma treatment and graft polymerization are the most popular techniques for surface hydrophilization. By using plasma treatment, WCA hysteresis is happened due to the change of the surface roughness and hydrophilic characteristic. Plasma treatment also generates the presence of polar groups on the membrane surface. Meanwhile, some studies showed that highly hydrophilic PP membranes with WCA less than 20° can be obtained by graft polymerization. The value of WCA is decreased with the increase of grafting degree.

6. Reference

[1] Arora P and Zhang Z 2004 Battery Separators *Chem. Rev.* **104** 4419-62
[2] Bae B, Chun B H and Kim D 2001 Surface characterization of microporous polypropylene membranes modified by plasma treatment *Polymer* **42** 7879-85
[3] Bhat N V and Upadhyay D J 2002 Plasma-induced surface modification and adhesion enhancement of polypropylene surface *J. Appl. Polym. Sci.* **86** 925-36
[4] Bhat N V, Upadhyay D J, Deshmukh R R and Gupta S K 2003 Investigation of Plasma-Induced Photochemical Reaction on a Polypropylene Surface *J. Phys. Chem. B* **107** 4550-9
[5] Boyd R D, Kenwright A M, Badyal J P S and Briggs D 1997 Atmospheric Nonequilibrium Plasma Treatment of Biaxially Oriented Polypropylene *Macromolecules* **30** 5429-36
[6] Briggs D, Brewis D M and Konieczko M B 1979 X-ray photoelectron spectroscopy studies of polymer surfaces *J. Mater. Sci.* **14** 1344-8
[7] Chaivan P, Pasaja N, Boonyawan D, Suanpoot P and Vilaithong T 2005 Low-temperature plasma treatment for hydrophobicity improvement of silk *Surf. Coat. Technol.* **193** 356-60
[8] Chung F H 1991 Unified theory and guidelines on adhesion *J. Appl. Polym. Sci.* **42** 1319-31
[9] Chung T-C 1998 Hydrophilic polypropylene membranes. Google Patents)
[10] Chung T C and Lee S H 1997 New hydrophilic polypropylene membranes; fabrication and evaluation *J. Appl. Polym. Sci.* **64** 567-75
[11] Ciszewski A, Gancarz I, Kunicki J and Bryjak M 2006 Plasma-modified polypropylene membranes as separators in high-power alkaline batteries *Surf. Coat. Technol.* **201** 3676-84
[12] Corn S, Vora K P, Strobel M and Lyons C S 1991 Enhancement of adhesion to polypropylene films by CF3Cl plasma treatment *J. Adhes. Sci. Technol.* **5** 239-45
[13] Dai Z-W, Wan L-S and Xu Z-K 2008 Surface glycosylation of polyacrylonitrile ultrafiltration membrane to improve its anti-fouling performance *J. Membr. Sci.* **325** 479-85
[14] Feng R, Wang C, Xu X, Yang F, Xu G and Jiang T 2011 Highly effective antifouling performance of N-vinyl-2-pyrrolidone modified polypropylene non-woven fabric membranes by ATRP method *J. Membr. Sci.* **369** 233-42
[15] Gérard E, Bessy E, Salvagnini C, Rerat V, Momtaz M, Hénard G, Marmey P, Verpoort T and Marchand-Brynaert J 2011 Surface modifications of polypropylene membranes used for blood filtration *Polymer* **52** 1223-33
[16] Goel N, Bhardwaj Y, Manoharan R, Kumar V, Dubey K, Chaudhari C and Sabharwal S 2009 Physicochemical and electrochemical characterization of battery separator prepared by radiation induced grafting of acrylic acid onto microporous polypropylene membranes *Lett* **3** 268-78
[17] Gomathi N, Sureshkumar A and Neogi S 2008 RF plasma-treated polymers for biomedical applications *Curr. Sci.* **94** 1478
[18] Guo H and Ulbricht M 2010 Surface modification of polypropylene microfiltration membrane via entrapment of an amphiphilic alkyl oligoethyleneglycolether *J. Membr. Sci.* **349** 312-20
[19] Hansen R H and Schonhorn H 1966 A new technique for preparing low surface energy polymers for adhesive bonding *J. Polym. Sci., Part B: Polym. Lett.* **4** 203-9
[20] Himma N F, Anisah S, Prasetya N and Wenten I G 2016a Advances in preparation, modification, and application of polypropylene membrane *J. Polym. Eng.* **36** 329-62
[21] Himma N F, Wardani A K and Wenten I G 2016b Preparation of Superhydrophobic Polypropylene Membrane Using Dip-Coating Method: The Effects of Solution and Process Parameters *Pol. Plast. Technol. Eng.*

[22] Hollahan J R, Stafford B B, Falb R D and Payne S T 1969 Attachment of amino groups to polymer surfaces by radiofrequency plasmas *J. Appl. Polym. Sci.* **13** 807-16

[23] Jin J, Zhang C, Jiang W, Luan S, Yang H, Yin J and Stagnaro P 2012 Melting grafting polypropylene with hydrophilic monomers for improving hemocompatibility *Colloids Surf., A* **407** 141-9

[24] Kato K, Uchida E, Kang E-T, Uyama Y and Ikada Y 2003 Polymer surface with graft chains *Prog. Polym. Sci.* **28** 209-59

[25] Kochkodan V M and Sharma V K 2012 Graft polymerization and plasma treatment of polymer membranes for fouling reduction: A review *Journal of Environmental Science and Health, Part A* **47** 1713-27

[26] Lanauze J A and Myers D L 1990 Ink adhesion on corona-treated polyethylene studied by chemical derivatization of surface functional groups *J. Appl. Polym. Sci.* **40** 595-611

[27] Le-Clech P, Chen V and Fane T A G 2006 Fouling in membrane bioreactors used in wastewater treatment *J. Membr. Sci.* **284** 17-53

[28] Lipnizki F, Hausmanns S, Ten P-K, Field R W and Laufenberg G 1999 Organophilic pervaporation: prospects and performance *Chem. Eng. J.* **73** 113-29

[29] Liston E M 1989 Plasma Treatment for Improved Bonding: A Review *J. Adhes.* **30** 199-218

[30] Liu Z-M, Xu Z-K, Wang J-Q, Wu J and Fu J-J 2004 Surface modification of polypropylene microfiltration membranes by graft polymerization of N-vinyl-2-pyrrolidone *Eur. Polym. J.* **40** 2077-87

[31] Meng J, Li J, Zhang Y and Ma S 2014 A novel controlled grafting chemistry fully regulated by light for membrane surface hydrophilization and functionalization *J. Membr. Sci.* **455** 405-14

[32] Mulder M 1996 Basic principles of membrane technology. Netherlands. Kluwer Academic Publishers)

[33] Nie F-Q, Xu Z-K, Huang X-J, Ye P and Wu J 2003 Acrylonitrile-Based Copolymer Membranes Containing Reactive Groups: Surface Modification by the Immobilization of Poly(ethylene glycol) for Improving Antifouling Property and Biocompatibility *Langmuir* **19** 9889-95

[34] Owens D K 1975 Mechanism of corona–induced self-adhesion of polyethylene film *J. Appl. Polym. Sci.* **19** 265-71

[35] Rajesh D and Mark J K 2003 A model for plasma modification of polypropylene using atmospheric pressure discharges *J. Phys. D: Appl. Phys.* **36** 666

[36] Shahidi S, Ghoranneviss M, Moazzenchi B, Rashidi A and Dorranian D 2007 Effect of using cold plasma on dyeing properties of polypropylene fabrics *Fibers Polym.* **8** 123-9

[37] Shim J K, Na H S, Lee Y M, Huh H and Nho Y C 2001 Surface modification of polypropylene membranes by γ-ray induced graft copolymerization and their solute permeation characteristics *J. Membr. Sci.* **190** 215-26

[38] Singh N, Chen Z, Tomer N, Wickramasinghe S R, Soice N and Husson S M 2008 Modification of regenerated cellulose ultrafiltration membranes by surface-initiated atom transfer radical polymerization *J. Membr. Sci.* **311** 225-34

[39] Slepička P, Vasina A, Kolská Z, Luxbacher T, Malinský P, Macková A and Švorčík V 2010 Argon plasma irradiation of polypropylene *Nucl. Instrum. Methods Phys. Res., Sect. B* **268** 2111-4

[40] Sykes J, Moloney A, Briggs D and Dahm R 1982 Surface analysis and pretreatment of plastics and metals *Appl. Sci. Pub.* 153

[41] Tsai C-Y, Juang R-S and Huang C 2011 Surface Modification of Polypropylene Membrane by RF Methane/Oxygen Mixture Plasma Treatment *Japanese J. Appl. Phys.* **50** 08KA2

[42] Ulbricht M 2006 Advanced functional polymer membranes *Polymer* **47** 2217-62

[43] Uyama Y, Kato K and Ikada Y 1998 *Grafting/Characterization Techniques/Kinetic Modeling,* ed H Galina, *et al.* (Berlin, Heidelberg: Springer Berlin Heidelberg) pp 1-39

[44] Venault A, Chang Y, Hsu H-H, Jhong J-F, Yang H-S, Wei T-C, Tung K-L, Higuchi A and Huang J 2013 Biofouling-resistance control of expanded poly(tetrafluoroethylene) membrane via atmospheric plasma-induced surface PEGylation *J. Membr. Sci.* **439** 48-57

[45] Wan L-S, Liu Z-M and Xu Z-K 2009 Surface engineering of macroporous polypropylene membranes *Soft Matter* **5** 1775-85
[46] Wang C-c, Yang F-l, Liu L-F, Fu Z-m and Xue Y 2009 Hydrophilic and antibacterial properties of polyvinyl alcohol/4-vinylpyridine graft polymer modified polypropylene non-woven fabric membranes *J. Membr. Sci.* **345** 223-32
[47] Wenten I G 1995 Mechanisms and control of fouling in crossflow microfiltration *Filtr. Sep.* **32** 252-3
[48] Xu Q, Yang J, Dai J, Yang Y, Chen X and Wang Y 2013 Hydrophilization of porous polypropylene membranes by atomic layer deposition of TiO2 for simultaneously improved permeability and selectivity *J. Membr. Sci.* **448** 215-22
[49] Yan M-G, Liu L-Q, Tang Z-Q, Huang L, Li W, Zhou J, Gu J-S, Wei X-W and Yu H-Y 2008 Plasma surface modification of polypropylene microfiltration membranes and fouling by BSA dispersion *Chem. Eng. J.* **145** 218-24
[50] Yang Q, Xu Z-K, Dai Z-W, Wang J-L and Ulbricht M 2005 Surface Modification of Polypropylene Microporous Membranes with a Novel Glycopolymer *Chem. Mater.* **17** 3050-8
[51] Yang Y-F, Li Y, Li Q-L, Wan L-S and Xu Z-K 2010 Surface hydrophilization of microporous polypropylene membrane by grafting zwitterionic polymer for anti-biofouling *J. Membr. Sci.* **362** 255-64
[52] Yang Y-F, Wan L-S and Xu Z-K 2009 Surface hydrophilization for polypropylene microporous membranes: A facile interfacial crosslinking approach *J. Membr. Sci.* **326** 372-81
[53] Yasuda H and Wang C R 1985 Plasma polymerization investigated by the substrate temperature dependence *J. Polym. Sci. : Polym. Chem. Ed.* **23** 87-106
[54] Yin J and Deng B 2015 Polymer-matrix nanocomposite membranes for water treatment *J. Membr. Sci.* **479** 256-75
[55] Yu H-Y, He X-C, Liu L-Q, Gu J-S and Wei X-W 2007a Surface modification of polypropylene microporous membrane to improve its antifouling characteristics in an SMBR: N2 plasma treatment *Water Res.* **41** 4703-9
[56] Yu H-Y, Liu L-Q, Tang Z-Q, Yan M-G, Gu J-S and Wei X-W 2008 Surface modification of polypropylene microporous membrane to improve its antifouling characteristics in an SMBR: Air plasma treatment *J. Membr. Sci.* **311** 216-24
[57] Yu H-Y, Xie Y-J, Hu M-X, Wang J-L, Wang S-Y and Xu Z-K 2005 Surface modification of polypropylene microporous membrane to improve its antifouling property in MBR: CO2 plasma treatment *J. Membr. Sci.* **254** 219-27
[58] Yu H-Y, Xu Z-K, Lei H, Hu M-X and Yang Q 2007b Photoinduced graft polymerization of acrylamide on polypropylene microporous membranes for the improvement of antifouling characteristics in a submerged membrane-bioreactor *Sep. Purif. Technol.* **53** 119-25
[59] Yun Y I, Kim K S, Uhm S-J, Khatua B B, Cho K, Kim J K and Park C E 2004 Aging behavior of oxygen plasma-treated polypropylene with different crystallinities *J. Adhes. Sci. Technol.* **18** 1279-91
[60] Zhang R, Liu Y, He M, Su Y, Zhao X, Elimelech M and Jiang Z 2016 Antifouling membranes for sustainable water purification: strategies and mechanisms *Chemical Society Reviews*
[61] Zhao J, Shi Q, Luan S, Song L, Yang H, Shi H, Jin J, Li X, Yin J and Stagnaro P 2011 Improved biocompatibility and antifouling property of polypropylene non-woven fabric membrane by surface grafting zwitterionic polymer *J. Membr. Sci.* **369** 5-12
[62] Zhao Y-H, Wee K-H and Bai R 2010 Highly hydrophilic and low-protein-fouling polypropylene membrane prepared by surface modification with sulfobetaine-based zwitterionic polymer through a combined surface polymerization method *J. Membr. Sci.* **362** 326-33
[63] Zheng Y, Liu H, Gurgel P V and Carbonell R G 2010 Polypropylene nonwoven fabrics with conformal grafting of poly(glycidyl methacrylate) for bioseparations *J. Membr. Sci.* **364** 362-71

[64] Zou L, Vidalis I, Steele D, Michelmore A, Low S P and Verberk J Q J C 2011 Surface hydrophilic modification of RO membranes by plasma polymerization for low organic fouling *J. Membr. Sci.* **369** 420-8

IOP Conf. Series: Materials Science and Engineering **214** (2017) 012015 doi:10.1088/1757-899X/214/1/012015

The Effect of Mo addition in stainless steels on the corrosion behavior in the nano fluids contain Al2O3 nano particles

Djoko Hadi Prajitno and Dani Gustaman Syarif
PSTNT-BATAN Jl. Tamansari 71 Bandung 40132

Email: djokohp@batan.go.id

Abstract. The effects of molybdenum in stainless steel on the corrosion behaviour in nano fluids contain nano particle Al2O3 in the de mineralized water have been investigated. Open circuit potential (OCP), Tafel polarization and electrochemical impedance spectroscopy (EIS) testing was performed in the demineralized water contain nano particle 0, 01% Al2O3 as Nano fluid. Metallurgical techniques such as X-ray diffraction (XRD) and optical microscopy were used to characterize the alloys before and after corrosion testing. According to the open circuit potential measurement of the stainless steel in the de mineralized water and nano fluid media, it is showed that stainless steel with Mo showed more negative OCP compared without Mo stainless steel. The results of the Tafel polarization technique show that corrosion currents stainless steels in the nano fluid decrease with the increasing of Mo content in the stainless steel. Surface morphology of the specimens by optical microscope examination showed that microstructure of stainless steel SS 304 and SS 316 alloys relatively unchanged before and after corrosion testing nano fluids. X ray diffraction examination of stainless steel with and without Mo after corrosion testing depict that γFe is major phase without oxide phase on the surface of alloy

1. Introduction

Nano fluids are a mixture of nano particles in a base fluid in the form colloidal. The properties of nano fluids are significantly improvement the heat transfer and conductivity characteristics from the original base fluid [1]. These nano particles usually a metal or metal oxide is increase conduction and convection coefficients behaviour of base fluid. Experimental studies on nanofluid single-phase heat transfer have been reported in the literature [2, 3]. Alumina-water and titania-water nanofluid are studies in turbulent convective heat transfer in tubes. Laminar convective heat transfer and viscous pressure loss of alumina–water and zirconia–water nanofluid are also having been studies. The studies of nanofluid for nuclear power plants application such as primary and secondary cooling system and emergency core cooling systems(ECCSs) were carried out by a number researcher [4,5] This studies were characterized by the improvement of critical heat flux (CHF) of nanofluid after injection in experiment of loss of coolant accidents (LOCAs). The benefits of nanofluid during simulation accident conditions show good results. It is clear that the nanofluid engineered ECCSs should be compatible with conventional systems during normal operations to make nanofluid technologies practical in Nuclear Power Plants [6].

Stainless steel shows very good corrosion resistance in various media because of its ability to form a thin protective passive film when its expose to environment [7]. The alloy 304 SS is austenitic steel with its properties non-magnetic, contain ii nickel, iron, carbon and at least 18% of chromium. The chromium element is responsible for the corrosion resistance of these alloys is attributed to the presence of a thin invisible passive film of Cr2O3 [8]. Tafel polarization technique has been used to study this corrosion behaviour of metal in various media. But the adopted scan rates in the literatures vary from 0, 5 to 115 mV/s, and there are few instances of research work about the effect of scan rate on the Tafel polarization curve. In this paper, the corrosion behaviour of stainless steel 304 in nanofluid solution will be studied to illustrate how the Tafel polarization curve is disturbed by the charging current, the corrosion of stainless steel in nanofluid media contain nanoparticle ZrO2 is infrequently studied. The present work is to investigate the corrosion behaviour of stainless steel SS 304 and 316 in nanofluid solution contain 0,01 gpl nanoparticle Al2O3, by open circuit potential and potentiodynamic measurements, electrochemical impedance spectroscopy (EIS). The surface morphology of the samples is observed by optical microscopy, scanning electron microscopy (SEM) and By X-ray diffraction.

2. Experimental

2.1. Material and nanofluid

The specimens used in this experiments were circular discs stainless steel having 1,6 Cm in diameter and 2 mm in thickness. The specimens is stainless steel 304 and 316 with the major chemical composition (in wt%) 17,29 % Cr, 7,17% Ni, 0,39% Si, 0,56% Mn, 0%Mo, 0,066% C balanced Fe and 16,39 % Cr, 8,54% Ni, 0,39% Si, 0,5% Mn, 2,14%Mo, 0,02% C balanced respectively. The specimen was wet abraded with 500, 800 and 1200 emery paper, degreased with detergent after that cleaned with ultrasonic cleaner in alcohol solution [8].

Aqua demineralized water (adm) and 0.01 gpl Al2O3 nanoparticle were used to produce nanofluid. The Al2O3 nanoparticles were supplied by the inframat Advanced Materials. The nanofluid were prepared by incorporating 0,01 gram of Al2O3 nanoparticles with an average particle size of 30-60 nm mixed into 1 L aqua demineralized water by magnetite stirrer for 30 minute. After magnetite stirrer then an ultrasonic vibrator operating at a frequency of 44 kHz was used to mixing for 20 min in order to the Al2O3 particles were dispersed and stabilized in demineralized water. Nanoparticle and nanofluid preparation were described in [9].The Al2O3 nanofluid was then used for corrosion testing. Characterization of nanoparticle Al2O3 by TEM shown in Figure1.

2.2. Electrochemical study

Electrochemical tests were performed using corrosion cell. Electrochemical experiments were conducted in three electrode cell, a stainless steels disk as the working electrode, a saturated calomel electrode (SCE) as reference electrode and platinum wire were used as an auxiliary electrode. The experiments were recorded using an apparatus PCI4, G300 Gamry Potentiostat Instrument [10] connected to a personal computer provide with DC 105 software from Gamry [11]. Tafel polarization curve were obtained by the polarization of working electrode from cathodic to anodic direction at a scan rate of 5 mVS-1 and the data were analyzed using software Gamry Echem Analyst version 5.66.

2.3. Sample characterization

Corroded and no corroded samples were analysed by x ray diffraction (XRD) for phase present analysed XRD measurements were performed by using a Pan Analytical diffractometer equipped with a Cukαα radiation source (1,5406A°). The scanning , 2 θ region ranged from 30 to100. Optical
Tables should have only horizontal rules and no vertical ones. Generally, only three rules should be used: one at the top of the table, one at the bottom, and one to separate the entries from the column headings. Table rules should be 0.5 points wide.

3. Results and discussion

3.1. Micrograph nanoparticle Al_2O_3

Figure 1 shows micrograph image of nanoparticle Al_2O_3 characterization by TEM. The image shows that the average of nanoparticle size approximately 5-20 nanometer.

IOP Conf. Series: Materials Science and Engineering **214** (2017) 012015 doi:10.1088/1757-899X/214/1/012015

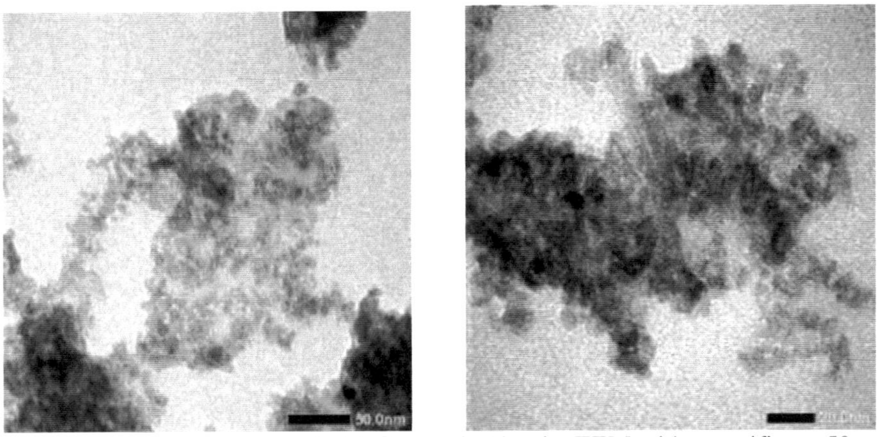

Figure 1. Micrograph nanoparticle Al_2O_3examination by TEM with magnificent 50 and 20 nm.

3.2 Open Circuit Potential (OCP)

Figure 2 shows the variation of open circuit potential (OCP) with immersion time of stainless steel SS 304 and SS 316 alloys in adm+0,01gpl Al2O3 nanoparticle at the room temperature for 600 seconds. Basically, at the first time, OCP decreases rapidly for 50S periods after that the OCP increase gradually with lower slope for long time immersion. The OCP value of SS 304 at the first time then decreases to -97 Vs (SCE) for 50s. After that the variation of OCP with immersion time are increases slowly to 600s finish. The OCP value of SS 316 start -45 then decreases to -123 Vs (SCE) for 75s, and then, the variation of OCP with time are increase slowly until 600s. This indicates that the open circuit potential of stainless steel in nanofluid is actively drop for first time due to equilibrium during 50s then the OCP of stainless steel increases slowly with time until 50s finish as a result the stainless steels corroded before stable protective film took place to the entire surface of alloys.

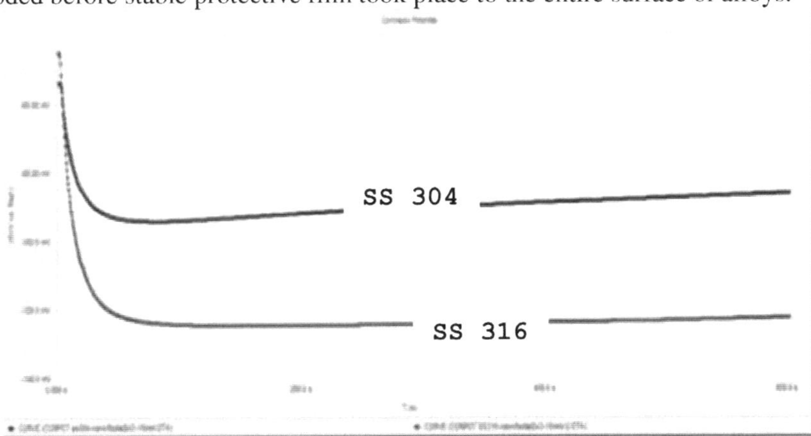

Figure 2. Variation of the open circuit potential as a function of time.

3. 3. Potentiodynamic polarization curves

Figure 3 depicted the cathodic and anodic polarization curves for SS 304 and 316 SS alloys immersed in nanofluid solution. The corrosion potential, polarization curves behavior and the plateau in the anodic and cathodic polarization of the SS 304 and 316 is dependent on the molybdenum content in the alloy [1].

The corrosion potential, corrosion current (icorr) , corrosion rate, anodic and cathodic Tafel slope were calculated by the extrapolation of the logarithmic section of the anodic and cathodic Tafel line to the point of intersection. With it of the software DC corrosion 105, the results of Tafel anodic, cathodic slope and corrosion current, corrosion rate for stainless steel with and without molybdenum are calculated and tabulated in table 1. In table 1 it can be seen that the increase of molybdenum in the alloy SS 316 makes the anodic Tafel slopes and corrosion rate will decreasing. Anodic Tafel slope value for SS 304 and 316 were 0,769 Vdec-1 and 0,327 Vdec-1, respectively. Cathodic Tafel slope value for SS 304 and 316 were 0,279 Vdec-1 and 0,171 Vdec-1, respectively. The distortion of the anodic slope can lead to the electrode process oxidation of stainless steel. The corrosion potential of SS 316 alloy in the nanofluid is -176 mv compared than for AISI 304 -148 mv. The higher corrosion potential of Stainless Steel 304 compared with SS 316 due to SS 304 more susceptible to corrosion take place. The corrosion current rate SS 304 and SS 316 alloys are 1.09 mA and 0.403 MPY, respectively. It can be concluded that addition Mo to Stainless Steel 316 improve corrosion rate du to Mo is stabilizer of passive film. Early study showed that corrosion SS 316 in chloride environment improved due to forming uniform passive layer [12].

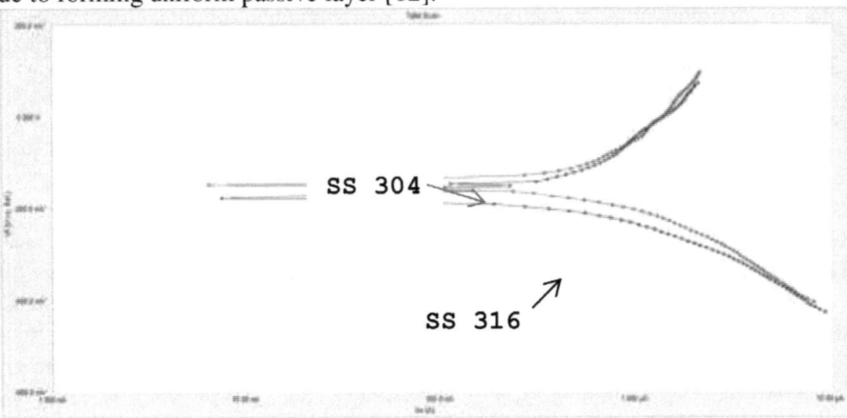

Figure 3. Tafel polarization curves obtained for Stainless steel SS 304 and 316 in nanofluid.

Table 1. Corrosion rate and Tafel slopes obtained from the polarization curves SS 304 and 316 in nanofluid

Stainless steel	Corpot. mV	Anodic Tafel slopes βa (mV/dec)	Cathodic Tafel slopes βc (mV/dec)	Icor μA	CR mpy
SS 304	-148	0.769	0.279	1.09	0.503
SS 316	-176	0.327	0.171	0.403	0.186

3.4. X-ray diffraction

The stainless steel electrode of SS 304 and 316 were potentially polarized in nanofluid solution at 25 ºC, after that time, the electrode was withdrawn and drying and finally examined by X-ray diffraction. Figure 4 a and b shows the diffractograms of SS 304 after polarization testing in nanofluid. XRD studies showed that the major phase for samples before polarization is associated with γ Fe phase as seen in Figure 4.a. After Tafel polarization testing, the peaks associated with γ Fe phase is still major phase in stainless steel. The scale of Fe2O3 associated with corrosion product of Fe cannot be seen in diffractograms. It can be concluded that corrosion stainless steel in adm + 0.01gpl nanoparticle ZrO2 is very low.

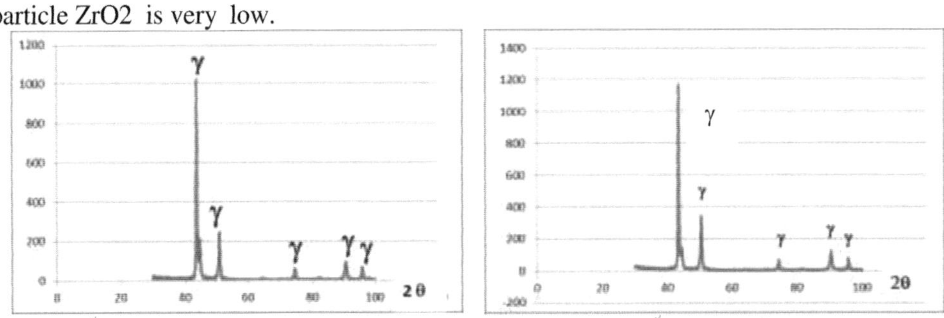

a **b**

Figure 4. XRD stainless steel 304 a) before b) after Tafel polarization in nanofluid.

3.5. Surface Morphology

Surface morphology of specimens stainless steel SS 304 and SS 316 after corrosion testing in nano fluid showed in Figure 5 a and b, respectively. In Figure 5a and 5b showed that the surface morphology specimen still smooth and free of corrosion product scale. From the data surface morphology stainless steel, it can be concluded that Nano fluid solution is not corrosive electrolyte.

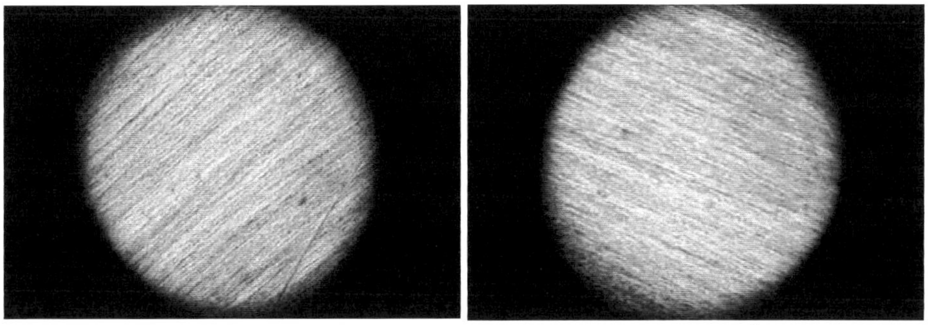

a **b**

Figure 5. Surface morphology of stainless steel (5a) SS 304 and (5b) SS 316 after corrosion test in the Nano fluids.

4. Conclusion

The corrosion behavior of the stainless steel SS 304 and 316 samples was studied in nanofluids contains 0.01 gpl nanoparticle alumina using the potentiodynamic polarization technique provided with XRD and SEM examinations. The results obtained can be summarized as follows:

The corrosion potential examination of the stainless steel SS 304 and SS 316 in nanofluid media containt 0.01gpl Al_2O_3 nanoparticle was actively corroded and the corrosion potential of SS 316 more negative compared with SS 304. The potentiodynamic polarization examination of the stainless steel SS 304 and SS 316 in nanofluid media contain 0.01gpl Al2O3 nanoparticle showed that the rate of corrosion of the stainless steel SS 304 and SS 316 in nanofluid was very low. The values of corrosion rate of these samples decreases in the order: SS 304I > SS 316. The increase in either Mo content in the stainless steels improves corrosion resistance.

5. References

[1] Djoko Hadi Prajitno 2015 Effects of scan rate on the corrosion behaviour SS 304 stainless steel in the nanofluid measured by Tafel polarization methods, AIP Conference Proceedings 1677, 070011

[2] Saidur R., Leong,K.Y.Leong, Mohammad,H.A., 2011. A review on applications and challenges of nanofluids, Renewable and Sustainable Energy Reviews ,**15**, 1646

[3] B.C. Pak, Y.I. Cho, 1998 Hydrodynamic and heat transfer study of dispersed fluids with submicron metallic oxide particles, Exp. Heat Transfer **11**, 151

[4] Ulzie Rea, Tom McKrell, Lin-wen Hu, Jacopo Buongiorno, 2009 International Journal of Heat and Mass Transfer **52**, 2042

[5] Pop MGM, Brian GL 2008 Nuclear power plant using nanoparticles in emergency systems and related method, U.S. Patent: US 2008/0212733A1.

[6] Benoît Ter-Ovanessian∗, Catherine Alemany-Dumont, Bernard Normand,, 2014 Electronic and transport properties of passive films grown ondifferent Ni-Cr binary alloys in relation to the pitting susceptibility Electrochimica Acta **133**, 373

[7] Steven J.Zinkle, et., all. 2009. Structural Materials for fusion and fission energy, Material Today v.**12** nov. 12

[8] Djoko Hadi Prajitno and dani G.S., 2017, Application Electrochemical Impedance Spectroscopy Methods to Evaluation Corrosion Behaviour of Stainless steels 304 in Nanofluids Media, Journal of Physics: Conf. Series **799,** 012007

[9] Syarif D G and Prajitno D H 2013 Journal of Materials Science and Engineering B **3 2** 122

IOP Conf. Series: Materials Science and Engineering **214** (2017) 012015 doi:10.1088/1757-899X/214/1/012015

[10] Gamry Potentiostat PCI4/G300, Electrochemical Measurement System 2010
[11] Gamry software DC 105 corrosion analysis system 2010
[12] J.M. Bastidas, C.L. Torres, E. Cano, J.L. Polo 2002 Corros. Sci. **44** 625

Acknowledgments
This work was in part my global research program in corrosion of metals in nanofluids media and financially supported by the DIPA Program Center for Applied nuclear science and technology National Nuclear Energy Agency

IOP Conf. Series: Materials Science and Engineering **214** (2017) 012016 doi:10.1088/1757-899X/214/1/012016

Microwave Synthesis of BCNO/SiO_2 Nanocomposite Material

I D Faryuni[1]*, F Ramdhani[1], J Sampurno[1], B W Nuryadin[2], F A Noor[3] and F Iskandar[3]

[1] Department of Physics, Faculty of Mathematics and Natural Sciences, Tanjungpura University, Pontianak,Indonesia
[2] Department of Physics, Faculty Sciences and Technology, UIN Sunan Gunung Djati Bandung, Cibiru-Bandung,Indonesia
[3]Physics of Electronic Materials Research Division, Faculty of Mathematics and Natural Sciences, Institut Teknologi Bandung, Bandung 40132, Indonesia

*irfana@physics.untan.ac.id

Abstract. In the present work, we report the preparation of BCNO/SiO_2 phosphor synthesized using a microwave-assisted method. This method allows a lower temperature and a shorter reaction time than simple heating (furnace). The phosphors were prepared from precursors containing, boric acid, urea, citric acid and SiO_2 nanoparticles. To These precursors, silica nanoparticles were added at various concentrations from 0 to 5 %wt. The emission wavelength produced by the phosphor was varied by varying the fraction mass of the silica that were added to the precursors. The results showed that higher photoluminescence (PL) intensity was produced by the BCNO/SiO_2 with 3 %wt silica addition. The novelty of this research is using microwave heating for BCNO/SiO_2 synthesis, which is usually conducted using a simple heating method.
Keywords:BCNO SiO_2, microwave synthesis

1. Introduction

Boron carbon oxynitride (BCNO) is a phosphoric material without rare earth metal, composed of boron (B), carbon (C), nitrogen (NO) and oxygen (O). BCNO can be an alternative fluorescent material for lighting applications[1]. BCNO can be prepared by several techniques, such as simple heating (furnace) [2][3][4], hydrothermal synthesis [5], electrospinning [6] and microwave heating[7]. Microwave-assisted BCNO synthesis was reported by Iwasaki[8].We have reported previously that BCNO/SiO_2 was successfully synthesized using a simple heating process [9].

In the present study, we investigated the utilization of microwave heating for synthesis of BCNO/SiO_2 nanoparticles, which has not been reported on previously. Using boric acid, urea and citric acid as raw materials, and the addition of fumed silica with microwave-assisted method allow a shorter reaction time and lower reaction temperature than electric-furnace heating.

2. Materials and synthesis

A schematic flow chart for the BCNO/SiO_2 synthesis via microwave heating method is shown in Fig.1. Compound of BCNO/SiO_2 was prepared using a microwave-assisted method. First, urea [$(NH_2)_2CO$ with mw of 60.07g/mol], boric acid [H_3BO_3 with mw of 61.83 g/mol] and citric acid [$C_6H_8O_7$ with mw of 192.124 g/mol] were used as the nitrogen sources, boron sources and carbon sources, respectively. The additional material consisted of SiO_2 nanoparticles (particle size of 40 nm) (Wacker

Chemicals Fumed Silica Co. Ltd. China). Precursor solutions were prepared by dissolving the urea, boric acid, citric acid and nanoparticle silica in 15 ml distilled water as the solvent, followed by stirring in a magnetic stirrer to obtain homogeneous solutions. The precursors were then heated in a commercial microwave with a maximum oven output of 1000 Watt for 120s.

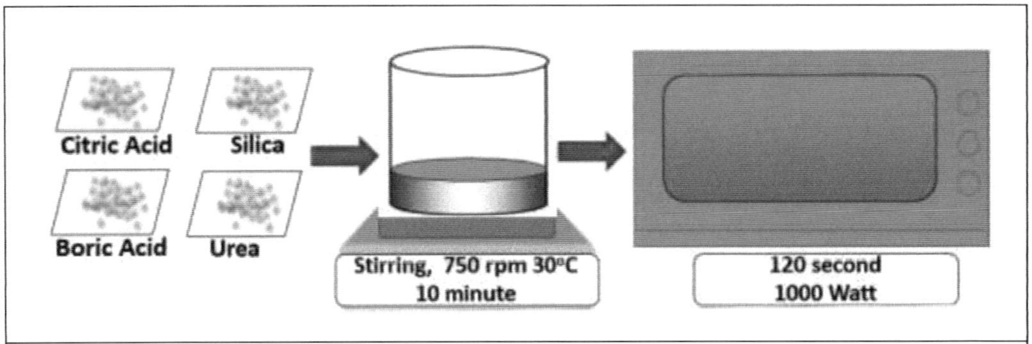

Figure 1. Microwave-assisted synthesis of BCNO/SiO_2 phosphor.

2.1 Measurement and Characterization

The photoluminescence (PL) spectrum of each sample was measured at room temperature using Cary Eclipse spectrofluorophotometer (Agilent, Australia). All PL analyses were performed with 365 nm excitation at room temperature.

3. Discussion

To investigate the effect of silica addition (0-5wt%) on the PL spectra, the microwave and the time were set constant at 120 second. Figure 2 shows the emission spectra of BCNO/SiO_2 and the PL spectra of the samples under excitation at 365 nm. The PL spectra of the sample with no silica addition has an emission peak at 488.029 nm with a PL intensity of 119.175 a.u., which is a greenish blue emission. The PL spectra of the sample with 3wt% silica addition has an emission peak at 461.059 nm with a PL intensity of 381.671 a.u, which is a blue emission. The PL spectra of the sample with 5 wt% silica addition has an emission peak at 486.059 nm with a PL intensity of 194.587 a.u, which is a greenish blue emission.

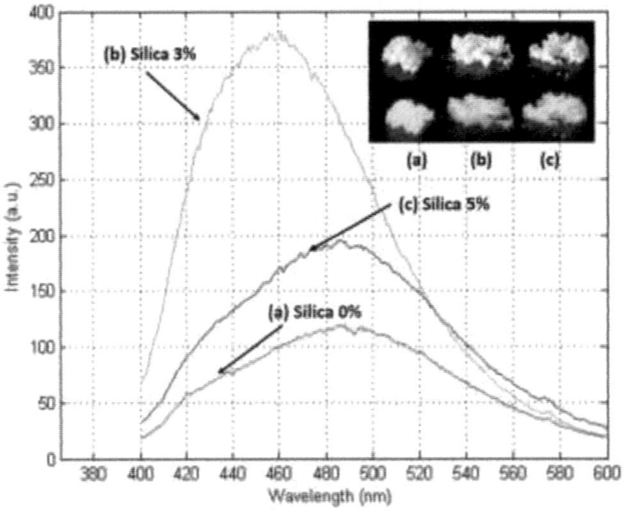

Figure 2. PL Spectra of BCNO/SiO_2 fabricated with microwave assisted method.

IOP Conf. Series: Materials Science and Engineering **214** (2017) 012016 doi:10.1088/1757-899X/214/1/012016

The PL spectra variation depends on the silica addition. The sample with 3wt% silica addition shows the highest intensity. The intensity is 3-fold larger than that of samples without SiO_2 nanoparticles. The PL spectra of the sample that is synthesized using the microwave method with 3 wt% silica has a higher intensity than the PL spectra of the sample with 0 wt% and 5 wt% silica. This phenomenon is consistent with our previous observations [8]. In our previous research, the PL spectra of the sample with 3 wt% silica that was synthesized using a simple heating method also has a higher intensity than the PL spectra of the sample with 0 wt% and 5 wt% silica. In our previous research the carbon source was PEG MW 20000, while in this report the carbon source is citric acid. The thermal properties of SiO_2 are suspected to be an important factor affecting the luminescence spectra and the uniformity of $BCNO/SiO_2$. In a crystallized nanocomposite, which comprised 3wt% SiO_2, more homogeneous photoluminescence is produced and more distributed heat happen. These phenomena are caused by SiO_2 which spreads to the entire of sample body and related with the nitrogen composition. On the other hand, while the fraction of SiO_2 was increased to 5wt%, the agglomeration of SiO_2 displacement happens. As the consequence, the luminescent centers were distributed with not homogeneous, the nitrogen compocition in the powder decreased and finally, in order to obtain the best luminescence properties, we believe that 3wt% is the optimum SiO_2 fraction for a $BCNO/SiO_2$ nanocomposite.

4. Conclusions

In this study, we have successfully synthesized $BCNO/SiO_2$ from boric acid, urea, citric acid and SiO_2 nanoparticles using a microwave-assisted method. It is found that the emission wavelength changes depending on silica addition. Samples with 3wt% SiO_2 shows the highest PL intensity, as previously reported for samples with 3wt% synthesized using a simple heating method. Shorter heating period and lower reaction temperature makes possible using microwave-assisted method, it could lead to efficient energy use.

References

[1] Kaihatsu, Y, Iskandar F, Widiyandari H, Wang W N and Okuyama K 2009 *Electrochem. Solid State Lett* **12** 3J33
[2] Wang W N, Ogi T, Kaihatsu Y, Iskandar F and Okuyama K 2011 *J. Mater. Chem* **21**5183
[3] Ogi T, Kaihatsu Y, Iskandar F, Wang W N and Okuyama K 2008 *Adv. Mater* **20**3235
[4] Liu X, Ye S, Qiao Y, Dong G, Zhang Q and Qiu J 2009 *Chem. Commun* **27**4073
[5] Xue Q, Zhang H, Zhu M, Wang Z, Pei Z, Huang Y, Huang Y, Song X, Zeng H and Zhi Chunyi 2016 *RSC Adv* **6**79090
[6] Suryamas A B, Munir M M, Ogi T, Khairurrijal andOkuyama K 2011 J. *Mater. Chem* **21** 12629
[7] Kang Y, Yuan B, Zhang D, Ma T, Huang X, Chu Z and Lai K 2016 *J. Lumin* **179** 501
[8] Iwasaki H, Ogi T, Iskandar F, Aishima K and Okuyama K 2015 *J. Lumin* **166** 148
[9] Faryuni I D, Nuryadin B W, Iskandar F, Abdullah M, Khairurrijal, Ogi T and Okuyama K2014 *J.Lumin* **148** 165

Acknowledgments

This work was supported by DRPM DIKTI (Hibah PEKERTI) Research Grant, Number 022/SP2H/PPM/DRPM/II/2016 in the fiscal years of2015-2016.

IOP Conf. Series: Materials Science and Engineering **214** (2017) 012017 doi:10.1088/1757-899X/214/1/012017

Transport and Retention Modelling of Iron Oxide Nanoparticles in Core Scale Porous Media for Electromagnetic Heating Well-Stimulation Optimization

R K Santoso[1,2,*], S Rachmat[1,2], W D K Putra[1,2], A H Resha[1,3] and H Hartowo[1,2]

[1]OGRINDO Research Consortium, Institut Teknologi Bandung, Indonesia
[2]Petroleum Engineering, Institut Teknologi Bandung, Indonesia
[3]Chemical Engineering, Institut Teknologi Bandung, Indonesia

* Email: ryan.kurniawan@students.itb.ac.id

Abstract. Understanding the transport and retention of iron oxide nanoparticles is critical in optimizing electromagnetic heating well stimulation. If the injected concentration or injection rate is too big, nanoparticles can build-up inside the pore throat, which can reduce the permeability of the reservoir. A numerical model has been created to describe the behavior of iron oxide nanoparticles in porous media. The model is coupling material balance equation and fluid flow in porous media equations. There are six parameters to be estimated through matching with experimental data: irreversible attachment rate, reversible attachment rate, irreversible attachment capacity, reversible attachment capacity, reversible detachment rate and permeability. All parameters were obtained directly through coreflooding result in previous study. We add Langmuir static isotherm test to limit the maximum adsorption capacity to provide a better estimation of concentration distribution. We use 1% NaCl solution as the base fluid and 45-50 mesh sand as the porous media. From the Langmuir static isotherm test, the maximum adsorption concentration is determined. Then, coreflooding is conducted using 10 ppm nanofluid and 12 cc/min injection rate. The proposed model is matched with the experimental data and its parameters are consistent with the maximum adsorption capacity provided from the test.

1. Introduction

Currently, nanoparticles have been extensively applied in various fields such as medicine, chemical, electronic and engineering. Nanoparticles, also called engineered nanomaterial, could be designed for certain purpose. Some of the nanoparticles types are highly responsive when they are induced with electromagnetic field. UV-visible, microwave and infra-red spectra may be used as sources to provoke translation, rotation and vibration motion of those magnetic nanoparticles [1]. Absorbed energies are then converted into heat. This process is known as electromagnetic heating assisted with nanoparticles.

The usage of nanoparticles to improve oil recovery has been popular with many methods and mechanisms applied to increase oil production. Some purposes such as stabilization of foam and emulsion, oil wettability alteration on the sand surface and interfacial tension (IFT) reduction could be conducted by nanoparticles support [2]. There are many developing nanoparticles application in oil exploration and production. Carbon dioxide in water foams may be generated by nano silica without surfactant addition, as reported by Espinosa et al. (2010) [3]. Huang et al. (2008) used nanoparticles to coat hydraulic fracture proppant to control fines migration without decreasing productivity [4]. Zhang et al. (2010) stated that oil-in-water and water-in-oil emulsions which are stabilized with different surface-coated silica nanoparticles of uniform size remain stable for several months without coalescence [5].

Although nanoparticles have many usages in oil production, one of the most critical issues is how to transport nanoparticles into the reservoir and fulfill the properties of colloidal transport in porous media. Emulsion properties depend on some factors like concentration of nanoparticle, fluid salinity, and initial injected volume ratio [6]. Aggregation, retention, adsorption and absorption phenomena can occur in porous media. Other than that, previously adsorbed nanoparticles may attach other particles injected afterwards [7]. Godinez and Darnault (2011) reported that the aggregation of nanoparticles and the electrostatic interaction between surfaces of porous media and nanoparticles are impacted by the solution ionic strength and pH [8].

Presenting those phenomena with equations will provide futher understanding and furthermore, can be applied to another cases without conducting the same experiment. Modifying computational flow design also gives a chance to modify the equations easily. In this paper, we modify material balance equation and couple it with fluid flow in porous media equation to describe nanoparticles flow behavior. Coreflooding is conducted to validate our proposed model.

2. Mathematical Model

Zhang et al. (2016) proposed a mechanistic model for nanoparticle retention in porous media. The model is an independent two-site model (ITSM) [9]. The ITSM is based on mass conservation, irreversible attachment, and reversible attachment. We proposed a modification of ITSM model by coupling it with diffusivity equation which provides continuously changing interstitial velocity, depends on pressure gradient instead of a constant interstitial velocity in ITSM.

2.1. Independent Two-Site Model

$$\frac{\partial c}{\partial t}+\frac{\rho_b}{\phi}\frac{\partial s_1}{\partial t}+\frac{\rho_b}{\phi}\frac{\partial s_2}{\partial t}=D\frac{\partial^2 c}{\partial x^2}-v\frac{\partial c}{\partial x} \tag{1}$$

$$\frac{\rho_b}{\phi}\frac{\partial s_1}{\partial t}=k_{irr}\left(1-\frac{s_1}{s_{1\max}}\right)c \tag{2}$$

$$\frac{\rho_b}{\phi}\frac{\partial s_2}{\partial t}=k_{ra}\left(1-\frac{s_2}{s_{2\max}}\right)c-\frac{\rho_b}{\phi}k_{rd}s_2 \tag{3}$$

Equation 1 is mass conservation equation, **Equation 2 and 3** are irreversible attachment and reversible attachment equation respectively where

c	: nanoparticle dispersion concentration, fraction
ρ_b	: bulk density, ML^{-3}
ϕ	: porosity, fraction
D	: hydrodynamic dispersion coefficient, L^2T^{-1}
v	: interstitial flow velocity, LT^{-1}
s_1	: irreversible attachment concentrations on solid surface, fraction
s_2	: reversible attachment concentrations on solid surface, fraction
$s_{1\max}$	: irreversible attachment capacity, fraction

$s_{2\max}$: reversible attachment capacity, fraction
k_{irr} : irreversible attachment rate coefficient, T^{-1}
k_{ra} : reversible attachment rate coefficient, T^{-1}
k_{rd} : reversible detachment rate coefficient, T^{-1}

2.2. *Fluid Flow in Porous Media*

$$\frac{\partial P}{\partial t} = \frac{k}{\phi \mu c_t} \frac{\partial^2 P}{\partial x^2} \tag{4}$$

where

P : pressure, $ML^{-1}T^{-2}$
ϕ : porosity, fraction
μ : dynamic viscosity, $ML^{-1}T^{-1}$
c_t : total compressibility, $M^{-1}LT^2$
k : permeability, L^2

2.3. *Boundary Condition and Constraint*

In the experiment setup explained in the next section, the boundary conditions are

1. injection condition at inner boundary (Neumann boundary)
$$\left.\frac{\partial P}{\partial x}\right|_{x=0} = -\frac{\mu Q B}{kA}$$
2. constant outlet pressure at outer boundary (Dirichlet boundary)
$$P|_{x=L} = P_{atmospheric}$$
3. constant injected concentration at inner boundary (Dirichlet boundary)
$$c|_{x=0} = c_{injection}$$
4. no concentration flux condition at outer boundary (Neumann boundary)
$$\left.\frac{\partial c}{\partial x}\right|_{x=L} = 0$$
5. initial pressure at all points in the core is confining pressure
$$P|_{t=0} = P_{confining}$$
6. initial concentration at all points in the core is zero
$$c|_{t=0} = 0$$
7. with the constraint
$$s_{2\max} < s_{Langmuir\_Test}$$

IOP Conf. Series: Materials Science and Engineering **214** (2017) 012017 doi:10.1088/1757-899X/214/1/012017

2.4. Stokes-Einstein Equation

In calculating the hydrodynamic dispersion coefficient, Stokes-Einstein equation [10] can be applied

$$D = \frac{k_B T}{6\pi\mu r} \tag{5}$$

where

k_B	: Boltzmann's constant, $ML^2T^{-2}\theta^{-1}$
T	: Temperature, θ
μ	: dynamic viscosity, $ML^{-1}T^{-1}$
r	: radius of nanoparticle, L

2.5. Darcy Equation

Darcy equation is a common equation used for fluid flow in porous media.

$$v = -\frac{k}{\mu}\phi\frac{\partial P}{\partial x} \tag{6}$$

where

v	: interstitial velocity, LT^{-1}
k	: permeability, L^2
μ	: dynamic viscosity, $ML^{-1}T^{-1}$
ϕ	: porosity, fraction
$\frac{\partial P}{\partial x}$	: pressure gradient, $ML^{-2}T^{-2}$

2.6. Calculation Algorithm

There are 6 parameters to be estimated. The parameter estimation is done by means of this algorithm

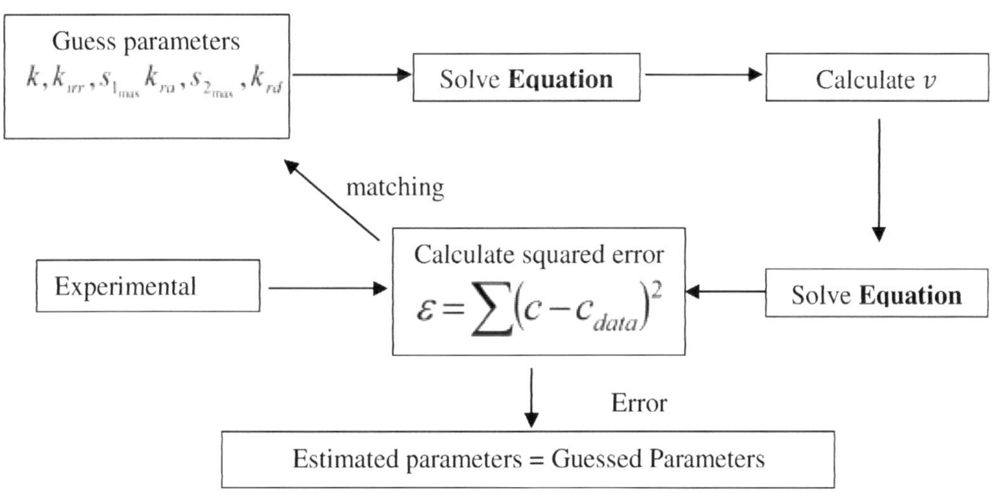

Figure 1. Parameter Estimation Algorithm.

3. Experimental Procedure

In order to obtain the five constant data (irreversible and reversible attachment rate, reversible detachment rate, irreversible attachment capacity and reversible attachment capacity) for the simulation purpose, we conducted coreflooding test using iron oxide nanoparticles in loose sand core. The coreflooding test mimics the real fluid flow in reservoir. Langmuir test was also conducted to provide the data for the boundary condition.

3.1. Materials and Tools

3.1.1. Materials

For making nanofluid, we use iron oxide nanoparticles from Sigma Aldrich, which known properties are <50 nm in size and 5 mg in weight. As a comparison, we use base fluid consists of NaCl powder for making 10 ppm brine solution. A loose sand core is also used as a representative to create a porous medium with average porosity of 38.4%.

3.1.2. Tools

Apparatus used include flooding setup, core holder (including wire net, rubber, and lids), laboratory glasswares, and digital scale.

3.2. Nanofluid Preparation

Aqua DM and NaCl were used to make 10 ppm synthetic brine. Then, 5 mg of iron oxide nanoparticles was gently added into 500 ml of synthetic brine to make 10 ppm nanofluid which is mixed with sonicator (Ultrasonic Cleaner, Krisbow 150179) for approximately 15-20 minutes.

Figure 2. Nanofluid used in the flooding experiment.

3.3. Core Preparation

3.3.1. Loose Sand Core Making

A Loose sand core was prepared from quartz. The wire net was installed near the core holder lids, thereafter sand was added and pressurized gradually using the core making device. The purpose of using wire net is to keep the sand from being carried away by the fluid during flooding experiment.

3.3.2. Pore Volume Determination

To calculate pore volume in this experiment, we first measured the mass of an empty set of core holder and the mass of a core holder with sand using the digital scale. From this procedure, we obtain the net mass of sand inside the core. Afterward, flooding was initiated using 10 ppm brine solution to saturate the entire core. The core was then briefly removed from hassler core holder to measure its mass. We will obtain the mass of fully saturated core.

Aside from the mass of each parameter, we also need to measure the dimension of the core to further obtain the bulk volume. Based on the measurement we have commenced, the loose sand core has a bulk volume of 17.73 cc. The results of core measurement are summarized in **Table 1**.

Table 1. Core Data

Rate	Mass of an empty core holder	Mass of a core holder + sand	Mass of a saturated core holder + sand	Net sand mass	Fluid mass	Porosity (fraction)
12 cc/min	111.1 g	148.4 g	155.2 g	37.3 g	6.8 g	0.38

Figure 3. Loose sand core.

3.4. Flooding Scenario

The flooding scenario was divided into 3 steps. The first step of the flooding was to saturate the loose sand core with 10 ppm brine solution using 1 cc/min injection rate. After the whole core was fully saturated, the mass was measured using the digital scale to calculate PV. The second step was injecting 10 ppm nanofluid with 12 cc/min injection rate for 15 cuvettes (about 4 cc/cuvette) or 8.8 PVs. Subsequently, post flush with 10 ppm brine solution and 12 cc/min injection rate for 15 cuvettes.

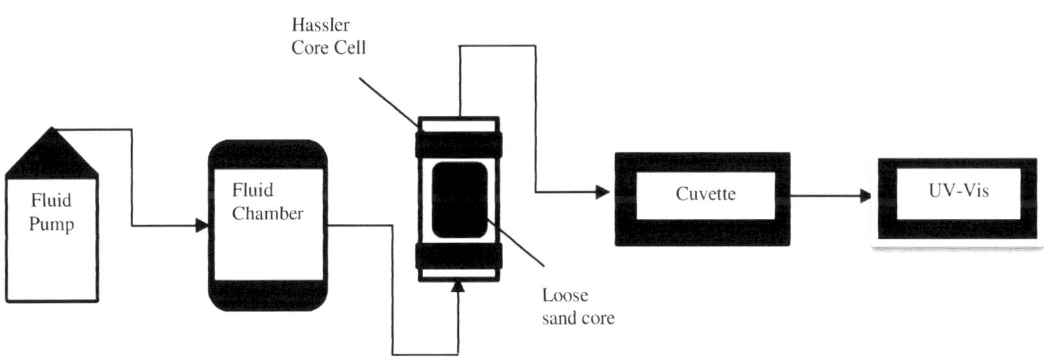

Figure 4. Schematic of flooding apparatus.

3.5. Langmuir Isothermal Test

In this section, we carried an experiment to generate Langmuir Isotherm plot. The first step was to create nanofluid samples comprised of 10 ppm brine with different nanoparticles concentration from 0 – 16 ppm and then measure them with UV – Vis spectrophotometer to obtain their initial concentration (c_0). Afterward, quartz and each sample were sonicated for 15 - 20 minutes. Each nanofluid sample should have a different concentration of nanoparticles (c_1) after sonication period. The adsorption concentration can be obtained by substracting c_0 with c_1.

4. Results and Discussion

The estimated parameters are shown in the table below

Table 2. Core Data

Parameter	Estimated value
k	$8\cdot10^{-21}$ m^2
k_{irr}	$9.9\cdot10^{-2}$ 1/s
$s_{1\max}$	$5\cdot10^{-5}$
k_{ra}	$8\cdot10^{-5}$ 1/s
$s_{2\max}$	$5\cdot10^{-8}$
k_{rd}	$4\cdot10^{-5}$ 1/s

Solving **Equation 1** and **Equation 4** with these estimated parameters and overlaying it with the experimental data result in **Figure 5** below. The proposed model is matched with the experimental data.

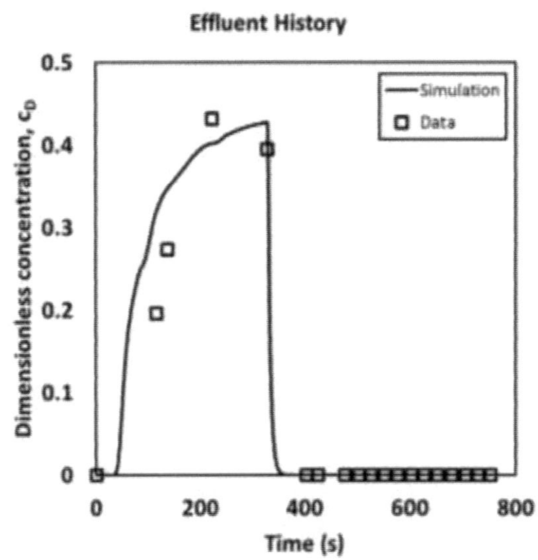

Figure 5. Effluent history simulation result overlayed with experimental data.

As we can see from the results, reversible detachment capacity ($s_{2\,\max}$) is smaller than irreversible attachment capacity ($s_{1\,\max}$), which we suspect is a notable sign of absorption phenomena.

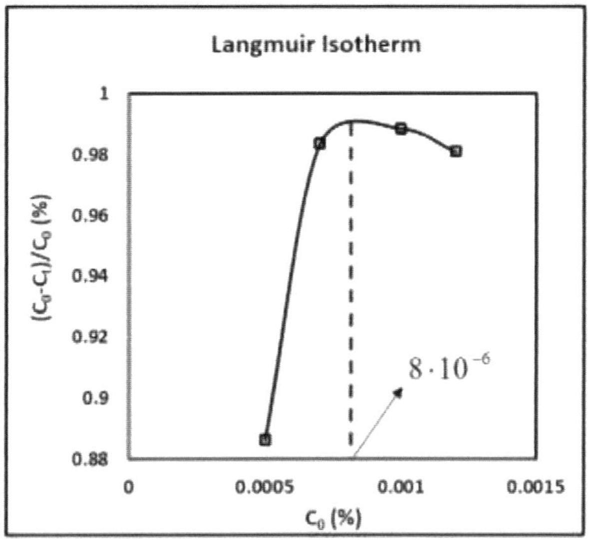

Figure 6. Langmuir isotherm plot for iron oxide shows maximum adsorption concentration of 8.10^{-6}.

Langmuir isotherm plot is conducted to confirm the estimated value of reversible detachment capacity. We can see the right part of Langmuir isotherm plot is decreasing. It is caused by agglomeration of the iron oxide nanoparticle. The maximum point before it decreases is the maximum adsorption concentration (MAC) which value is $8 \cdot 10^{-6}$. This result provides a more powerful sense of $s_{2\,\max}$ value. Langmuir isotherm test is conducted in static condition while $s_{2\,\max}$ is in dynamic condition, which means the value of $s_{2\,\max}$ should be less than MAC. While the estimated value of $s_{2\,\max}$ is less than $8 \cdot 10^{-6}$, the estimated value is confirmed to be correct.

5. Conclusion

From the simulation results, it can be concluded that the combined model to predict the transport and retention of iron oxide nanoparticles in porous media has been successfully developed. The model is developed using material balance (ITSM) equation and coupled with fluid flow in porous media equation.

Langmuir adsorption test result is used as a new constraint to increase the accuracy of prediction. The value of $s_{2\,\max}$ inside the ITSM model should be less than concentration obtained from the MAC test.

6. References

[1] Angayarkanni, S. A. & Philip, J 2015 *Adv. Colloid Interface Sci* **225** 146–176
[2] Zaid, H. M., Yahya, N., Rasyada, N. & Latiff, A 2013 *Journal of Nano Research* **21** 103–108
[3] Espinosa, D., Caldelas, F., Johnston, K., Bryant, S. L., Huh, C 2010 *Nanoparticle-Stabilized Supercritical CO2 foams for potential mobility control applications* SPE 129925, presented at SPE Improved Oil Recovery Symposium, Tulsa, Oklahoma, USA, 24-28 April 2010

[4] Huang, T., Crews, J. B., Willingham, J.R 2008 *Using nanoparticle technology to control formation fines migration* SPE 115384, presented at SPE Annual Technical Conference and Exhibition held in Denver, Colorado, USA, 21-24 September 2008

[5] Zhang, T., Davidson, D., Bryant, S. L., Huh, C 2010 *Nanoparticle-stabilized emulsions for application in enhanced oil recovery* SPE 129885, presented at SPE Improved Oil Recovery Symposium, Tulsa, Oklahoma, USA, 24-28 April 2010

[6] Solovitch, N., Lablle, J., Rose, J., Chaurand, P., Borschneck, D., Wiesner, M. R. and Bottero, J. Y 2010 *Environ. Sci. Technol* **44** 4897–4902

[7] Li, S., Torsæter, O. 2015 *The Impact of Nanoparticles Adsorption and Transport on Wettability Alteration of Water Wet Berea Sandstone* SPE 176256, presented at SPE/IATMI Asia Pacific Oil & Gas Conference and Exhibition, Nusa Dua, Bali, Indonesia, 20-22 October 2015.

[8] Godinez, I. G. and Darnault, C. J. G 2011 *Water Res* **45** 839–851

[9] Zhang, T., Murphy, M., Yu, H., Huh, C., Bryant, S.L 2016 *Transp Porous Med* **115** (2) 387–406

[10] Fullstone, G., Wood, J., Holcombe, M. and Battaglia, G. 2015 *Scientific Reports* **5** (10649)

Acknowledgement
We would like to thank DIKTI 2016 for funding this research.

IOP Conf. Series: Materials Science and Engineering **214** (2017) 012018 doi:10.1088/1757-899X/214/1/012018

Synthesize and Analysis of $Li_{1.33}Mn_{1.67}O_4$ as Adsorbent for Lithium Extraction of Lumpur Sidoarjo

L Noerochim[1], W P Kistiyanto[1], D Susanti[1], H Purwaningsih[1], and A Widodo[2]

[1]Department of Materials and Metallurgical Engineering, Institut Teknologi Sepuluh Nopember, Surabaya 60111, Indonesia
[2]Department of Geophysical Engineering, Institut Teknologi Sepuluh Nopember, Surabaya 60111, Indonesia

Email: lukman@mat-eng.its.ac.id

Abstract. $Li_{1.33}Mn_{1.67}O_4$ (LMO) was successfully synthesized by solid state reaction method at temperature variation of 500, 600, and 700 °C. As-prepared samples are immersed in 1 lt 0.5M HCL for 24 hours to create vacant site for lithium ion. Lithium extraction process is employed by soaking the adsorbent LMO in 1 lt of Lumpur Sidoarjo for 24 hrs. SEM images of adsorbent with variation of calcination temperature show irregular shape with particles size in the range of 15-52 μm. The calcination temperature effects on the size of the grain and crystallinity of adsorbent. LMO 700 sample has the highest capacity of lithium adsorption with capacity of 29.8 mg/g. It is attributed to the high crystalline of LMO 700 sample compared with other samples.

1. Introduction

Recently, lithium world consumption level has increased very significantly. The high demand for lithium is evidenced by soaring consumption in 2015. Industry analysts predict as much 36,000 ton lithium are used throughout the year. This number was up 15% compared to 2013 [1]. However, the increasing number of lithium does not have any effect in Indonesia economic growth because there is no a single company engaged in the production of lithium. Lithium sources in nature are found in seawater or brines, mineral rock and the geothermal fluid [2]. Sidoarjo mud (Lumpur Sidoarjo) is one of many geothermal fluids containing lithium, approximately 6 ppm [3]. Many researches are carried out to find the best way of extracting lithium. One such way is by adsorption method used in seawater for lithium extraction while only few researches work on geothermal fluid such as Lumpur Sidoarjo [4]. Lithium manganese oxide is one the best adsorbent which is cheap, safe and non-toxic. Previous research results show that lithium manganese oxide is capable of absorbing lithium up to 6.6 mg/g [5, 6]. In this work, lithium manganese oxide are synthesized by solid state reaction method with a mole ratio of Li / Mn 0.8 namely $Li_{1.33}Mn_{1.67}O_4$. Lithium carbonate and manganese oxide is used as a reactant. All precursors are calcined at various temperatures of 500, 600 and 700 °C. Each as-prepared adsorbent were analyzed systematically based on its ability to absorb lithium contained in Sidoarjo mud.

2. Experimental Procedures

$Li_{1.33}Mn_{1.67}O_4$ (LMO) was synthesized from $LiCO_3$ and MnO_2 by solid-state reaction process. The two reagents were dissolved and mixed well with 0.8 mole ratio in ethanol. The mixture was calcined at various temperature of of 500, 600 and 700 °C for 5 hrs namely LMO 500, LMO 600, and LMO 700, respectively. After cooling down to room temperature, as-prepared LMO powder was dissolved in 0.5 M HCl solution for 24 hrs to elute lithium and to form a vacant site in the crystal structure of LMO. This process will produce an ion sieve type adsorbent to adsorb lithium ion in geothermal fluid of Sidoarjo mud. The change in the spinel structure of each sample before and after the acid treatment was examined using X-ray diffraction (XRD, Philips Xpert with a Cu-Kα radiation tube), the morphology of particles were observed using the scanning electron microscopy (SEM, Philips FEI). The concentration of Li in the supernatant solution were determined through inductively coupled plasma atomic emission spectrophotometer (ICP-AES, 138 Ultrace, Jobin-Yvon) and the extractabilities of the lithium ions were calculated accordingly. Lithium adsorption and desorption of LMO adsorbent was started with preparation of membrane reservoir made of polypropylene (Kimtex®). The membrane reservoir was subsequently filled with the powder of LMO as much as 0.1 grams. After the introducing LMO adsorbent, all side of membrane reservoir was sealed.. The filled-membrane reservoir was placed in the 1L of 0.5M HCl aqueous solution for 24 hrs. At the 2nd day, the filled-membrane reservoir was dipped into 1L of the extractive Sidoarjo mud solution for 24 hrs. Desorption of lithium was performed in the 1L of 0.5M HCl aqueous solution for 24 hrs. Amount of lithium in the solution were measured by inductively coupled plasma atomic emission spectrophotometer (ICP-AES, 138 Ultrace, Jobin-Yvon).

3. Results And Discussion

Figure 1 presents that all XRD peaks of LMO 500, LMO 600 and LMO 700 are indexed as $Li_{1.33}Mn_{1.67}O_4$ (LMO) with spinel structure (JCPDS No. 01-088-0460). It shows that LMO is successfully synthesized by solid state reaction process at different temperature. However, XRD peaks of LMO 700 has extra peaks attributing to Li_2MnO_3 with JCPDS card no. 0-018-0737.

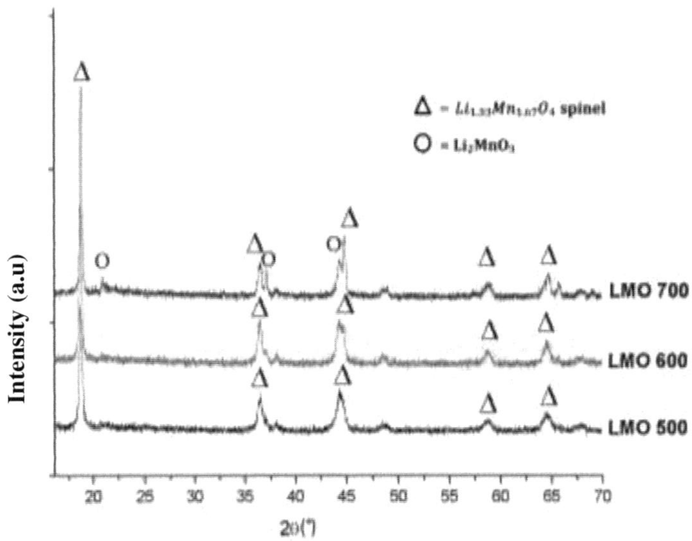

Figure 1. XRD patterns of each LMO samples.

Figure 2 shows the SEM images of as-prepared LMO. As-prepared LMO powders have irregular shape with particles size varying between 15 μm - 52 μm. There is no much different morphology from LMO 600 to LMO 700 sample.

Figure 2. SEM images of (a) LMO 500, (b) LMO 600, and (c) LMO 700.

XRD test after acid treatment for LMO adsorbent as shown in Figure 3. Generally, the shape of the diffraction pattern of LMO 500, 600 and 700 after acid treatment is not much difference with the diffraction pattern before acid treatment. The only difference is the peaks slightly shifted toward the right side to higher 2 theta values. This indicates that the Li ion extraction reaction proceed topotactically, involving a decrease of lattice constant of the spinel structure.

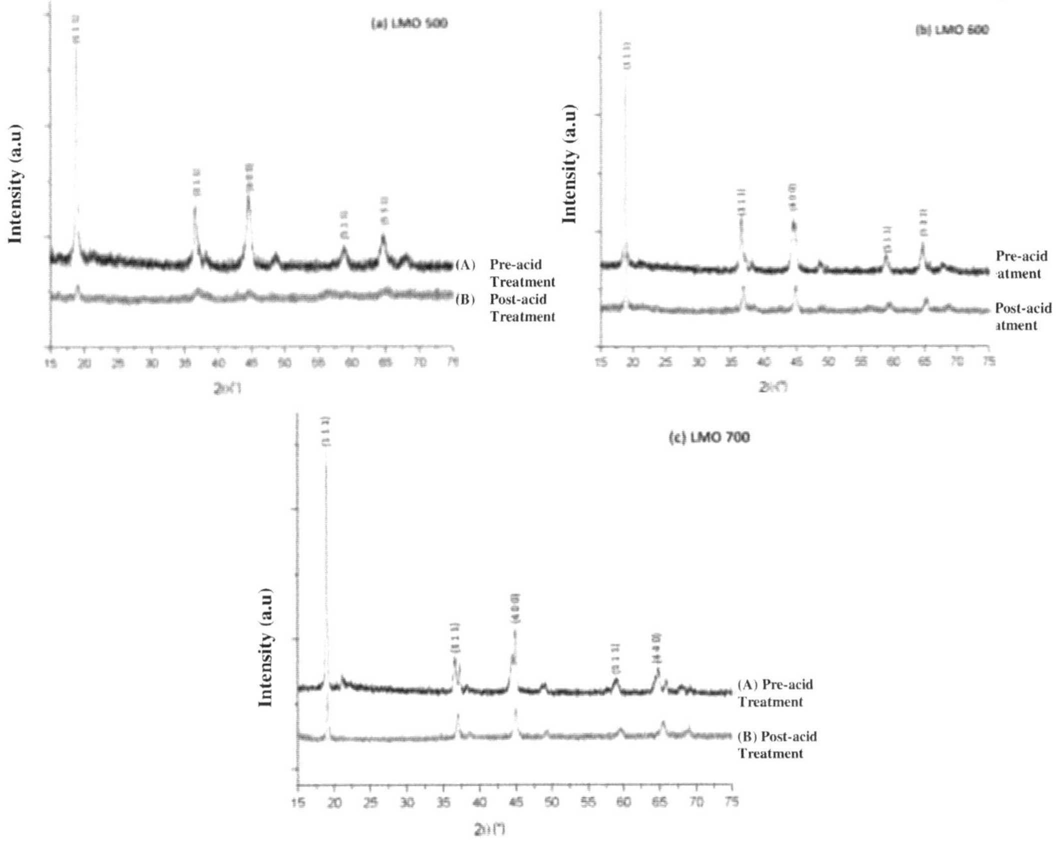

Figure 3. XRD pattern of all samples LMO before and after acid treatment (a) LMO 500, (b) LMO 600, and (c) LMO 700.

IOP Conf. Series: Materials Science and Engineering **214** (2017) 012018 doi:10.1088/1757-899X/214/1/012018

Diffraction pattern of LMO 500, 600 and 700 were in the shift of each peak to the right side. This shift is due to the ion exchange and redox mechanism at the acid treatment process. The position of each peak shifted to the right after the acid treatment can be indicated that there are changes in lattice parameters [7].

Lithium uptake
Table 1 shows the results of the ICP test for all samples of LMO. LMO 500, 600 and 700 have successfully adsorbed the lithium from Sidoarjo mud. Lithium uptake efficiency of $Li_{1.33}Mn_{1.67}O_4$ (LMO) adsorbent is calculated by equation below.

$$\text{Efficiency} = \frac{Lithium\ Post\text{-}adsorption}{Lithium\ Pre\text{-}adsorption} \text{ x } 100\% \quad (1)$$

Table 1. Lithium uptake efficiency of $Li_{1.33}Mn_{1.67}O_4$ (LMO) adsorbent

Adsorbent	Pre-adsorption (mg/g)	Post-adsorption (mg/g)	Efficiency (%)
LMO 500	29.5	27.2	92.2
LMO 600	29.5	24.8	89.85
LMO 700	29.8	27.6	92.62

The result shows that the increasing the temperature will result in the greater ability of the adsorbent to absorb lithium. LMO 700 has the highest of absorption with 29.8 mg/g and efficiency of 92.62 % (Table 1). It could be attributed to the increasing number of sites that can attract lithium ions in the crystal structure of the adsorbent which has greater temperature calcination. The higher temperature calcinations is employed, the more active sites are created. Other researches with temperature calcinations of range 300°C–800°C also exhibit similar result [8].

4. Conclusion
$Li_{1.33}Mn_{1.67}O_4$ (LMO) adsorbent was successfully prepared through the solid-state reaction process. Lithium adsorption process from Lumpur Sidoarjo (Sidoarjo mud) was performed by using the LMO-based adsorbents. LMO 500, 600 and 700 have stable spinel crystal structure. Adsorbent LMO 700 with the crystal structure of spinel has the highest ability to absorb Li. It can be attributed to the increasing number of active sites that can attract lithium ions in the crystal structure of the adsorbent which has greater calcinations temperature.

References
[1] Jaskula. Brian W 2013 “Lithium”. U.S. Geological Survey Mineral Commodity Summaries, 94-95
[2] Bakane, Piyush A 2013 “Overview Of Extraction Of Mineral/Metals With The Help Of Geothermal Fluid”. *Proceedings of The Thirty-Eighth Workshop on Geothermal Reservoir Engineering*
[3] Takinawa, Wataru 2011 “The Mechanism of Overpressure Generation in The LUSI Mud Volcano”. *E-Proceeding Symposium On Future Lusi*
[4] Chung, K.-S., Lee, J.-C., Kim, W.-K., Kim, S.B. and Cho, K.Y 2008 *J. Membrane Sci* **325** 503-508
[5] L. Noerochim, G. A. Satriawangsa, D. Susanti, A. Widodo 2016 *J. Eng. and Tech. Sci* **48** 2
[6] L. Noerochim, G. P. A. Sapputra, A. Widodo 2016 *AIP Conf. Proceed* **1725** 020054
[7] Feng, Q., Miyai, Y., Kanoh H., And Ooi, K., 1992 *Langmuir* **8** 1861
[8] Jie, Feng., Din, Yaijing., Guo, Yunpeng., Li, Xiaohong., Li, Wenying 2013 *Fuel* **109** 110-115

IOP Conf. Series: Materials Science and Engineering **214** (2017) 012019 doi:10.1088/1757-899X/214/1/012019

Systemic Assessment of Calcium and Phosphorus Level after Implantation of Porous Iron in Rats

S F Siallagan[1] , F Amelia[1], N D Utami[1], M F Ulum[1], A Boediono[1], S Estuningsih[1], H Hermawan[2,*] and D Noviana[1,*]

[1]Faculty of Veterinary Medicine, Bogor Agricultural University, Bogor, Indonesia
[2]Department of Mining, Metallurgical and Materials Engineering & CHU de Quebec Research Centre, Laval University, Quebec City, Canada

* Email: deni@ipb.ac.id

Abstract. One of important aspects in bone healing process is physiological level of calcium (Ca), and phosphorus (P) that can be altered by implantation of biodegradable porous iron. Therefore, this study aims to investigate the concentration of Ca, P and Ca/P ratio in the peripheral blood during the implantation period up to 4 months. Forty adult male Sprague Dawley rats were used and divided into 3 groups receiving different pore size of iron implants (pore size 450, 580, 800μm) and one group of sham. The implants (5x2x0.5mm) were inserted into flat bone defects at latero-medial of femoral bone. Blood sample was taken from ventral tail artery before and after 4 month of implantation. Calcium and P concentrations in the blood were determined by BA-88A Semi-Auto Chemistry Analyzer. Results showed that concentration of Ca and P are slightly higher after implantation than before implantation, except for the 450μm group. The Ca/P ratio before and after implantation was increased in the sham group, and decreased in the 450 and 800μm groups. Concentration of Ca, P and Ca/P ratio insignificantly change between before and 4 months after surgery in some groups.

1. Introduction

Porous biodegradable metal made of pure iron has been considered as ideal material for developing bone scaffolds. Porous structure of sintered iron foams allowed increase in degradation rate [1]. Although iron utilizing as bone material implant is still debated due to its cytotoxicity [2], but iron as bone metal implant has proven non toxic to rats [3]. Iron is considered an essential mineral as the central element in oxygent transport and utilization [4]. Study about Porous iron as bone metal implantation already done in mice. This study showed that the porous iron implant have minimal inflammation effect through peripheral white blood cell examination [5].

Iron as biodegradable metal frequently combined with other material and also increased the number of porous surface of the implant to increase degradation rate [6]. Materials that commonly combine with iron are poly (lactic-co-glycolic acid), hydroxyapatite (HA), tricalcium phosphate (TCP) and biphasic calcium phosphate (BCP) [6,7]. Increasement of the degradation rate of iron implant is expected to accelerate the bone healing processes. Bone scaffold frequently use in bone disorders to restore the damaged bone [8].

In human, 99% of calcium and 85% of phosphor are found in bone [9]. Calcium, together with phosphorus, constitute the main component of bones [10] as hydroxyapatite. One of important aspects in bone healing/growth is the in-vivo physiological level of calcium (Ca) and phosphorus (P) that can be altered due to the implantation of a scaffold. The skeleton is the body's principal reservoir of calcium and phosphorus. Contrary to its appearance, bone is a dynamic tissue, and calcium and phosphate are continuously being deposited and released [11].

Several studies of porous iron implant had been done, but only a few study discussed the correlation between the porous iron implant and systemic of bone mineral such as Ca and P. Previous study has been discussed about this correlation just in 1 month observation [3]. Therefore, this study aims to investigate the concentration of Ca, P and Ca/P ratio in the peripheral blood during the implantation of porous iron in rats in longe period 4 month observation.

2. Materials and Methods

2.1Specimens preparation

Pure porous iron specimens (Alantum, Korea) with pore size of 450, 580 and 800 μm were cut from commercial product to 5 x 2 x 0.5 mm of size without any prior treatments. Specimens were sterilized using an autoclave with high pressure saturated steam at 121^{0}C, and followed by UV ray for 1 hour.

2.2Implantation process

This study was approved by Animal Care and Use Committee, Bogor Agricultural University (12-2015 RSHP FKH IPB). Forty adult male Sprague Dawley rats were used and divided into 3 groups receiving different pore size implants (450, 580, and 800μm) and one group of sham as control. After 12 hours of fasting, the animals received premedication (Atropine®, Indofarma, Indonesia). All rats were implanted under general anesthesia ketamin (Ketamil®, Illium, Australia) and xylazin (Illium xyla®, Illium, Australia) influenced. After anaesthetized, femoral hair was shaved and the skin was disinfected with 70% alcohol and 10% iodine before implantation surgery. The rats were place in left lateral recumbence on the operation table. The implantation surgery procedure was started with skin incision in the femoral area, the muscle was then retracted until the femur bone reached. The implants (5 x 2 x 0.5mm) were inserted into flat bone defects at latero-medial of femoral bone of the rats. Sham treatment was done to all the procedure, but inserted the material implant. Femoral muscle and skin were sutured using absorbable 4/0 polyglactin suture (Hinglact, HiCare, India). To prevent infection, rats were treated with general antibiotic amoxicillin and clavulanic acid (claneksi® dry syrup, sanbe, Indonesia) for 5 days post surgery, orally.

2.3Blood examination

The peripheral blood was collected under general anesthesia influenced. Blood sample was taken from ventral tail artery before at 0 and 4 months after implantation. Calcium and phosphorus concentrations in the serum were determined by BA-88A Semi-Auto Chemistry Analyzer (Mindray Bio-Medical Electronics, China).

2.4Data analysis

Data were analyzed by one-way analysis of variance (ANOVA) with a post hoc Duncan test using SPSS software (SPSS Inc, USA) at the 95% confidence level, then expressed as mean±standard deviation.

3. Result

Figure 1 shows that concentration of Ca and P is slightly higher after implantation compared to before implantation, except for the 450 μm group. The significant changes in Ca and P concentration are showed in 800μm group. Significant changes concentration also can be seen between control and 800 μm groups. The Ca/P ratio before and after implantation was insignificantly increased in the sham group, and insignificantly decreased in the 450 and 800 μm groups.

IOP Conf. Series: Materials Science and Engineering **214** (2017) 012019 doi:10.1088/1757-899X/214/1/012019

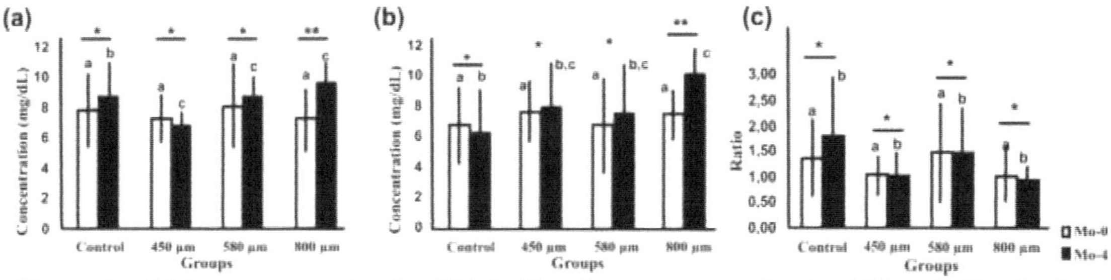

Figure 1.a Calcium concentration (mg/dL), b. Phosphorus concentration (mg/dL), c. Ca/P ratio after and before implantation of porous metal iron at femoral bone of rats. Different letters in superscript following values indicate statistical significant ($P < 0.05$), *represent $P > 0.05$ indicate statistic insignificant, ** represent $P < 0.05$ indicate statistical significant.

4. Discussion

Calcium-phosphorus interaction is important. Fig 1a and b show an increasement in Ca and P that indicate the mineralization process of bone healing [12] that regulated by parathyroid hormone (PTH) and vitamin D [9]. Iron, calcium and phosphorus deficient of rats have provent decreased bone mass and increased fragility [13]. Fig 1.a. shows an increase of calcium concentration except the group 450 µm. Bone mineralization is the final step in bone formation. Bone defect will stimulate PTH to increase calcium in extracellular fluid [9]. The higher concentration of calcium and phosphorus then will increase the number and function of osteoclast, then induce the osteoblast proliferation [14, 15]. Fig 1.a also shows that the concentration of Ca insignificantly decrease in group 450 µm. The smaller pore size lead to the greater number of porous implant surface. Porous surface plays important role in cell growth and proliferation due to wider surface contact area and interconnectivity within the pores which lead to greater cell spreading [16] thus expected to reduce the requirement of calcium in mineralization processes. There is no direct interaction between iron degradation product of implant to calcium in bone remodelling.

Phosphorus exists in two forms within the body such as 10% inorganic (phosphate) which is routinely measured for laboratory purposes and 90% organic (e.g nucleic acids, phospholipids, ATP) [9]. Phosphorus is important in maintaining the structure of bones and teeth, maintaining of cell membranes, and supplying of energy. Most inorganic phosphorus is deposited in bone [9, 12]. Bone healing process will increased phosphate consumption that will increase the concentration of phosphorus in the fourth month after implantation except the control group (Fig 1.b). Insignificantly decreasing of phosphorus concentration in control group can be affected by a slight decrease in erythrocytes due to implantation process. Other groups have an indirect iron supplement from the material implant, that play role in oxygen transport and erythrocytes metabolism [11]. Also, there is no direct interaction between iron degradation product of implant to phosphorus in bone remodelling.

The result of Ca/P ratio is depend on the concentration of Ca and P. In all group Ca/P ratio insignificantly different at 0 and 4 months after implantation processes (Fig 1.c). Although insignificanly different but calcium and phosphorus consentration should be observed carefully. Nitric oxide synthase (NOS) is induced during fracture healing in rats and humans. Nitric oxide has pleiotropic effects in bone cell response, and potently decrease resorption through decreasing osteoclast formation and activity [17]. Nitric oxide synthase is expressed at low level in their tissue of origin and their activity is mainly regulated by changes in free intracellular Ca^{2+} concentration. Rapid increase in intracellular calcium concentration could be occur due to phospholipase C activity and IP3 signaling resulting in release of calcium from intracellular stores in many cases [18]. As a free radical, smaller level of NOS will increase blood flow and induce bone healing but higher level of NO will increase oxidative stress probability and increase the activity of osteoclast [2]. Therefore, Ca, P, and Ca/P ratio examination with other examination such as blood examination should be investigated to determine the body condition during bone healing processes.

5. Conclusion

Bone healing involves complex interactions between mechanically stable environment, cell, osteo-conductive matrix with blood supply that affected the level of Ca and P. Concentration of Ca and P, and also Ca/P ratio insignificantly change between at 0 and 4 months after surgery in some groups. Iron supplement from the material implant play role in oxygen transport and erythrocytes metabolism, but there is no direct interaction between porous iron implant degradation product to Ca and P concentration. The concentration of Ca and P were affected by the porous size of iron implant, The smaller pore size will increase the surface contact area between the iron implant and surrounding cells.

6. References

[1] Oriňáková R, Oriňák A, Bučková L M, Giretová M, Medvecký L, Labbanczová E, Kupková M, Hrubovčáková M, Kovaľ K 2013 Iron Based Degradable Foam Structures for Potential Orthopedic Applications *Int. J Electrochem Sci* **8** 12451 – 12465

[2] Jia P, Xu Y J, Zhang Z L, Li K, Li B, Zhang W, Yang H 2012 Ferric Ion Could Facilitate Osteoclast Differentiation and Bone Resorption through the Production of Reactive Oxygen Species *J Orthop Res* **30** 1843–1852

[3] Ulum M F, Paramitha D, Siallagan S F, Sariningrum A, Maharani A D K, Harjo F A, Nashrulloh M F, Laily R, Rahmadani D, Wulansari R, Maheswari H, Hermawan H, Noviana D 2016 Local and systemic analysis of porous iron implantation in femoral bone of rats *10th World Biomaterials Congress* 2970

[4] Wang J and Pantopoulos K 2011 Regulation of cellular iron metabolism *Biochem J* **434** 365–381

[5] Paramitha D, Noviana D, Estuningsih S, Ulum M F, Nasution A K, Hermawan 2015 *Proc Int Conf on AIP Conf Proceedings* 1677

[6] Yosop A H, Daud N M, Nur H, Kadir M R A, Hermawan H 2015 Controlling the degradation kinetics of porous iron by poly(lactic-co-glycolic acid) infiltration for use as temporary medical implants *Scientific Reports* **5** 11194

[7] Ulum M F, Arafat A, Noviana D, Yusop A H, Nasution A K, Abdul K M R, Hermawan H 2014 In vitro and in vivo degradation evaluation of novel iron-bioceramic composites for bone implant applications *Mater Sci Eng C Mater Biol Appl* **1** 336-44

[8] Bose S, Roy M, Bandyopadhyay A 2012 Recent advances in bone tissue engineering scaffolds *Trends Biotechnol* **30** 546-554

[9] Grossman J 2014 Disorder of Fluids and Electrolyte Balance *Porth's Pathophysiology* ed 9th ed Grossman S C and Porth C M Chapt 39 1019-1059

[10] Christy L 2014 Structure and Function of the Musculoskeletal System *Pathophysiology The Biologic Basis for Disease in adult and children* ed 7th, ed K L McCance, S E Huether, V L Brasher, N S Rote (Missouri: elsevier) chapt 43 pp 1515-16

[11] Bhagavan N V and Ha C-E 2015 *Essential of Medical Biochemistry* 2nd ed (Elsevier: San Diego) pp: 511-516

[12] Schenck P A 2010 Electrolyte Disorders : Ca-P and Mg *Textbook of Veterinary Internal Medicine* Vol 1, ed Ettinger S J, Feldman E C (Missouri: Elsevier) Chapt 79 Sect 2 pp 308-313

[13] Medeiros D M, Plattner A, Jennings D, and Stoecker B 2002 Bone Morphology, Strength and Density Are Compromised in Iron-Deficient Rats and Exacerbated by Calcium Restriction. *J. Nutr* **132** 3135–3141

[14] Boonrungsiman S, Gentleman E, Carzaniga R,Evans N S, McComb D W, Porter A E, Stevens M M 2012 The role of intracellular calcium phosphate in osteoblast-mediated bone apatite formation *PNAS* **109** (35) 14170-14175

[15] Park J-W, Suh J-Y,Chung H-J 2008 Effects of calcium ion incorporation on osteoblast gene expression in MC3T3-E1 cells cultured on microstructured titanium surfaces *J Biomed Mater Res* **86**A 117–126

[16] Matassi F, Botti A, Sirleo L, Carulli C, and Innocenti M 2013 Porous metal for orthopedics implants *Clin Cases Miner Bone Metab* **10** (2) 111-115

IOP Conf. Series: Materials Science and Engineering **214** (2017) 012019 doi:10.1088/1757-899X/214/1/012019

[17] Fan X, Roy E, Zhu L, Murphy T C, Ackert-Bicknell C, Hart C M, Rosen C, Nanes M S, Rubin J 2004 Nitric oxide regulates receptor activator of nuclear factor-kappaB ligand and osteoprotegerin expression in bone marrow stromal cells Endocrinology **145** (2) 751-759
[18] Rubin J, Rubin C, Jacobs C R 2006 Molecular pathways mediating mechanical signaling in bone *Gene* **367** 1-16

Acknowledments
The authors acknowledge the support of Indonesian Ministry of Research, Technology and Higher Education in International research collaboration and scientific publication No 079/SP2H /LT/DPRM/II/ 2016. This work is part of an international research collaboration between Bogor Agricultural University, Indonesia and Laval University, Canada.

IOP Conf. Series: Materials Science and Engineering **214** (2017) 012020 doi:10.1088/1757-899X/214/1/012020

Release Profile of *Andrographis paniculata* Leaf Extract Nanocapsule as α-Glucosidase Inhibitors

K Zahrani[1], F Imansari[1], T S Utami[1] and R Arbianti[1]

Bioprocess Engineering, Department of Chemical Engineering, Faculty of Engineering Universitas Indonesia, UI Campus, Depok 16424

arbianti@che.ui.ac.id

Abstract. *Andrographis paniculata* is one of 13 leading commodities Indonesian medicinal plants through the Ditjen POM. Andrographolide as main active compound has been shown to have many pharmacological activities, one of which is as α-glucosidase enzyme inhibitors which has clinical potential as an antitumor, antiviral, antidiabetic, and immunoregulator agents. This study aims to do nanoencapsulation of *Andrographis paniculatar* leaf extract to increase its active compound bioavailability and get a release profile through synthetic fluids media simulation. Nanoencapsulation with ionic gelation method result the encapsulation efficiency and loading capacity values of 73.47% and 46.29% at 2%: 1% of chitosan:STPP ratio. The maximum α-glucosidase inhibition of 37.17% was obtained at 16% concentration. Burst release at gastric pH conditions indicate that most of the drug (in this study is an *Andrographis paniculata* leaf extract) adsorbed on the surface of the nanoparticles an indicates that the kind of nanoparticle formed is nanosphere.

1. Introduction

Andrographis paniculata is one of 13 leading commodities of Indonesian medicinal plants by the government [1]. The active ingredient contained in Andrographis paniculata is diterpene lactones and flavonoids [2]. Diterpene lactone found in Andrographis paniculata is andrographolide, 14-deoxy-11,12-didehidro andrographolide, neo andrographolide, and 14-deoksi andrographolide, while the two flavonoids were identified 5,7,2 ', 3'-tetrametoksiflavanon and 5-hydroxy-7 , 2 ', 3'-trimetoksiflavon [3]. Andrographolide is the most active compounds and are most commonly found in *Andrographis paniculata.* Andrographolide have the ability as an antidiabetic, hepatoprotective, cardioprotective, anti-inflammatory, antiangiogenic, antithrombotic, and antiviral agents [4]. Diterpene compounds have the ability as an inhibitor of α-glucosidase enzyme, wherein α-glucosidase enzyme inhibitors has clinical potential that can serve among other things as an antitumor, antiviral, antidiabetic, and agents immunoregulator [5][6].

Various pharmacological activity of *Andrographis paniculata* makes this plant has a very good potential to be used to cure various diseases, including diabetes and hepatitis B. However, andrographolide as an active compound primarily on *Andrographis paniculata* is unstable in acidic conditions and alkaline digestive extreme, and also has a very short biological half-life (2 hours), so it need a method to deliver an active compound into the body. The application of nanotechnology is one method that can help this problem. Nanotechnology can increase the surface area of active compound,

IOP Conf. Series: Materials Science and Engineering **214** (2017) 012020 doi:10.1088/1757-899X/214/1/012020

increasing solubility, improving bioavailability in oral administration, and protect drug from fast degradation so it can sustain the release of active compounds in the long term.

The polymer can be selected to facilitate the preparation of nanoparticles are in the form of a water-soluble polymer. One of the water-soluble polymer is chitosan. Chitosan has the properties of biocompatible, biodegradable, non-toxic and not expensive [7]. In addition, chitosan is a polysaccharide second in terms of its availability in nature and included as polyelectrolitecationic [8]. The usage of chitosan as encapsulator has been widely used before, including the encapsulation of vitamin C [9] and encapsulation of tea [10]. However, using chitosan as encapsulator also has the disadvantage of mechanical properties of chitosan are fragile and should be stabilized by polyanion cross-linker. Polyanion cross-linker that most widely used is sodium tripolyphosphate (TPP) because not toxic and has multivalen. In addition, the use of tripolyphosphate for the formation of Chitosan gel can improve the mechanical properties of the particles that are formed due to the negative charge density tripolyphosphate has a high interaction with polycation Chitosan will be greater [11]. The existence of cross-linker to make chitosan formed is still not decayed so that the active substances contained cannot be separated before reaching the target when passing through the highly acidic pH.

Based on this background, the study aims to do nanoencapsulation of *Andrographis paniculata* extract with chitosan-STPP and see its benefits as an inhibitor of α-glucosidase enzyme. Inhibition tes of α-glucosidase enzyme performed in in-vitro test with p-nitrophenyl-α-D-glucopiranoside substrate. In this case it is important to review the release profile of *Andrographis paniculata* leaf extract that has nanoencapsulated with a coating of chitosan in order to determine the performance of the drug in the body. In this study, the release profile test performed by in-vitro using synthetic fluids media of Simulated Gastric Fluid (SGF) pH 1.2 which shows the condition of the gastric and Simulated Intestinal Fluid (SIF) pH 7.4 which shows the condition of the intestine.

2. Methods

2.1. Instruments

Main instruments used consist of a beaker glass (Pyrex Iwaki), analytical massbalance (Kern, 870), magnetic stirrer (ColeParmer), syringe (Terumo), sentrifuge (Hanil), digital incubator (Boekel133730Model), sonicator (Ultrasonic Cleaner ElmaS30H), and UV Vis Spectrophotometry (Labomed,Inc.).

2.2. Materials

Materials used consist of *Andrographis paniculata* dry leaf, chitosan, sodium tripolyphosphate (STPP), tween80, acetic acid (Merck), α-glucosidase enzymes (Sigma Aldrich), the p-nitrophenyl-α-D-glucopiranoside substrate (SigmaAldrich), bovine serum albumin (Merck), andrographolide standard (Sigma Aldrich), ethanol, methanol, cloroform (Merck), KH_2PO_4 (Merck), NaOH (Merck), Na_2CO_3 (Merck), dimethyl sulfokside (DMSO) (Merck), and aquades.

2.3. Andrographis paniculata Extract

Extraction using ethanol 70% as solvent with sonication method. 50 g of *Andrographis paniculata* dry leaves is dissolved with a ratio of 1:10 (w / v) (do two circulation). Extraction with sonicator lasted for 60 minutes at a frequency of 40 kHz and a temperature of 30^0 C. Filtrate obtained was concentrated using a vacuum rotary evaporator to form a thick extract.

2.4. Nanoencapsulation

One gram of chitosan dissolved in 50 mL of acetic acid 1% (b/v), then 0.15 extract added and stirred with magentic stirer until homogene. In other beaker glass, 0.25 g of STPP dissolved in 25 mL of aquades, and 200 μL Tween 80 0.1% (v/v) was added to the solution of STPP and stirred. Solution of chitosan-extract then added with syringe into STPP solution and stirred for 30 minutes. Then, centrifuge at 10,000 rpm speed for 15 minutes. Sediment formed then dried with freeze dryer.

Calculation of loading capacity is done by dry nanocapsule formed, and calculation of encapsulation efficiency is done by the andrographolide level on nanocapsule using High Performance Liquid Chromatography (HPLC). Particle size and morphology are identifed using Field Emission Scanning Electron Microscopy (FE-SEM).

2.5. Solution Prepare

Buffer phosphate with pH 6,8 made by KH_2PO_4 0,1 M solution and NaOH 0,1 N solution. Substrate solution made by 0,0315 g of p-nitrophenyl-α-D-glucopiranoside dissolved in 20 mL buffer phosphate. Enzyme carrier solution made by 6 mg of BSA dissolved in buffer phosphate until 6 mL, then enzyme solution made by dissolve α-glucosidase in BSA solution until get 0,3 u/mL, 0,15 u/mL, 0,075 u/mL, and 0,0375 u/mL concentration. Na_2CO_3 solution to end enzymatic reaction made by dissolve 1,06 g of Na_2CO_3 in 100 mL buffer phosphate. Sample of crude extract, nanocapsule of *Andrographis paniculata* leaf extract, and andrographolide standard dissolved by DMSO then buffer phosphate until get concentration of 0,5%, 1%, 2%, 4%, 8%, 12%, 16%, and 18%.

2.6. α-Glucosidase Inhibition [12]

Sample testing of crude extract, nanocapsule, and andrographolide standard done by adding 10 μL of sample with various concentrations into 490 μL of buffer phosphate and 250 μL of substrate solution. Solution then incubated for 5 minutesat 37°C. 250 μL of enzyme solution 0,0375 u/mL was added, then the solution was incubated for 15 minutes. Reaction end by adding 2000 μL of Na_2CO_3. Absorbance measured at wavelength 400 nm. Sample control testing done by adding 10 μL of sample with various concentrations into 490 μL of buffer phosphate and 250 μL of substrate solution. Solution then incubated for 5 minutesat 37°C. 2000 μL Na_2CO_3 was added, then the solution was incubated for 15 minutes. Then 250 μL of enzyme solution 0.0375 u/mL was added. Absorbance measured at wavelength 400 nm.

2.7. Inhibition Percentage Calculation

Inhibiton calculated with the equation of:

$$\% \, \boldsymbol{Inhibition} = \frac{A_B - A_S}{A_B} \tag{1}$$

where A_B is blank absorbance minus blank control absorbance, and A_S is sample absorbance minus sample control absorbance.

2.8. Making of Synthetic Fluid Media

Buffer media with pH 1.2 was prepared by dissolving 4.0824 g KH_2PO_4 with aquades to 300 mL, then 0.8 g NaOH dissolved with aquades to 200 mL. Buffer media with pH 7.4 was prepared by dissolving 1.7 mL of 37% HCl to 100 ml with aquades, then as much as 6 g KCl dissolved with aquades to 400 mL.

2.9. Release Profile

A total of 1 g nanocapsule dissolved in 50 mL of buffer pH 1.2, and then incubated at 37°C. The data was undertaken every 1 hour by taking 10 ml solution and centrifuge at speeds 2000 rpm for 10 minutes. Then, supernatant was taken to measure its absorbance. Replacing the media buffer from pH 1.2 to pH 7.4 is done on the 3rd hours by filter media solution pH 1.2 with paper Whatmann paper No. 42 by vacuum pump. Nanocapsule solids were filtered and diluted with media buffer pH 7.4. The data taken were previously as procedure.

3. Result and Discussion

3.1. Andrographis paniculata Extraction

Andrographis paniculata extract is performed three times to produce an average yield of 16% which is almost the same as Yonanda study [13] that is equal to 15.88% and vary by study Mathew, et al. [14] using methanol and generate value yield of 39.8%. This is because the solvent with low viscosity, easier to pour. The use of percolation methods by Dewi [15] produces a yield of 17.09%.

3.2. Nanoencapsulation

Dry nanocapsule leaf extract produced is as much 1.88 g. Calculation result the loading capacity and the encapsulation efficiency that high enough of 46.29% and 73,47%.This is due to the magnitude of the chitosan concentration that cause chitosan molecules will be increasingly bound to each other which leads into the increasing of intermolecular crosslinking bonds [16].The bonds formed at these conditions are strong and stable so produce particles that hard and not easily broken. The resulting particle size ranged from 557.9 nm to 892.6 nm. This particle size meets the definition of nanoparticles between 10-1000 nm [14], but not yet to meet the range of nanoparticles for drug of <100 nm [11].

The morphology of the particles shown in Figure 1 are identified by FE-SEM 1,000x zoom in and 2,000x zoom in, result particles that not spherical perfect with a smooth surface and aglomerated each other but still visible form of a circle. This is due to the use of high concentration of chitosan that cause increased viscosity solution that produces a strong form of the particle surface with the interaction by TPP, so the particles result will form the more spherical [17].

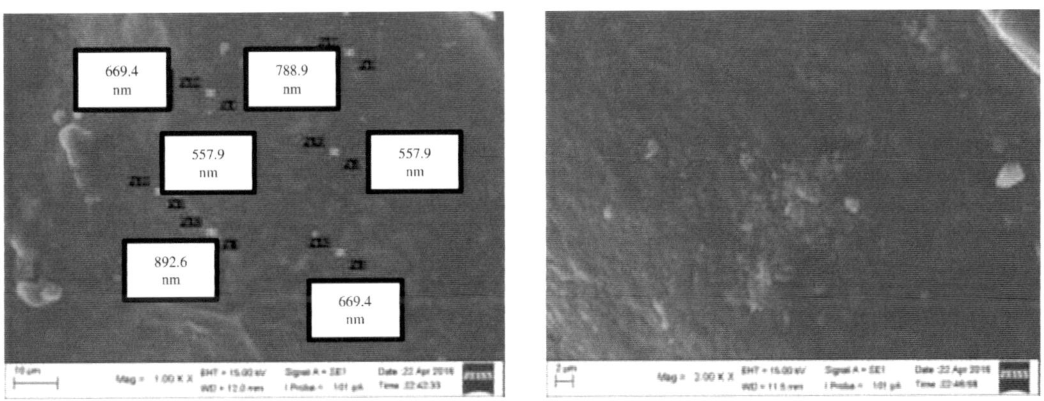

(a) **(b)**

Figure 1. SEM result of *Andrographis paniculata*-Chitosan nanoparticle in 1,000x zoom in (a) and 2,000x zoom in (b).

3.3. α-Glucosidase Inhibition

α-glucosidase enzyme inhibiton of sample shows in the Figure 2. The maximum percent inhibition and optimum concentration needed from each sample are different each other. Crude extract result maximum inhibition of 33.17% at 12% concentration. Nanocapsule result higher maximum inhibiton of 37.17%, however the concentration required to get that value is 16%.

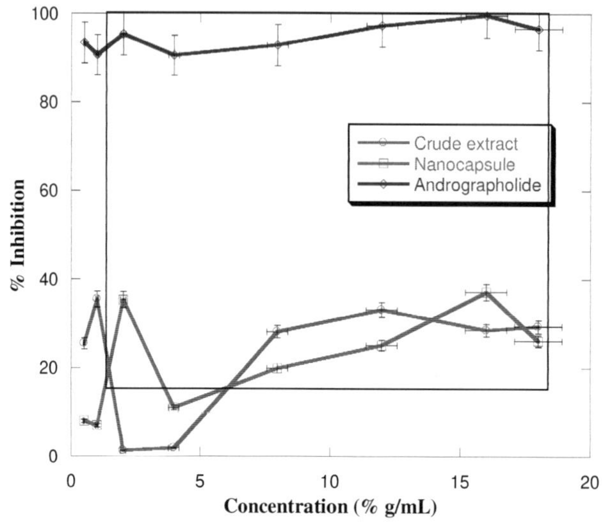

Figure 2. Inhibition of α-glucosidase by AP extract, AP nanocapsule, and andrographolide standard.

3.4. Release Profile

In Figure 3 shows the percent cumulative release of 10.45% on hour-7. This shows that the concentration ratio of chitosan: STPP produce a strong cross-linking between them and a strong matrix so difficult split and make the release (release) drug lasts longer. Characteristics of chitosan unstable under acidic conditions led to the electrostatic repulsion between molecules on the surface of chitosan nanoparticles that produces the drug adsorbed diffusion, thereby making high levels of drug release from the matrix [18][19]. Burst release indicates that some drugs (in this study is an extract of *Andrographis paniculata*) adsorbed on the surface of the nanoparticles [20] and it also indicates that the kind of nanoparticle formed is nanosphere [21]. Sustained release profile caused by the properties of chitosan as a main ingredient matrix properties insoluble at pH below 6.0, but has resistance at pH above that value, and the hydrogen bonds between the molecules of chitosan nanoparticles create stronger which causes diffusion into the molecule becomes more difficult , followed by release of the drug lasts longer [18].

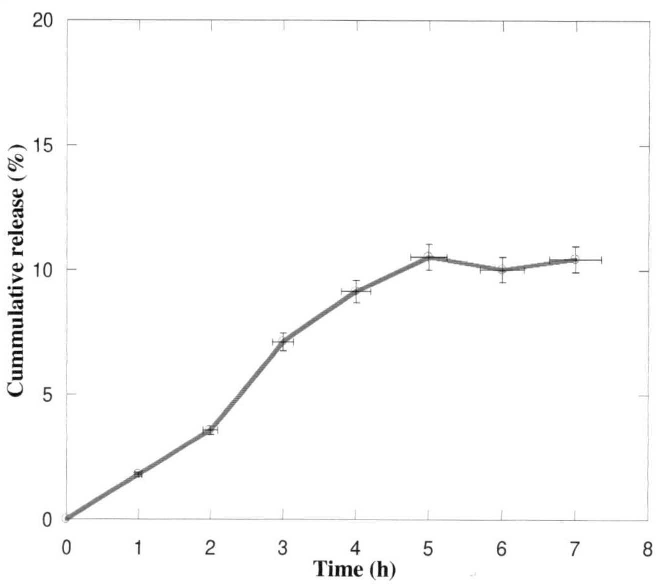

Figure 3. Release profile of *Andrographis paniculata* nanoparticle.

4. Conclusion

Nanoencapsulation of *Andrographis paniculata* can take the problem of andrographolide short half-life by maintain the drug release into the organ target by percent cumulative release of 10.45% on hour-7 in order to inhibits α-Glucosidase activity. Inhibition test of α-glucosidase activity by *Andrographis paniculata* nanocapsule result the maximum inhibition of 37.17% with 16% concentration.

5. References

[1] Herdiani, Elvina 2012 *Potensi Tanaman Obat Indonesia*. [online] Available at http://www.bbpp-lembang.info/index.php/arsip/artikel/artikel-pertanian/585-potensi-tanaman-obat-indonesia [accessed September, 11th 2015]

[2] Akbar, Shahid 2011 *Andrographis paniculata*: A Review of Pharmacological Activities and Clinical Effects.*Alternative Medicine Review* **16** (1)

[3] Chao, Wen-Wan & Lin, Bi-Fong 2012 Hepatoprotective Diterpenoids Isolated from *Andrographis paniculata.Scientific Research* **3** 136-143

[4] Hidalgo, Maria A., Hancke, Juan L., Bertoglio, Juan C., & Burgos, Rafael A 2013 Andrographolide a New Potential Drug for the Long Term Treatment of Rheumatoid Arthritis Disease. *Intech.*

[5] Dai, Gui-Fu., Xu, Hai-Wei., Wang, Jun-Feng., Liu, Feng-Wu., & Liu, Hong-Min 2006 Studies on the Novel α-Glucosidase Inhibitory Activity and Structure-Activity Relationships for Andrographolide Analogues.*Bioorganic & Medicinal Chemistry Letters* **16** 2710-2713

[6] Xu, Hai-Wei., Dai, Gui-Fu., Liu, Gai-Zhi., Wang, Jun-Feng., dan Liu, Hong-Min 2007 Synthesis of Andrographolide Derivatives: A New Family of α-Glucosidase Inhibitors. *Bioorganic & Medicinal Chemistry* **15** 4247-4255

[7] Tiyaboonchai, W 2003 Chitosan nanoparticles: Apromising system for drug delivery. *NaresuanUniversity Journal* **11** (3) 51–66

[8] Wu, Y., Yang, W., Wang, C., Hu, J., Fu, S 2005 Chitosan nanoparticles as a novel delivery system for ammonium glycyrrhizinate. *International journal of pharmaceutics* **295** (1-2) 235-45

[9] Desai, K. G. H. and Park, H. J. 2005 Encapsulation of vitamin C in tripolyphosphate crosslinked chitosan microspheres by spray drying. *J. Microencapsulation* **22** 179-192

[10] Hu, B., Pan, C., Sun, Y., Hou, Z., Ye, H., Hu, B. And Zeng, X. X 2008 Optimization of fabrication parameters to produce chitosan tripolyphosphate nanoparticles for delivery of tea catechins. *J. Agric. Food Chem.* **56** 7451-7458

[11] Aini, Desna Qurratul 2015 Uji Pelepasan Terkendali Ekstrak Daun Keji Beling (Strobilanthes cripus) untuk Sediaan Obat Antihiperkolesterolemia. Thesis. Depok: Universitas Indonesia.

[12] Aisyah, Nurul. 2014. Ekstraksi Akar Ilalang (Imperata cylindrica) serta Aktivitas Inhibisinya terhadap α-Glukosidase yang Berfungsi dalam Pembentukan Virus Hepatitis B. *National Seminar Prosiding Chemical Engineering & Process*

[13] Yonanda, Rona Jutama. 2011. Formulasi Ekstrak Sambiloto (Andrographis paniculata) dan Brotowali (Tinospora crispa) sebagai Inhibitor α- Glukosidase dan Analisis Sidik Jari Menggunakan Teknik Kromatografi. Institut Pertanian Bogor, Bogor, 9-10

[14] Mathew,Arhewoh., Falodun, Abiodun., Okhamafe, Augustine O.,Bao, Yang., Sheng, Qui 2011 Ultrasonic Assisted Extraction and radical scavenging activity of some selected medicinal plants. *Journal of Pharmacy Research* **4**(2) 408-410

[15] Dewi, Veronica. 2013. Enkapsulasi Ekstrak Sambiloto (Andrographis paniculata) dengan Kasein Susu sebagai Sediaan Antidiabetes. Thesis. Depok: Universitas Indonesia.

[16] Nanda, Rabindra K., Patil, Subodh S., and Navathar, Dipak A 2012 Chitosan Nanoparticles Loaded with Thiocolchicoside. *Der Pharma Chemica* **4**(4) 1619-1625

[17] Ko, J.A., *et al.* 2002 Preparation and Characterization of Chitosan Microparticles Intended for Controlled Drug Delivery. *International Journal of Pharmaceutics* **249** 165-174

[18] Dhanasingh, Selvakumar & Nallaperumal, Shunmuga Kumar 2010 Chitosan/Casein Microparticles: Preparation, Characterization and Drug Release Studies. *World Academy of Science, Engineering and Technology* **4** 08-23

[19] Dounighi, Mohammadpour., et al. 2012 Preparation and In Vitro Characterization of Chitosan Nanoparticles Containing *Mesobuthus eupeus* Scorpion Venom as an Antigen Delivery System. *The Journal of Venomous Animals and Toxins Including Tropical Diseases* **18** (1) 44-52

[20] Nesalin, J. Adlin Jino & Smith, A. Anton. 2012. Preparation and Evaluation of Chitosan Nanoparticles Containing Zidovudine. *Asian Journal of Pharmaceutical Sciences* **7** (1) 80-84

[21] Kurniawan, Edi. 2012. Preparasi dan Karakterisasi Nanopartikel Sambung Silang Kitosan-Natrium Tripolifosfat dalam Gel Verapamil Hidroklorida. Thesis. Depok: Universitas Indonesia

IOP Conf. Series: Materials Science and Engineering **214** (2017) 012021 doi:10.1088/1757-899X/214/1/012021

Synthesis, Properties and Application of Glucose Coated Fe_3O_4 Nanoparticles Prepared by Co-precipitation Method

Ayu Y Sari[1,*], Eko A S[1], Candra K[1], Denny P Hasibuan[2], M Ginting[1], P Sebayang[1], and P Simamora[2]

[1] Research Center for Physics – LIPI, South Tangerang, Banten, Indonesia
[2] Department of Physics, Universitas Negeri Medan, Medan, North Sumatera, Indonesia

*ayuy001@lipi.go.id

Abstract. Synthesis of glucose coated Fe_3O_4 magnetic nanoparticles have been successfully prepared with co-precipitation method. Raw material of natural iron-sand was obtained from Buaya River, Deliserdang, Indonesia. The milled iron-sand was dissolved in HCl (37 mole %), and stirred in 300 rpm at 70°C for 90 minutes. Glucose was added to the filtered powder with varied content of 0.01, 0.02, and 0.03 mole, and precipitated by NH_3 (25 mole%). After drying process, the final product subsequently was glucose coated magnetite (Fe_3O_4) nanoparticles. The characterizations performed were true density measurement, FTIR, VSM, XRD, BET, and adsorbent performance by AAS. The FTIR analysis showed that M-O (bending) with M=Fe (stretching vibration) with υ = 570.92 and 401.19 cm^{-1}. While glucose coated well on nanoparticle Fe_3O_4, proved by functional groups C=O (stretching), M-O (stretching) and C-H (bending) with υ = 1404.17, 570.92, and 2368.58 cm^{-1}, respectively. Single phase of magnetite (Fe_3O_4) structure was determined from XRD analysis with cubic spinel structure and lattice parameter of 8.396 Å. The optimum conditions, obtained on the Fe_3O_4 nanoparticles with 0.01 mole of glucose addition, which has true density value of 4.57 g/cm^3, magnetic saturation, M_s = 35,41 emu/g, coercivity, H_{cJ} = 83.58 Oe, average particle size = 12.3 nm and surface area = 124.88 m^2/g. This type magnetic nanoparticles of glucose-coated Fe_3O_4 was capable to adsorbed 93.78 % of ion Pb. Therefore, the glucose-coated Fe_3O_4 nanoparticle is a potential candidate to be used as heavy metal removal from wastewater.

1. Introduction

Nanoparticle is a particle with nanometer size (1-100 nm), and having different properties from the original [1]. Nanoparticle, such as nano magnetic, can be prepared from the natural resource iron-sand. Iron-sand is abundant in Indonesia and it is needed to be added-value for the competitiveness of economy. Iron-sand has four phases: maghemite (γ-Fe_2O_3), hematite (α-Fe_2O_3), goethite (FeO.OH) and magnetite (Fe_3O_4). Among them, only maghemite (γ-Fe_2O_3) and magnetite (Fe_3O_4) phases have a good magnetic properties. In addition, magnetite (Fe_3O_4) is also amphoteric and has high absorption. Magnetite contains one ion of Fe^{2+} and two ions of Fe^{3+} in spinel structure, and has a black color [2]. There are several methods to treatment iron-sand, such as co-precipitation, to obtain Fe_3O_4 products. The interests of this mechanism are low temperatures required, easily control the particle size, and short time needed[3].

There are some applications of Fe_3O_4 magnetic nanoparticle, such as : ferrofluid [4], Radar Absorber Material (RAM), biosensors, biotechnology, biomedicine [5], catalyst [6], magnetic resonance imaging (MRI) [7], magnetic gel, and data storage [8]. In addition, magnetite is a promising materials as magnetic adsorbents due to its capability of adsorption in aqueous solutions [4]. Furthermore, to avoid corrosion and coagulation mass, the modification of Fe_3O_4 require a coating substance from polymer and organic material[9]. One of organic coating material for Fe_3O_4 is glucose and it can be used as a particle size reduction agent [10]. However, there are limited study of glucose coated Fe_3O_4 especially in the application as adsorbent material to remove heavy metal elements. Therefore, in this study we investigated the effects of glucose addition as a coating agent. The synthesis of Fe_3O_4 magnetic nanoparticles from iron-sand Sungai Buaya – Deliserdang then coated with glucose, are conducted by co-precipitation method.

2. Experimental Method

Iron-sand from sungai Buaya - Deliserdang was used as raw material in the manufacture of Fe_3O_4 magnetic nanoparticles. The process were performed from iron-sand milling with a planetary ball mill (PBM) for 15 hours. Then, a milled powder was dissolved using HCl (Merck 37%, 12M) and stirred using a magnetic stirrer (300 rpm) at temperature 70 °C for 90 minutes. The solution was then filtered using Whatman filter paper 40 (d = 110 mm and size of pore = 8 μm). The varied of glucose, 0.01, 0.02, and 0.03 mole, was combined into filtered powder. Then, the mixtures were precipitated by NH_3 solution (Merck 6.5 M). Furthermore, the solutions are heated at temperature 70°C for 90 minutes and stirred at 300 rpm speed using a magnetic stirrer, resulting a blackish brown sediment. Moreover, the sediment was washed repeatedly with distilled water until normal pH solution is obtained, and then dried at 100 °C (for 5 hours) and resulting a coated Fe_3O_4 nanoparticles.

Characterizations were carried out by true density (ASTM D. 1217-15), analysis of functional groups using the FTIR, magnetic properties using the VSM, microstructures using the XRD, surface area and particle size using the BET, and adsorption of heavy metal ions Cu and Pb using the AAS.

3. Results and Discussion

The true density value of Fe_3O_4 magnetic nanoparticle with variation of glucose composition (0, 0.01, 0.02 and 0.03 mole) as a coated are shown in Figure 1. From the result show that the highest true density, about 4.57 g/cm3, is obtained by the addition of 0.01 mole of glucose. It is indicated that the densification reach 88.4%, and from literature, the theoretical value of density is 5.17 [11]. The decrease in density in other samples occurs due to the existence of the cavity and the non-homogenous size of grains produced.

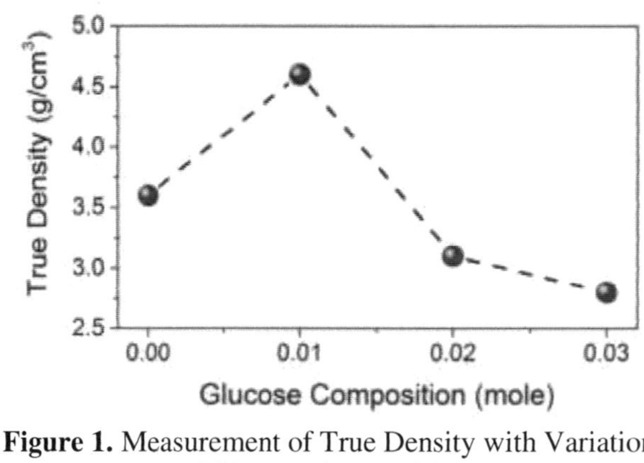

Figure 1. Measurement of True Density with Variation of Glucose Composition.

The analysis results of the FTIR spectroscopy from the samples are shown in Figure 2 and Table 1. The Fe_3O_4 measurement give only M-O and C-H bond. For M-O bond has wave number (υ) = 570.92 and 401.19 cm^{-1} with the stretching vibration, where M is the identification of Fe. While the O-H bond, with the bending and stretching vibration, occurs at υ = 1627.9 and 3387.0 cm^{-1} and it describe the presence of water content. On the other hand, Fe_3O_4 magnetic nanoparticle with variation of glucose coated give a M-O, C-O, C=O, O-H and C-H bonds. In addition, M-O, C=O and C-H bond are indicating of glucose existence. For Fe_3O_4 with 0.01 mol glucose has υ = 1026.1 and 1404.2 cm^{-1} as C-O bond and C = O , respectively with the stretching vibration. And as C-H bond give at υ = 2368.6, and 2931.8 cm^{-1} with bending vibration. The stretching vibration is only change the interatomic distance along bond axis, while bending vibration can change the angle between two bonds.

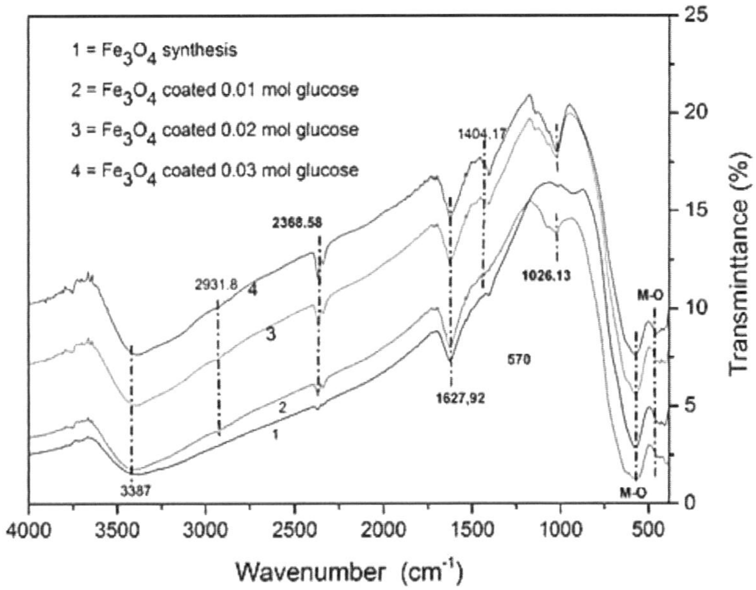

Figure 2. Results of the FTIR Spectroscopy of Magnetic Nanoparticles 1). Sample A, 2). Sample B, 3). Sample C, and 4). Sample D.

Table 1. Bond, Wave Number and Vibration of Fe_3O_4 and Fe_3O_4 Coating Glucose

NO	Bond	Wave number, cm^{-1}				Vibration
		A	B	C	D	
1	M-O	401,2	393,5	455,2	401,2	Stretching
2	M-O	570,9	570,9	578,6	570,9	Stretching
3	C-O		1026,1	1026,1	1026,1	Stretching
4	C=O	-	1404,2	1404,2	1404,2	Stretching
5	O-H	1627,9	1627,9	1620,2	1620,2	Bending
6	C-H	-	2368,6	2368,6	2368,6	Bending
7	C-H	-	2931,8	2931,8	2931,8	Bending
8	O-H	3387,0	3387,0	3387,0	3387,0	Stretching

A = Fe_3O_4, C = Fe_3O_4 + 0.02 mol glucose and
B = Fe_3O_4 + 0.01 mol glucose, D = Fe_3O_4 + 0.03 mol glucose.

The magnetic properties of samples A (Fe_3O_4), B (Fe_3O_4 + 0.01 mole of glucose), C (Fe_3O_4 + 0.02 mole of glucose) and D (Fe_3O_4 + 0.03 mole of glucose) were measured using the VSM, the results are shown in Figure 3 and Table 2.

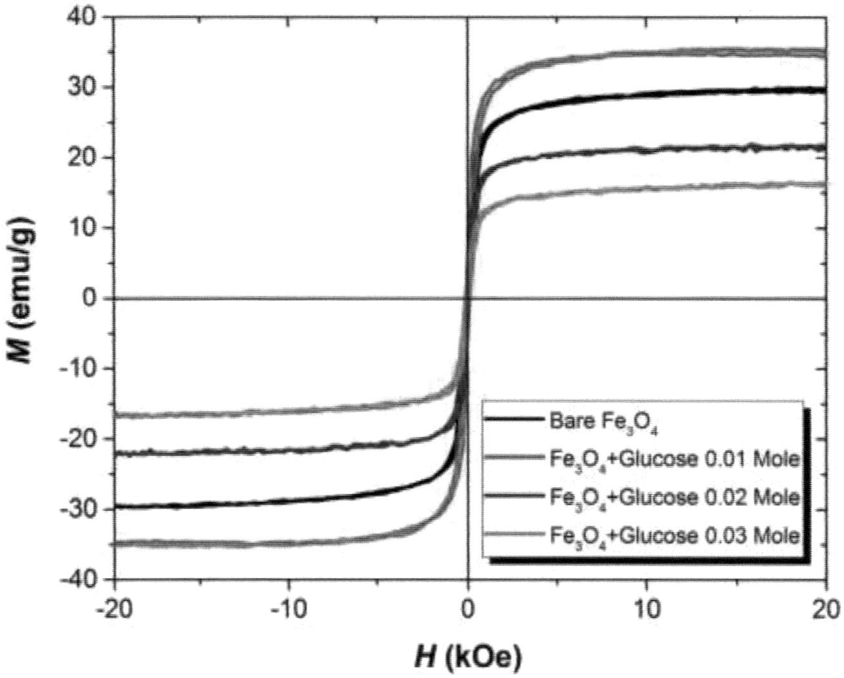

Figure 3. Hysteresis Curve of Sample A, B, C and D.

From Figure 3, it can be seen that all samples show a soft magnetic behaviour since the coercivity (H_{CJ}) is smaller than 126 Oe (the boundary conditions of soft magnetic). The lowest coercivity (H_{CJ}) was obtained in sample B, about 83.58 Oe. It also has a high saturation magnetization (σs), around 35.41 emu/g, resulting a superparamagnetic properties. According to the reference, saturation magnetization (σs) value for Fe_3O_4 coated with glucose is 29.55 emu/g [10].

Table 2. H_{CJ}, M_s and M_r Values of Nanoparticle Magnetic Fe_3O_4 and Fe_3O_4 Glucose Coated

Sample	Magnetic Nanoparticle	H_{CJ} (Oe)	σ_s(emu/g)	σ_r (emu/g)
A	Fe_3O_4	92.80	30.01	4.20
B	Fe_3O_4 + 0.01 mol glukosa	83.58	35.41	4.62
C	Fe_3O_4 + 0.02 mol glukosa	92.05	16.48	2.47
D	Fe_3O_4 + 0.03 mol glukosa	88.24	21.88	3.19

Analysis of XRD pattern for sample A and B are depicted in Figure 4. From the Figure 4, it can be seen that both samples only have a single phase magnetite (Fe_3O_4) with a cubic spinel structure and the lattice parameter is 8.123 Å. A cubic spinel structure can be determined by planes (hkl) for each peaks in XRD pattern, which are (111), (202), (311), (222), (400), (422), (333) and (404) [12]. Within the main peak 2θ = 35.56 ° and hkl (311), the crystal size of sample B =14.7 nm and sample A = 32.3 nm were obtained using scherrer equation [7].

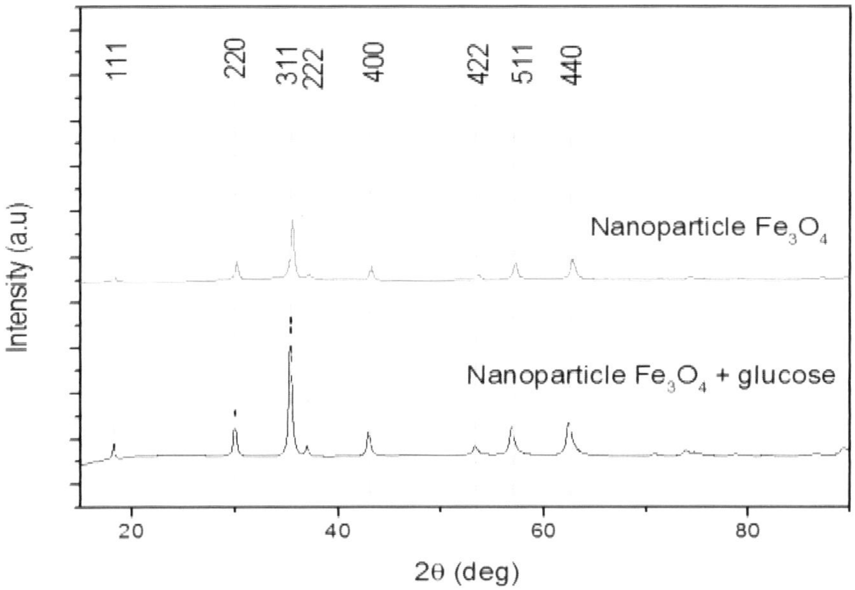

Figure 4. XRD Result of Nanoparticle Magnetic, a). Sample A, and b).Sample B.

From the measurement of surface area and pore size of Fe_3O_4 magnetic nanoparticles + 0.01 mole of glucose (sample B) conducted using the BET, 124.88 m^2/g and 4.8 nm were obtained. The average particle diameter given value 12.3 nm, which the calculation can be obtained from BET data measurement. While the surface area and pore size from reffrence [13] using glucose-coated Fe_3O_4 (Aldrich) are 35.7 m^2/g and 13.4 nm.

The sample B compound has been attempted for the adsorbent of heavy metal ions Cu and Pb using the AAS and the results are shown in Table 3. Apparently, Fe_3O_4 magnetic nanoparticles with 0.01 mole of glucose was capable of adsorbing heavy metal ions Pb in the amount of 93.78%.

Table 3. Analysis Measurement of AAS on sample B, for ion Cu and Pb

Heavy Metal	Concentration (ppm)		Adsorption (%)
	Initial	Final	
Cu	277	32.2	88.37
Pb	706	43.9	93.78

4. Conclusion

Fe3O4 magnetic nanoparticles and glucose coated Fe_3O_4 have successfully been synthesized using the co-precipitation method from iron-sand sungai Buaya - Deliserdang as a raw material. Fe_3O_4 are well coated by glucose, with presence of glucose functional group in the C=O, M-O and C-H bonds with stretching, stretching and bending vibration. The results of the XRD analysis show that there is only single phase magnetite (Fe_3O_4) with a cubic spinel structure and the lattice parameter of 8.123 A. The optimum condition was obtained in the composition of Fe_3O_4 coated with 0.01 mole of glucose. This composition resulting value of true density = 4.57 g/cm3, Ms = 35.41 emu/g, HCJ = 83.58 Oe, average particle size = 12.3 nm and surface area = 124.88 m^2/g. Fe_3O_4 magnetic nanoparticles coated with 0.01 mole of glucose is capable of adsorbing metal ion Pb up to 93.78%.

5. References

[1] Ghosh Chaudhuri R and Paria S 2012 Core/Shell Nanoparticles: Classes, Properties, Synthesis Mechanisms, Characterization, and Applications *Chem. Rev.* **112** 2373–433

[2] Gubin S P, Koksharov Y A, Khomutov G B and Yurkov G Y 2005 Magnetic nanoparticles: preparation, structure and properties *Russ. Chem. Rev.* **74** 489–520

[3] Meng H, Zhang Z, Zhao F, Qiu T and Yang J 2013 Orthogonal optimization design for preparation of Fe3O4 nanoparticles via chemical coprecipitation *Appl. Surf. Sci.* **280** 679–85

[4] Oxana V K, H V Rasika D, and Boris I Kharisov 2014 Magnetc adsorbents based on micro- and nano- structured materials. *RSC Adv.* **5** 6695–6719

[5] Lu A-H, Salabas E L and Schüth F 2007 Magnetic Nanoparticles: Synthesis, Protection, Functionalization, and Application *Angew. Chem. Int. Ed.* **46** 1222–44

[6] Kolen'ko Y V, Bañobre-López M, Rodríguez-Abreu C, Carbó-Argibay E, Sailsman A, Piñeiro-Redondo Y, Cerqueira M F, Petrovykh D Y, Kovnir K, Lebedev O I and Rivas J 2014 Large-Scale Synthesis of Colloidal Fe_3O_4 Nanoparticles Exhibiting High Heating Efficiency in Magnetic Hyperthermia *J. Phys. Chem. C* **118** 8691–701

[7] Akbarzadeh A, Samiei M and Davaran S 2012 Magnetic nanoparticles: preparation, physical properties, and applications in biomedicine *Nanoscale Res. Lett.* **7** 144

[8] Suh S K, Yuet K, Hwang D K, Bong K W, Doyle P S and Hatton T A 2012 Synthesis of Nonspherical Superparamagnetic Particles: *In Situ* Coprecipitation of Magnetic Nanoparticles in Microgels Prepared by Stop-Flow Lithography *J. Am. Chem. Soc.* **134** 7337–43

[9] Junejo Y, Baykal A and Sözeri H 2013 Simple hydrothermal synthesis of Fe3O4-PEG nanocomposite *Open Chem.* **11**

[10] Sun X, Zheng C, Zhang F, Yang Y, Wu G, Yu A and Guan N 2009 Size-Controlled Synthesis of Magnetite (Fe_3O_4) Nanoparticles Coated with Glucose and Gluconic Acid from a Single Fe(III) Precursor by a Sucrose Bifunctional Hydrothermal Method *J. Phys. Chem. C* **113** 16002–8

[11] Khayat Sarkar Z and Khayat Sarkar F 2011 Synthesis and Magnetic Properties Investigations of Fe3O4 Nanoparticles *Int J Nanosci Nanotechnol* **7** 197–200

[12] Silva V A J, Andrade P L, Silva M P C, Bustamante D. A, De Los Santos Valladares L and Albino Aguiar J 2013 Synthesis and characterization of Fe3O4 nanoparticles coated with fucan polysaccharides *J. Magn. Magn. Mater.* **343** 138–43

[13] Hou C, Wang Y, Zhu H and Wei H 2016 Preparation of mesopourous Fe3O4 nanoparticle with template reagent: Tannic acid and the catalytic performance *J. Taiwan Inst. Chem. Eng.* **60** 438–44

Acknowledgment

We are grateful to DIPA program from research center for physics – LIPI 2016 for funding this research program.

IOP Conf. Series: Materials Science and Engineering **214** (2017) 012022 doi:10.1088/1757-899X/214/1/012022

Effect of Mixing Dyes and Solvent in Electrolyte Toward Characterization of Dye Sensitized Solar Cell Using Natural Dyes as The Sensitizer

Nurrisma Puspitasari[1], Nurul Amalia Silviyanti S[2], Gatut Yudoyono[1] ,Endarko[1]

[1]Physics Department Faculty of Mathematics and Natural Science, Institut Teknologi Sepuluh Nopember Surabaya
[2]Abdurrahman Saleh University (UNARS) Situbondo

nurrisma@physics.its.ac.id, nurrismapuspitasari@gmail.com

Abstract. Dye Sensitized Solar Cell (DSSC) using natural dyes (chlorophyll, curcumin from turmeric extract, and anthocyanin from mangosteen extract) have been successfully fabricated for determining the effect of variation natural dyes, mixing dyes and acetonitrile in electrolyte toward characterization of DSSC. DSSC consists of five parts namely ITO (Indium Tin Oxide) as a substrate; TiO_2 as semiconductor materials; natural dyes as an electron donor; electrolyte as electron transfer; and carbon as a catalyst that can convert light energy into electric energy. Two types of gel electrolyte based on PEG that mixed with liquid electrolyte have utilized for analyzing the lifetime of DSSC. Type I used distilled water as a solvent whilst type II used acetonitrile as a solvent with addition of concentration of KI and iodine. The main purpose of study was to investigate influence of solvent in electrolyte, variation of natural dyes and mixing dyes toward an efficiency that resulted by DSSC. The result showed that electrolyte type II is generally better than type I with efficiency 0,0556 and 0,0456 %, respectively. An efficiency values which resulted from a variation of mixed three natural dyes showed the greatest efficiency compared to mixed two natural dyes and one dye, with an efficiency value can be achieved at 0,0194 % for chlorophyll; 0,111 % for turmeric; 0,0105 % for mangosteen; 0,0244% (mangosteen and chlorophyll); 0,0117 % (turmeric and mangosteen); 0,0158 % (turmeric and chlorophyll); and 0.0566 % (mixed three natural dyes).

1. Introduction

Solar energy makes comfortable for humans and the environment and has been widely used as energy source in solar cells. The solar cell is a device that can convert light energy into electrical energy [1]. The use of solar cells as a source of electrical energy is still limited due to constraints of the high cost of the main ingredient of the cell, namely silicon. However, after passing through the times, solar cell technology also developed. These developments have come to the solar cells developed by Gratzel. These cells are often also called Gratzel cells or Dye-Sensitized Solar Cells (DSSC) [2].

The Dye Sensitized Solar Cell is a photoelectrochemical solar cell, it uses a liquid electrolyte or other ion-conducting phase as a charge transport medium [3]. DSSC usually consists of five parts namely TiO_2 as semiconductor materials for photoanode; ITO (Indium Tin Oxide) as a substrate; dyes as an electron donor; carbon as a catalyst that can convert light energy into electric energy; and electrolyte as electron transfer. The electrolyte is one of the key components of DSSC that provide internal electric ion conductivity that spreads in the mesoporous TiO_2 layer and an important factor for determining the performance of the cells [4]. Therefore, various attempts have been made to study the various types of electrolytes. In liquid electrolytes, ionic conductivity is generated is high. However, in use in a DSSC, leakage and evaporation of the solvent liquid cannot be avoided so that the stability of the low electrolyte which causes low efficiency [5]. To overcome these problems, gel electrolyte have been a good candidate for the application of DSSC. In the gel electrolyte, polymer is used as a container to trap liquid solvents, organic solvents and inorganic salts such as propylene, ethylene carbonate and iodide salts of lithium, sodium and potassium. In the resulting gel electrolyte stability and conductivity high enough when compared with liquid electrolyte. Therefore, it will do further research on the influence of the electrolyte material such as a polymer or solvent of the electrolyte gel material used in DSSC

2. Experimental

2.1 Materials

The materials used in this research: Alfalfa chlorophyll (K-Link), curcumin from turmeric extract, and anthocyanin from mangosteen extract as the dyes solution, graphite carbon (from pencil), ITO (Indium Tin Oxide) that has been cleaned by ultrasonic cleaner, ethanol, HCl (Merck), $TiCl_3$ (Merck), NH_4OH (Merck), HNO_3 (Merck), Triton X-100 (Merck), aquades, PEG (polyethylene glycol), chloroform, iodine (Merck), acetonitrile (VWR), and KI (Potassium Iodide).

2.2 Synthesis of TiO_2 Nanoparticles

The precursor solution to synthesize of nanocrystalline TiO_2 anatase containing a mixture of 50 mL aquades, 20 mL HCl (Merck) 2M 37% and 20 mL $TiCl_3$ (Merck) was magnetically stirred for half hour at room temperature. Furthermore, a solution of 50 mL of NH_4OH (Merck) 25% was added dropwise to the precursor solution and was then stirred again for 60 minutes and precipitated for 1 day at room temperature and sealed. After the filtering the sediment then dried using a furnace with a temperature at 400 ^{o}C for 3 hours [6].

2.3 Preparation of Dye Sensitized Solar Cell

The paste of TiO_2 made from 1 ml distilled water, 0,7 ml of acetic acid ((Merck) and 0,7 ml of Triton X-100 (Merck) and PEG 1000 [7]. TiO_2 layer is deposited on the ITO glass by Doctor-Blade method for DSSC photoanode (working electrode). These photoanode are soaked in natural dyes solution for 24 hours at room temperature and was then taken out from the dye solution and rinsed in 70% ethanol. The counter electrode made of a ITO (Indium Tin Oxide) glass coated with graphite. Two types of the gel electrolyte fabricated by adding a polymer in a liquid electrolyte. Type I consists of a mixture of 7 g PEG (Merck), 25 mL cloroform (Merck), and liquid electrolyte (liqid electrolyte : 0,8 g KI dissolved in 10 ml of aquades and 5 drops of iodine solution (Merck). The precursor solution was magnetically stirred for one hour at 80° C until homogeneous and become a gel [8]. Type II consists of a mixture of 7 g PEG (Merck), 25 mL cloroform (Merck), and liquid electrolyte (liqid electrolyte: 3 g KI dissolved in 10 mL acetonitrile and 3 mL iodine solution (Merck)). The precursor solution was magnetically stirred for one hour at 80° C until homogeneous and become a gel. The counter electrode made of a ITO glass coated with graphite. Subsequently, the working electrode, electrolyte, and counter electrode arranged in thesandwich for fabricating of DSSC.

2.4 Measurements

The synthesized anatase TiO_2 nanoparticles were characterized by XRD (*X-Ray Diffraction*) whereas the absorbance spectrum of extracted natural dyes were characterized by usng UV-Vis Spectrophotometer 110 (Ocean Optics). The result of characterization was done by measuring dyes' absorbance to visible light spectrum with the wavelength ranging from 300 – 700 nm. The performance of DSSC (Dye Sensitized Solar Cell) was tested by characterizing the current and voltage

(I-V) with the power density 125 W/m^2. Fill factor of DSSC was determined acording to equation (1) and the energy conversion was determined acording to equation (2) [3]:

$$FF = \frac{V_{MPP} I_{MPP}}{V_{OC} I_{SC}} \quad (1)$$

Where *FF* is the fill factor (value should be between 0 and 1), *Vmpp* is the voltage maximum power point (V), *Impp* is The current maximum power point (Ampere)

$$\eta = \frac{P_{\max}}{P_{light}} x100\% \quad (2)$$

Where η is efficiency (%), Pmax is the maximum power (Watt)

3. Result and Discussion

3.1 Characterization of Dye Absorbance

The characterization of dye sensitizer has been done by using the dye absorbance UV-Vis Spectrophotometer (UV-Vis 1100 Spectrophotometer, Ocean Optics). The result of UV-Vis absorption spectra of Chlorophyll Sensitizer is shown in Figure 1. Based on the results of UV-Vis absorption in Figure 1, All of natural dyes has the absorbance spectrum 300 – 700 nm. The absorbance peak of chlorophyll sensitizer located at 400 nm. The absorbance peak of turmeric sensitizer located at 435 nm. The absorbance peak of mangosteen sensitizer located at 429 nm. The absorbance peak of mangosteen-chlorophyll sensitizer located at 300 nm and 418 nm. The absorbance peak of turmeric-mangosteen sensitizer located at 300 nm and 410 nm. The absorbance peak of turmeric-chlorophyll sensitizer located at 300 nm and 422 nm. The absorbance peak of mixing dyes (chlorophyll-turmeric-mangosteen) sensitizer located at 300 nm and 432 nm.

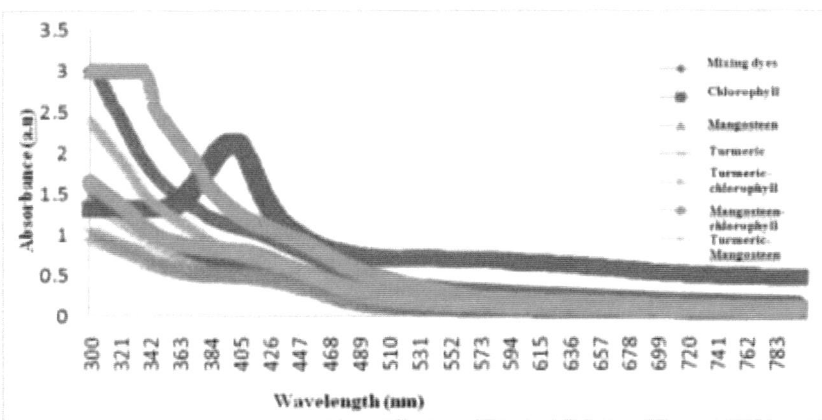

Figure 1. Absorption spectrum of natural dyes.

3.2 Characterization of I-V Curves

The characterization of current and voltage need to be done to determine the performance of all types of DSSC samples that have been fabricated in this paper. In DSSC type I (using gel electrolyte type I) are summarized in Table 1. It can be concluded from Table 1 that the highest maximum power and

IOP Conf. Series: Materials Science and Engineering **214** (2017) 012022 doi:10.1088/1757-899X/214/1/012022

the highest efficiency were achieved for the DSSC using mixing dyes (chlorophyll-turmeric-mangosteen) with the values obtained at 8548,8 nW and 0,0456%, respectively. Meanwhile, the performance of the DSSC using electrolyte type II are summarized in Table 2. The best result was achieved for the DSSC using mixing dyes (chlorophyll-turmeric-mangosteen) with the highest efficiency at 0,0566%.

Table 1. Photochemical parameters of the DSSC by natural dyes using electrolyte type I

Nama Sel	V_{oc} (mV)	I_{sc} (µA)	P_{max} (nW)	*FF* (%)	η (%)
Chlorophyll	426	20	1720,5	20,193	0,0092
Turmeric	460	18,9	1780	20,474	0,0095
Mangosteen	377	27,4	2811,9	27,221	0,0150
Mangosteen-Chlorophyll	376	53	4360,5	21,881	0,0232
Turmeric-Mangosteen	504	62,2	5795,5	18,487	0,0309
Turmeric-Chlorophyll	563	64	7046	19,555	0,0376
Chlorophyll-Turmeric-Mangosteen	574	69	8548,8	21,586	0,0456

Table 2. Photochemical parameters of the DSSC by natural dyes using electrolyte type II

Cell Name	V_{oc} (mV)	I_{sc} (µA)	P_{max} (nW)	*FF* (%)	η (%)
Chlorophyll	295	43	3634	28,648	0,0194
Turmeric	320	24	2076	27,031	0,0111
Mangosteen	450	30	1958,8	14,510	0,0105
Mangosteen-Chlorophyll	530	44	4488	19,245	0,0244
Turmeric-Mangosteen	340	20	2200,5	32,360	0,0117
Turmeric-Chlorophyll	387	30	2964	25,529	0,0158
Chlorophyll-Turmeric-Mangosteen	496	60	10605	35,635	0,0566

The results of the characterization of the current and voltage are obtained, it is known that when the light illuminated DSSC, TiO_2 nanoparticles that have been created, capable of generating current and voltage response was quite good. It is proved that the dyes used in this study has a good potential for DSSC application. The voltage generated at the DSSC is affected by the conditions of material arranged at DSSC, namely semiconductors and electrolytes. In theory, the voltage arising from the difference, or the difference between the Fermi energy level of the semiconductor and the potential energy of the redox electrolyte [9]. The resulting current in DSSC is influenced by the quantity and activity of electrons located in the cell. The more electrons, the greater the current generated. In the first type of this DSSC, the current generated from single DSSC dye is smaller than the current generated from a two mixing dyes and three mixing dyes. According to UV-Vis data, absorbance values on a single dye bigger than two mixing dyes and three mixing dyes but the mixing dyes produce wide wavelength range than single dye. This is consistent with research of A.Z Abidin et al and Bandara, which in their study that the results DSSC written by a mixing of more than one dye can improve the efficiency of the solar cells [10-11]. Where the dye is meant to capture photons from the light source has the ability to absorb light of a wavelength range that is generated between each dye was done mixing. So that more photons are absorbed by the dye and cause electrons generated more and more. However, the current results obtained in this study is the first type DSSC, the results obtained current is small, so the optimization was done by replacing the type of electrolyte. In this second type of DSSC, optimization is done by adding the KI and iodine concentration in the gel electrolyte and also replace the previous solvent distilled into acetonitrile. Acetonitrile as a solvent increase the efficiency of DSSC because acetonitrile is an organic solvent which has many advantages compared to distilled water, such as low viscosity and rapid diffusion of ions. With this second optimization, the voltage generated is not much different, but as a whole, the current generated has risen 20 – 60 µA. This is due to the addition concentration of KI and Iodine. With the addition of this concentration, the ion - ion produced was also growing. Given that the ions in the electrolyte serves as an electron transfer from the carbon electrode to the dye was oxidized on the working electrode. The more ions contained in the electrolyte, the faster the electrons recombine [12]. The Characterization of current and voltage DSSC type I shown in Figure 2. The Characterization of current and voltage DSSC type I shown in Figure 3.

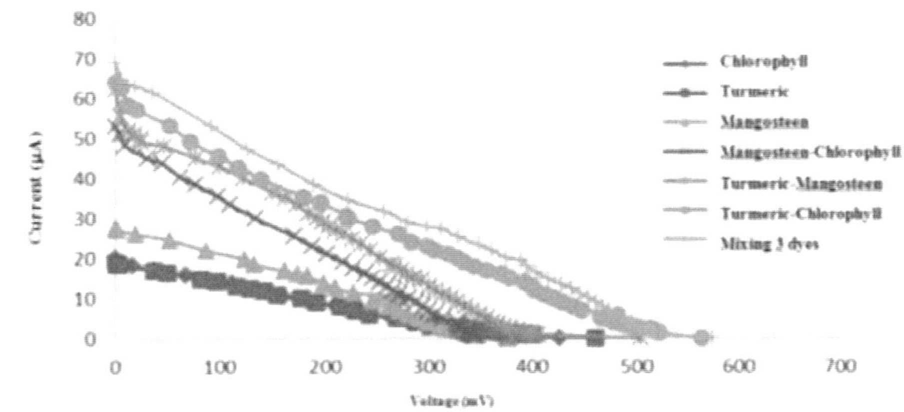

Figure 2. The Characterization of Current-Voltage DSSC type I.

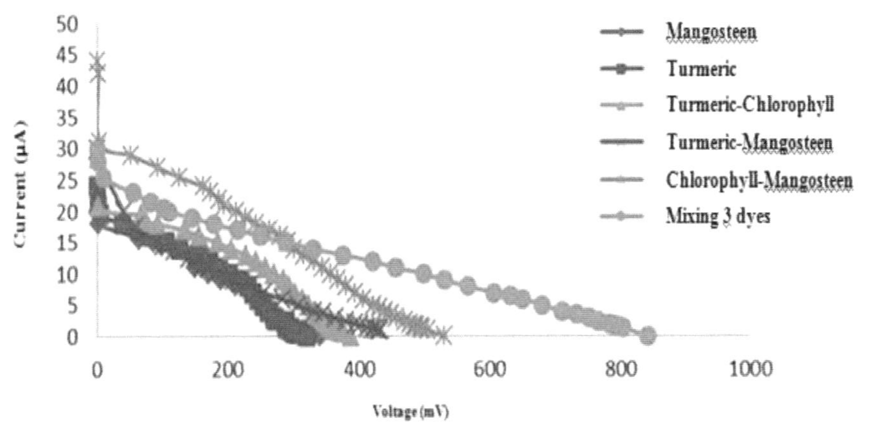

Figure 3. The Characterization of Current-Voltage DSSC type II.

4. Conclusion

The features of the DSSC have successfully been fabricated with natural dyes sensitizers containing chlorophyll, turmeric, mangosteen, chlorophyll-mangosteen, chlorophyll-turmeric, turmeric-mangosteen and chlorophyll-mangosteen-turmeric. DSSC with all natural dyes were optimezed using two types of gel electrolyte. DSSC type I and type II with mixing three dyes generate greater efficiency compared to sensitizer single dye and mixing two dyes. Meanwhile, DSSC type II using acetonitrille solvent generate greater than DSSC type I using aquades solvent.

5. References

[1] O'Regan & Gratzel, M 1991 A low-cost, high efficiency solar cell based on DS Colloidal TiO2 films Nature Vol. **353** Issue 6346,737

[2] Janne Halme 2002 *Dye-Sensitized Nanostructured and Organic Photovoltaic Cells: Technical review and Preliminary Test* Master's thesis submitted in partial fulfillment of the requirements for the degree of Master of Science in Technology Helsinki University of Technology

[3] Green, A. Martin 1982 Solar Cells Operating Principles Technology and System Application Prentice Hall Inc. Evylewood Cliffs

[4] Wang Y 2009 Recent research progress on polymer electrolytes for dye-sensitized solar cells Sol Energy Mater Sol Cells; **93**:1167–75

[5] Freitas JN, Longo C, Nogueira AF, Paoli MA. 2008 *Solar module using dye-sensitized solar cells with a polymer electrolyte* Sol Energy Mater Sol Cells; **92**:1110–4

[6] Nur Widaryanti, H., n.d. 2010, Pembentukan Nanopartikel TiO2 Fasa Anatase dan Rutile dengan Metode Bervariasi, ITS Surabaya.

[7] Kook, Lee Jin, Jeong Bo-Hwa, Jang Sung-il, Kim Young-Guen, Jang Yong-Wook, Lee Su-Bin, Kim Mi-Ra 2009 *Preparations of TiO2 pastes and its application to light-scattering layer for dye-sensitized solar cells* Journal of Industrial and Engineering Chemistry **15** 724-729.

[8] Maddu, A., Zyhri, M., Irmasyah 2010 Penggunaan Ekstrak Antosianin Kol Merah Sebagai Fotosensitizer Pada Sel Surya TiO_2 Nanokristal Tersensitisasi Dye Jurnal Matematika, Sains dan Teknologi MST Volume 11.

[9] Grätzel M. 2003 Review dye-sensitized solar cells. Journal of Photochemistry and Photobiology Photochemistry Reviews **4** 145–153.

[10] Abidin, A. Z. ,Noezar,I 2012 Identification of Potential Dyes and Developing Methods to Improve Dye-sensitized Solar Cell's Efficiency Regional Symposium on Chemical Engineering

[11] Bandara, J., Weerasinghe, H. 2006 *Design of high-efficiency solid-state dye-sensitized solar cells using coupled dye mixtures* Solar Energy Materials and Solar Cells **90**, 864–871.

[12] Careem,et al. 2014 *Polyvinyl alcohol gel polymer electrolyte based Dye Sensitized Solar Cells*, University of Malaya 50603 Kuala lumpur Malaysia

6. Acknowledgments

The author would like to thank to Research Institutions and Community Service Institut Teknologi Sepuluh Nopember Surabaya (LPPM-ITS), and Centre of Excellence of Alternative Energy ITS for assistantship in absorption spectrum measurement is gratefully acknowledged.

IOP Conf. Series: Materials Science and Engineering **214** (2017) 012023 doi:10.1088/1757-899X/214/1/012023

Formation Mechanism of Titanium Silicon Carbide: The Effect of Different Composition of Starting Materials

Solihin[1], Anggoro Tri Mursito[1], Zhengming Sun[3]

[1] Indonesian Institute of Science Research Center for Geotechnology, Komplek LIPI, Jl Sangkuriang, Bandung 40135, West Java, Indonesia

[2] National Institute of Advanced Industrial Science and Technolog (AIST), Moriyama-ku, Nagoya, Aichi 463-8560, Japan

* Email: solihin@lipi.go.id

Abstract. Titanium silicon carbide (Ti_3SiC_2) is a kind of ceramic that has physical property value similar with metal. Ti_3SiC_2 has been synthesized through various methods based on solid state reaction. Although Ti_3SiC_2 has been synthesized through various methods by using various starting materials consisting titanium (Ti), silicon (Si), and carbon (C) the mechanism of Ti_3SiC_2 formation through sintering has not fully understood. The aim of this research is to reveal the mechanism happening during sintering. Two composition of starting material was used, 2Ti/2Si/3TiC and 5Ti/2Si/3C. The analysis through XRD and SEM-EDS shows that the formation of intermediate phases, TiC and Ti_5Si_3, takes place prior to the formation of Ti_3SiC_2. In other words, Ti_3SiC_2 can only be formed through solid state reaction between TiC and Ti_5Si_3. Since TiC has already available in the system 2Ti/2Si/3TiC, the phase purity of Ti_3SiC_2 in 2Ti/2Si/3TiC is always higher than that of 5Ti/2Si/3C.

1. Introduction

Titanium silicon carbide (Ti_3SiC_2) is a ceramic compound that behaves like metal. Unlike ordinary ceramics, the hardness of Ti_3SiC_2 is extremely low and its thermal/electrical conductivity is relatively high. The detail properties of Ti_3SiC_2 had been reported by Barsoum et al [1]. Since it was found, Ti_3SiC_2 has been synthesized through various methods such as chemical reaction (CR) [2],chemical vapor deposition (CVD) [3], reactive sintering (RS) [4], hot isostatic pressing (HIP) [5,6,7], arc melting (AM) [8] solid-state synthesis (SSS) [9,10] and pulse discharge sintering (PDS) [11,12,13]. Various starting materials combinations have been used to synthesize Ti_3SiC_2. Some of them are Ti/Si/C [4,5,7,8] Ti/SiC/C[1,6,9] Ti/Si/TiC [11], Ti/SiC/TiC [12] and Ti/TiS2/TiC [13].

The utilization of various starting materials results in the formation of intermediate compounds as well as Ti_3SiC_2. The melting of elemental powder (Ti/Si/C) at 1200-1400 °C for 100 hours results in the formation Ti_3SiC_2, TiC, $TiSi_2$, and $Ti_5Si_3C_x$ [4,5,7,8]. The excess of silicon in the process that used elemental powder as starting material leads to the formation of $TiSi_2$ [4]. Depending on process temperature, other combination of other starting materials such as Ti/Si/TiC also results in the formation of TiC, TiS_2 as well as Ti_3SiC_2 [11]. The phase purity of Ti_3SiC_2 sample depends on the combination of starting materials. By using Ti/Si/TiC as starting material, the phase purity of Ti_3SiC_2 was more than 99 %, whereas by using the elemental powder (Ti/Si/C), the phase purity is lower than that of Ti/Si/TiC. The previous works deal with Ti_3SiC_2 synthesis usually focuses on the characterization of Ti_3SiC_2 after being sintered at single temperature. There is no firm explanation on the mechanism of Ti_3SiC_2 formation and the actual reactions takes place at lower temperature prior to the formation of Ti_3SiC_2. Therefore, the aim of this research is to reveal the mechanism of Ti_3SiC_2 formation through calcinations at various temperatures. A series of sintering experiment from low to high temperature has been done to reveal the intermediate phases in the sample sintered at various temperatures. Based on previous research [4,5,7,8,11], we concluded that only two kinds of starting material that can produce Ti_3SiC_2 at relatively lower temperature, that is, the starting powders with composition and ratio 2Ti/2Si/3TiC and 5Ti/2Si/3C. Therefore, in this experiment we used similar composition and atomic ratio. The technique chosen was Plasma Discharge Sintering (PDS) technique, since it can synthesize highest phase purity of Ti_3SiC_2 at lowest temperature and shortest time [11].

2. Experimental Procedure

Commercial Ti, Si and TiC powders at mol ratio of 2Ti/2Si/3TiC were mixed by tubular shaker for 24 hours in argon atmosphere. The average particle size and purity of the powders are 10 μm and 99.9 % for Ti, 2-5 μm and 99 % for TiC, 10 μm and 99 % for Si, respectively. The mixed powder were compacted and sintered in vacuum using PDS technique at 700 – 1300 °C with a constant mechanical pressure of 50 MPa. After being sintered, the sample was ground and polished to remove the graphite layer on the surface of the sample. X-ray diffractometer, with CuKα radiation at 30 kV and 40 mA, was used to analyze the sample. The back scattering electron (BSE) image of each sample was observed by using Scanning Electron Microscope (SEM). The phase in the BSE image was analyzed through Energy-Dispersive X-ray Spectroscopy (EDS). Another starting powder consisting of commercial Ti, Si and C powders at mol ratio of 5Ti/2Si/3C was also prepared by using the same procedure with that of 2Ti/2Si/3TiC.

3. Result and Discussion

The XRD patterns for 2Ti/2Si/3TiC samples sintered at various temperatures are summarized in figure 1. The XRD pattern of sample sintered at 700 °C shows that this sample still contains only starting materials, which indicates Ti, Si and TiC has not reacted at this temperature. But in the sample sintered at 800 °C, a new phase, Ti_5Si_3, which is an indication that a reaction has taken place, can be found. The formation of Ti_5Si_3 becomes rapid at 900-1000 °C. It is indicated by the increasing of the diffraction intensity corresponding to Ti_5Si_3 and the decreasing of diffraction intensity corresponding to Ti and Si. Ti_5Si_3 is one the phases in binary phase diagram of Ti-Si, as well as TiSi, Ti_3Si, Ti_5Si_4, and $TiSi_2$ [14]. Although there are various phases in binary phase diagram of Ti-Si, the reaction takes place between Ti and Si always results in the formation of single phase Ti_5Si_3 [15,16,17]. Therefore, based on these facts, the formation of Ti_5Si_3 can be written as follows.

$$5Ti + 3Si \rightarrow Ti_5Si_3 \quad (1)$$

The new set of diffraction intensity corresponding to Ti_3SiC_2 starts to appear at 1000 °C . The 2θs of the diffraction intensity corresponding to Ti_3SiC_2 in our sample are 39.59, 40.81 and 42.57. These are almost the same with 2θs of Ti_3SiC_2 in ICDD database (39.58, 40.86, and 42.57). Other researcher that deals with Ti_3SiC_2 synthesis also reported that 2θs of their Ti_3SiC_2 sample are almost similar with that of our sample [18,19,20,21].

Meanwhile, the diffraction intensity corresponding to TiC decrease with an increasing of temperature, and then they finally disappear at 1200 °C. The disappearance of the diffraction intensity corresponding to TiC obviously indicates that a reaction involving TiC has taken place. It is interesting to see that as the diffraction intensity corresponding to TiC and Ti_5Si_3 decrease and finally disappear, the intensity of the diffraction intensity corresponding Ti_3SiC_2 is getting higher. Thus, the formation of Ti_3SiC_2 must be the cumulative result of solid state reaction between TiC and Ti_5Si_3.

Figure 2 presents the back scattering electron (BSE) image of mixed and sintered 2Ti/2Si/3TiC powder at 900, 1100 and 1300 °C. The selection of these temperatures for SEM-EDS evaluation was based on XRD profile of sintered sample, presented in Figure1, which shows that the lowest temperature where Ti_3SiC_2 can be formed, one among temperatures where all product can be found, and one among temperatures where the sample consisting only Ti_3SiC_2 are 900, 1100, and 1300 °C, respectively. At 900 °C, three phases can be distinguished and identified through Energy-Dispersive X-ray Spectroscopy (EDS), they are Si, TiC and Ti_5Si_3. Meanwhile, at 1100 °C, Si has disappeared and Ti_3SiC_2 can be found, as well as TiC and Ti_5Si_3. Finally at 1300 °C in the entire image, other phases have disappeared, leaving Ti_3SiC_2 as the only phase in the sample. These result is agree with the XRD result that shows the formation of Ti_5Si_3 at 800-900 °C, the disappearance of Si and the formation of Ti_3SiC_2 at 1000 °C, and Ti_3SiC_2 as single phase at 1300 °C.

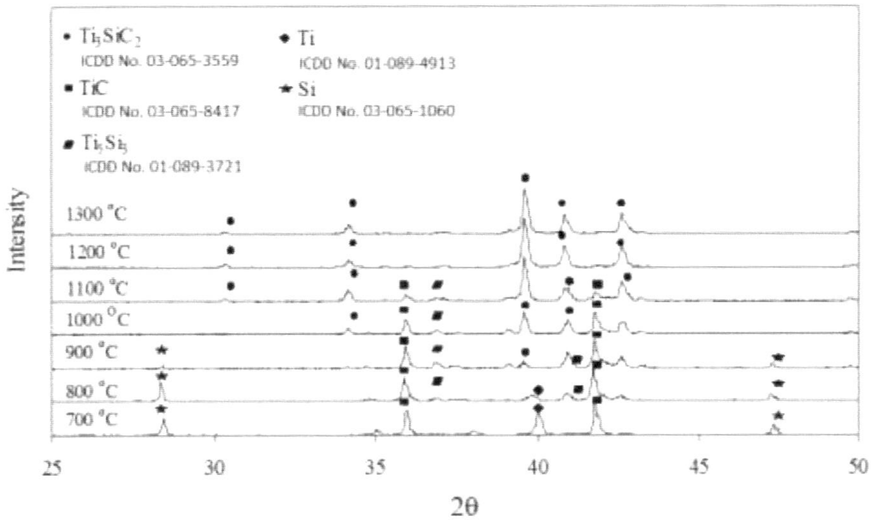

Figure 1. XRD profile of mixed 2Ti/2Si/3TiC sintered at 700-300°C for 15 minutes.

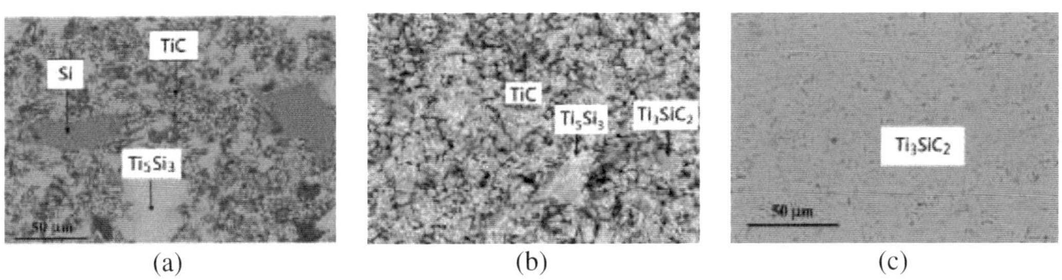

Figure 2. BSE image of mixed 2Ti/2Si/3TiC sintered at (a) 900 °C, (b) 1100 °C, (c)1300 °C for 15 minutes.

XRD patterns for the samples sintered from the powder mixture of 5Ti/2Si/3C are shown in figure 3. It was found that except the intermediate compounds of Ti_5Si_3, another intermediate compound, Ti_8C_5, was formed during sintering. The diffraction intensity of Ti_8C_5, increases with an increasing in temperature until it disappears at 1000 °C. As Ti_8C_5 disappears at 1000 °C, TiC has been formed and it still can be found until 1200 °C. The formation of TiC was at the same time with the disappearance of C and Ti_5C_8. Therefore, the reaction of TiC formation can be written as follows.

$$8Ti + 5C \rightarrow Ti_8C_5 \quad (2)$$

$$Ti_8C_5 + 3C \rightarrow 8TiC \quad (3)$$

Meanwhile, diffraction intensity corresponding to Ti_3SiC_2 can be found significantly for the first time in the XRD pattern of sample sintered at 1000 and it continues to exist until 1200 °C. At these temperatures, silicon and titanium has been reacted to form Ti_5Si_3 and TiC has also been formed through mechanism as shown in equation (2) and (3). Thus, the reaction takes place during sintering should be happen between TiC and Ti_5Si_3.

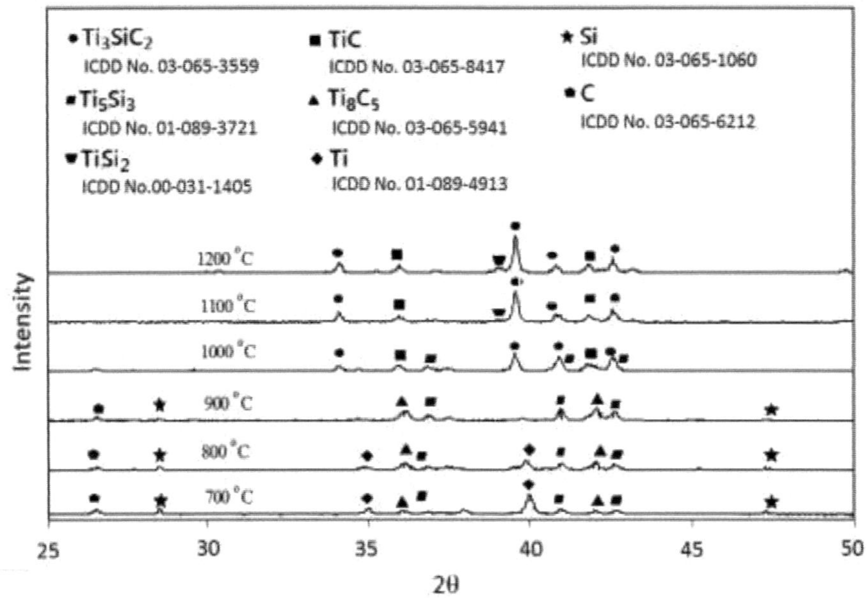

Figure 3. XRD profile of the mixed 5Ti/2Si/3C sintered at temperature 700–1200 °C for 15 minutes.

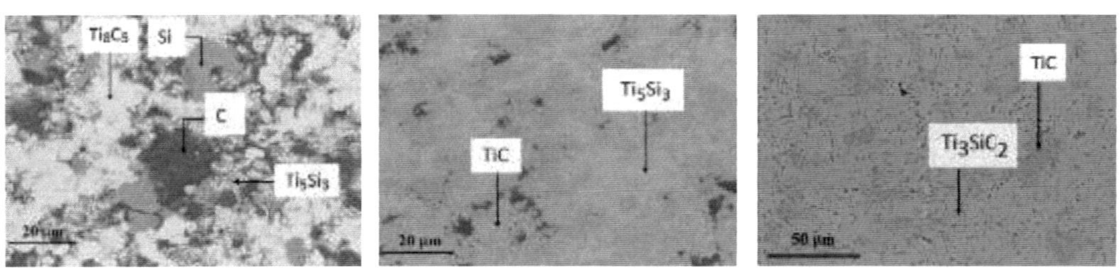

(a) (b) (a)

Figure 4. BSE image of mixed 5Ti/2Si/3C sintered (a) 700 °C, (b) 1000 °C, (b) 1200 °C for 15 minutes.

Figure 4 shows the BSE image of mixed 5Ti/2Si/3C powder sintered at 700, 1000, and 1200 °C. The phases within this image were also identified through Energy-Dispersive X-ray Spectroscopy (EDS). At 700 °C the phases that can be found are C, Si, Ti_8C_5, and Ti_5Si_3. Obviously, the reactions among Ti, Si and C have taken place, resulted in the formation of Ti_8C_5 and Ti_5Si_3. The SEM-EDS analysis results quite similar with the result obtained through XRD analysis that shows the formation of Ti_8C_5 and Ti_5Si_3. The SEM-EDS analysis to the sample sintered at 1000 °C is also quite similar with the analysis obtained through XRD, which shows that Ti, C and Si have been reacted to form Ti_8C_5 and Ti_5Si_3. Finally at 1200 °C, Ti_3SiC_2 becomes the matrix phase in the image with TiC as dispersed phase. This result is also similar with the result obtained through XRD which shows that at this temperature the sintered powder consist of Ti_3SiC_2 with TiC.

The fraction of Ti_3SiC_2 to TiC is the representation of phase purity of Ti_3SiC_2 in the sintered sample. It was calculated by using integrated intensity ratio method, in which the integrated intensity of Ti_3SiC_2 was compared to the sum of the integrated intensity of Ti_3SiC_2 (104) and TiC (200). The result of calculation is presented in Figure 5. The figure clearly shows that at any temperature, the phase purity of 2Ti/2Si/3TiC is always higher than that of 5Ti/2Si/3C. The previous XRD and SEM-EDS analysis can be summarized that the formation of TiC and Ti_5Si_3 takes place prior to the formation of TSC. In other words, TiC and Ti_5Si_3 are the necessary reactant for Ti_3SiC_2 formation. Since in the 2Ti/2Si/3TiC, the TiC was already available in the starting material, there was no energy needed to form TiC in 2Ti/2Si/3TiC. Therefore, the rapidity of Ti_3SiC_2 formation in 2Ti/2Si/3TiC is higher than that of 5Ti/2Si/3C.

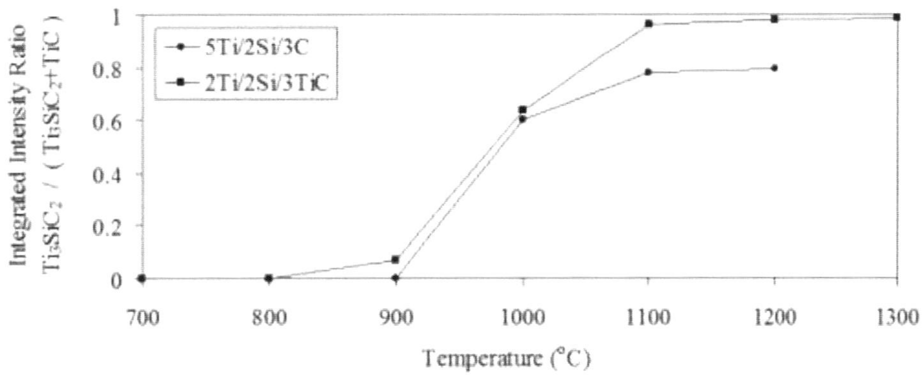

Figure 5. Integrated intensity ratio for Ti_3SiC_2 in 2Ti/2Si/3TiC and 5Ti/2Si/3C sintered at temperatures of 700-1200 °C for 15 minutes.

4. Summary

Ti_3SiC_2 can be formed through plasma discharge sintering method by using molar composition and ratio Ti/2Si/3TiC or 5Ti/2Si/3C. XRD and SEM-EDS analysis shows that formation of certain intermediate phases takes place prior to the formation of Ti_3SiC_2. Some intermediate phases are Ti_8C_5 and Ti_5Si_3, and TiC, Ti_8C_5 gradually converts to TiC at 1000 °C, whereas Ti_5Si_3 did not convert to another intermediate phase. The analysis through XRD and SEM-EDS combined with the calculation fraction of Ti_3SiC_2 to TiC shows that TiC and Ti_5Si_3 is needed in the formation of Ti_3SiC_2, and therefore, the sample that already posses those phases TiC (2Ti/2Si/3TiC) can produce more Ti_3SiC_2 than that of 5Ti/2Si/3C.

5. Refference

[1] Barsoum M W and El-Raghy T 1996 *J. Am. Ceramic Soc.* **79** 1953
[2] Jeitschko W and Nowotny H 1967 *Monath. Chem.* **98** 329
[3] Goto T and Hirai T 1987 *Mater. Res. Bull.* **22** 1195
[4] Radhakrishnan J, William JJ, and Akinc M 1999 *J. of Alloy and Compound* **85** 285
[5] Lis J, Miyamoto Y, Pampuch R, and Tanihata K 1995 *Material Letters* **22** 163
[6] Hong X, Okano T, Iseki T, and Yano T 1995 *J. of Material Science Letters* **30** 3087
[7] Li J F, Sato F, and Watanabe R 1999 *J. of Material Science Letters* 1595
[8] Arunajatesan S, and Carim AH 1995 *J. Am. Ceramic Soc.* **78** 667
[9] Goesmann F, Wenzel R, and Fetzer R S 1998 *J. Am. Ceram. Soc.* **81** 3025
[10] Racault C, Langlais F, and Naslain R 1994 *J. of Mat. Sci.* **29** 3384
[11] Sun Z M, Zhang Z F, Hashimoto H and Abe T 2002 *Mater. Trans.* **43** 428
[12] Zhang Z F, Sun Z M, Hashimoto H, and Abe T 2001 *Scripta Materialia* **45** 1461
[13] Zhang Z F, Sun Z M, Hashimoto H, and Abe T 2002 *Mat Res Innovat.* **5** 185
[14] Massalski T B, Okamoto H, Subramanian P R, Kacprzak L 1990 *Binary Alloy Phase Diagrams*, ASM Internasional, Materials Park, OH
[15] Vishnyakov V, Lu J, Eklund P, Hultman L, Colligon J 2013 *Vacuum,* (93), 56-59
[16] Kachelmyer C R, Khomenko IO, Rogachev AS, Varma A 1997 *J. Mater. Res.* **12** 12

[17] Sabooni S, Karimzadeh F, Abbasi M H 2012 *Bull. Mater. Sci.* **35** 439
[18] Zou Y, Sun Z M, Tada S, Hashimoto H 2006 *Materials Transactions* **47** 2987
[19] Tang K, Wang C, Huang Y, Xia J 2001 *J. European Ceramic Soc.* **21** 617
[20] Tang K Wang C, Xu X, Huang Y 2002 *Materials Letters* **55** 50
[21] Wang L, Jiang W, Chen l, Bai B 2004 J. Mater. Res. **19** 3004

IOP Conf. Series: Materials Science and Engineering **214** (2017) 012024 doi:10.1088/1757-899X/214/1/012024

Enhanced Salt-removal Percentage in Capacitive Deionization of NaCl Solutions with Modified Activated Carbon Electrodes by HNO_3

Diani Ainun Nisa and Endarko

Department of Physics, Institut Teknologi Sepuluh Nopember
Jl. Arif Rachman Hakim, Sukolilo Surabaya (60111), Indonesia

endarko@physics.its.ac.id

Abstract. Carbon electrodes for desalination system have successfully been synthesized with and/or without modified activated carbon by chemically activated using HNO_3. The freezing-thawing method was used to synthesize the carbon electrodes. In this study, 5 cycles of freezing-thawing were applied in the synthesized carbon electrodes (1 cycle is 12 hours for freezing and 6 hours for thawing). Electrochemical properties of the synthesized carbon electrodes with and/or without modified activated carbon were characterized and observed by cyclic voltammetry (CV) and Electrical impedance spectroscopy (EIS). The salt-removal percentage experiments were conducted to evaluate the performance of capacitive deionization (CDI) cell using the two pairs of carbon electrodes with each pair consisting of two parallel electrodes that separated by a spacer. The result showed that the salt removal percentage of the carbon electrodes with modified activated carbon has greater than the carbon electrodes without modified activated carbon, with reduction level at 55.7 and 24.8%, respectively.

1. Introduction

Global warming is tied to our huge demand for energy and the consequent burning of fossil fuels. Furthermore, energy and water are also related. Energy is needed to deliver water and water is needed to generate energy. Indeed, one does not site a nuclear power plant in the middle of the desert but rather near a major water body such as a river, lake or the sea. By the same token, we are likely, not able to produce energy such as bio-fuels without a sufficient supply of water [7]. On the other hand, we need large quantities of energy to desalinate seawater for potable use. The scarcity of clean water is often hit several parts of the world due to the dry season and pollution of water resources. The challenge to solve this problem is to convert seawater and brackish water into clean water. Researchers have been conducted to meet the needs of fresh water using bracpowerkish water and sea water as a source of fresh water. Desalination process for getting fresh water from seawater has been growing rapidly in several developed countries [3].

Many methods that have been developed for desalinating of seawater such as Multi-Stage Flash (MSF) and Reverse Osmosis (RO). According to energy consume RO system is suitable for processing with low energy while MSF system needs high energy. Despite energy consume in the RO is low, but the cost production for RO desalination system is still expensive. For the reason, capacitive deionization (CDI) is one of the technologies that can be applied to the desalination process of brackish water and seawater [8].

IOP Conf. Series: Materials Science and Engineering **214** (2017) 012024 doi:10.1088/1757-899X/214/1/012024

Capacitive deionization (CDI) is a method for removing salt from aqueous solutions through an electrochemically controlled by an external power supply. The significant factor of an electrode in the CDI cell is when the electrode has a high specific surface area regarding the electrosorption process quantitative and attractive for water treatment [1]. The performance of CDI largely depends on the physical and structural properties of the electrode materials. An ideal electrode should have excellent electrical conductivity for effective charge holding, a high specific surface area for ionic accumulation, suitable pore size, and networks for easy ionic motions as well as good electrolyte wettability. Currently, most of the electrodes for CDI are made of different carbon materials including activated carbons, carbon aerogel, ordered mesoporous carbon and carbon nanotubes [4].

In the present work, the carbon electrodes will be synthesized using freezing-thawing method for CDI cell with or without modified activated carbon. The electrochemical properties and surface morphology are measured and observed to show the performance of the carbon electrode. Furthermore, the molecular bonds contained in active carbon with and without modified are also analyzed. Desalination experiments to determine the desalting performance of CDI cell will also be studied using NaCl solution.

2. Material and Methods

2.1 Materials

The materials used in this study namely nitric acid (60% HNO_3), polyvinyl alcohol (PVA) (Merck Co.), aquades, graphite sheet as a current collector, and activated carbon.

2.2 Fabrication of carbon electrode

Carbon electrodes were prepared with unmodified carbon and modified carbon (adding nitric acid 60% to the mixture) using the freezing-thawing method. 1 g of polyvinyl alcohol (PVA) and 100 mL distilled water at 100 °C were placed on a magnetic stirrer and was then stirred for 1 hour. Subsequently, the mixture was added with 20 g of activated carbon powder and was then stirred for 20 mins. Afterward, the paste was coated onto graphite sheet with a size of 6×8 cm^2 and was then dried it at room temperature. Furthermore, the dried carbon electrodes were placed in the freezer at -14°C for freezing process (12 h). Two variations of the time for thawing process (at 30°C) namely 6 and 12 h were used in this study. The carbon electrodes were synthesized by a freezing-thawing method with 3, 4 and 5 cycles.

2.3 Characterization of carbon electrode

Scanning Electron Microscopy (SEM) FEI-Inspect S25-EDAX was used to characterize the morphology of the carbon electrodes. Hereafter, Cyclic Voltammetry (CV) and Electrical Impedance Spectroscopy (EIS) (Autolab PG-Stat 302 Methrom) were used to investigate the electrochemical properties of the carbon electrodes. EIS was performed using three electrodes method in 0.5 M KCl solution. CV measurement was conducted with the voltage spans from -0.5 – 0.5 V (vs. Ag/AgCl) at 10 mV/s for 10 cycles whereas EIS is performed with frequency in the range of 0.01-10 Hz at 10 mV peak to peak.

2.4 Desalination System of Capacitive Deionization

Desalting performance has been optimized by passing a solution of 0.0025 M NaCl. The conductivity and salinity measurements were then conducted using a salinity meter and conductivity meter in the initial condition of the salt solution and after it passed from the CDI system. Desalination efficiency can be calculated according to the following equation [7]:

$$Desalination\ efficiency = \frac{C_0 - C_t}{C_0} \times 100\% \quad (1)$$

where C_0 and C_t are initials for initial conductivity and conductivity at each duration t.

3. Result and Discussion

3.1 Carbon Analysis by FTIR Test

The Activated carbon adsorption capacity is not only determined by the pore structure but also influenced by the chemical nature of the surface. Properties of activated carbon surface chemistry can be selectively modified with the aim to further enhance the adsorption capacity. Results of activated carbon modified by HNO_3 can be seen through the test FTIR.

Fourier transform infrared spectroscopy (FTIR) is a technique which is used to obtain an infrared spectrum of absorption or emission of a solid, liquid or gas. An FTIR spectrometer simultaneously collects high spectral resolution data over a wide spectral range. This confers a significant advantage over a dispersive spectrometer which measures intensity over a narrow range of wavelengths at a time.

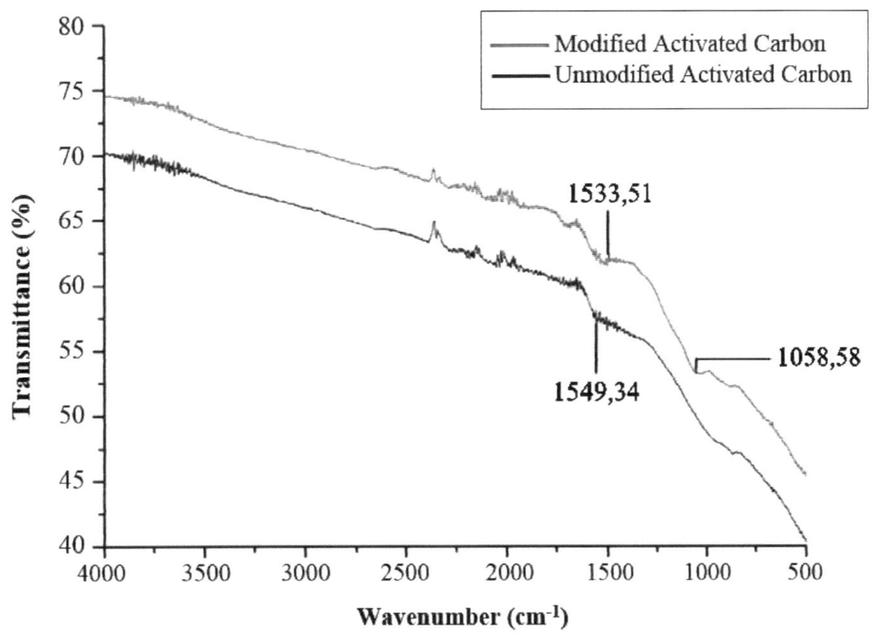

Figure 1. The frequency spectrum of modified and unmodified carbon by FTIR test.

FTIR testing aims to determine the molecular bonds (force element) which is in active carbon. On activated carbon modified by an additional functional group that is the wave number 1058.58 cm^{-1}. The wave numbers indicate a functional group ether (C-O). So that the activated carbon modification has two functional groups which aromatic groups (C=C) and the functional group ether (C-O). The addition of oxygen-containing groups can increase uptake is possible to increase the value of the capacitive electrochemical. Based on research conducted by Gokce and Over in 2014 increased the carboxylic group can increase the adsorption methylene liquid blue and phenol for carboxylic group adding electrostatic interactions on the surface of the carbon decisive role in the adsorption process [6].

IOP Conf. Series: Materials Science and Engineering **214** (2017) 012024 doi:10.1088/1757-899X/214/1/012024

3.2 Carbon Electrode Morphology

Scanning Electron Microscopy (SEM), also known as SEM analysis or SEM microscopy, is used very effectively in microanalysis and failure analysis of solid inorganic materials. Scanning electron microscopy is performed at high magnifications, generates high-resolution images and precisely measures very small features and objects. The composition of the carbon electrodes was measured and characterized by Scanning Electron Microscopy coupled with Energy Dispersive X-ray (SEM/EDX). SEM cannot determine the composition of the observed carbon electrode.

Figure 2 shows the composition of the carbon electrode structure which measured by SEM/EDX. It can be analyzed that the cycle variation does not influence the composition of the carbon electrodes.

Figure 2. SEM image of modified electrode was fabricated for 5 Cycles.

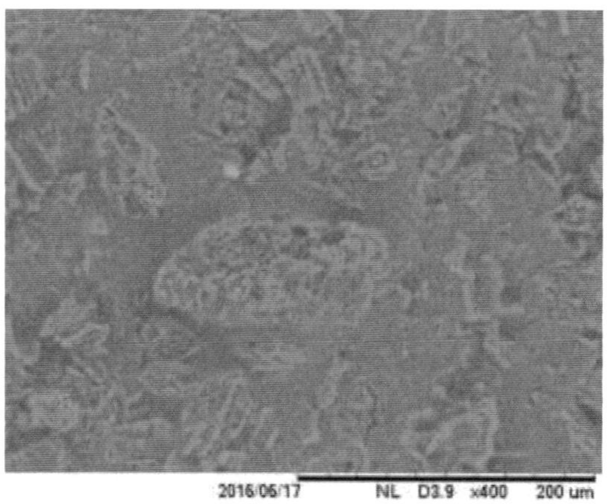

Figure 3. SEM image of unmodified electrode was fabricated for 5 Cycles.

Freezing-thawing process on the fabrication of carbon electrodes resulted in gelation process, so that made a strong electrode and insoluble in water. In the freezing process, the solvent will freeze and form crystals along microphase of solute carbon [6] PVA which is not frozen because of the concentration of the liquid becomes very small so that the concentration of gel increases triggered gel formation. When it freezes, solvents crystal is grown to touch the outer portion of the crystal to another, so that after melting the interconnection system of porous appears in gels. Gelation can be occurred at any phase of the process of freezing – thawing [7]. When melted, the surface tension between the liquid solvent and the gel phase causing carbon coarse pores arranged. Conversion between spongy cryogels and non-spongy relies on freezing, the concentration and composition of solute [6]

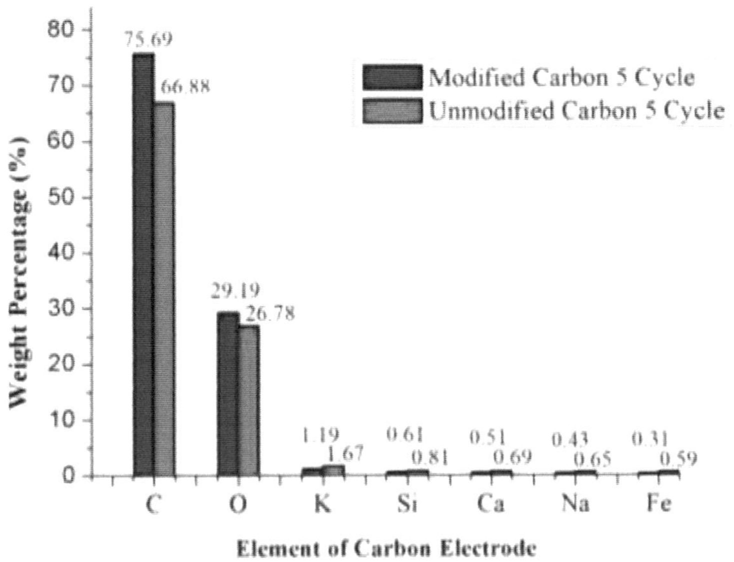

Figure 4. Elements of carbon electrode in unit (%) measured by SEM/EDX.

Figure 3 shows the elements of the carbon electrode in the unit (%). The result showed that around 70% the electrode is formed by carbon and 20% from oxygen and less than 1% from K, Si, Ca, Na and Fe.

3. 3 Electrochemical Properties of Carbon Electrode

Besides the physical characterization, the carbon electrodes are also characterized by the electrochemical property. The electrochemical properties were characterized by CV and EIS. Cyclic voltammetry or CV is a type of potentiodynamic electrochemical measurement. In a cyclic voltammetry experiment, the working electrode potential is ramped linearly versus time. Unlike in linear sweep voltammetry, after the set potential is reached in a CV experiment, the working electrode's potential is ramped in the opposite direction to return to the initial potential. These cycles of ramps in potential may be repeated as many times as desired. The current at the working electrode is plotted versus the applied voltage (i.e., the working electrode's potential) to give the cyclic voltammogram trace. Cyclic voltammetry is generally used to study the electrochemical properties of an analyte in solution. CV measurement was performed with the voltage ranging from 0.5 – 0.5 V (vs. Ag/AgCl) at a scan rate of 10 mV/s for 10 cycles whereas EIS is carried out in the frequency range of 0.01-100.000 Hz at a signal amplitude of 10 mV peak to peak. CV test results for carbon electrode modification and without modification is shown in Figure 4.

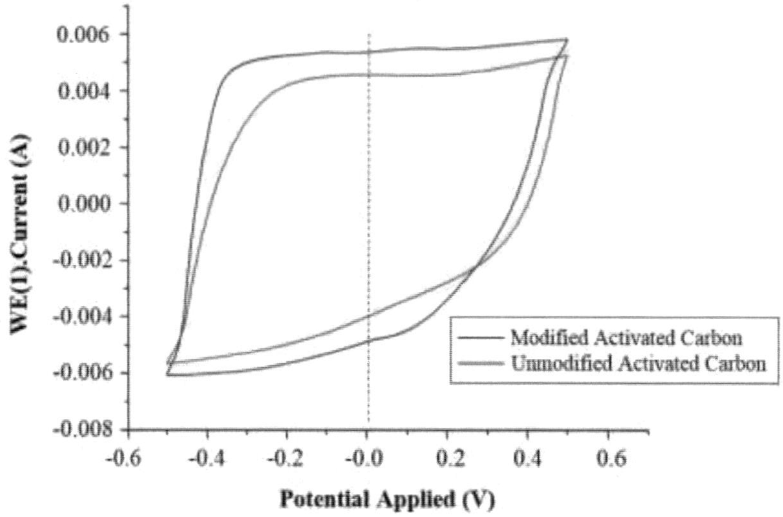

Figure 5. CV curve of carbon electrodes for unmodified carbon and modified carbon for 5 cycles.

Based on data from CV curves in Fig. 4, value of capacitance of carbon electrodes can be calculated using equation as follows [7] :

$$C = \int \frac{dq}{dV} = I \int \frac{dt}{dV} = \frac{I}{v} \tag{2}$$

where q is a charge (C), V is the potential difference (V), I is intensity (A), t is time (s) and v is scan rate (V/s). Meanwhile, the value of specific capacitance can be calculated with equation (2) divided by mass of the carbon powder. From Figure 4 shows that the modified carbon electrodes have the largest area compared to the unmodified carbon electrode. The specific capacitance for both the carbon electrodes with and/or without modification which are synthesized by the freezing-thawing at 5 cycles can be calculated at 26.78 and 22.77 F/g, respectively.
Voltammogram curve of the carbon electrode is ideally a rectangular-shaped [5] which shows the flow reduction and oxidation are clear and stable. The CV curve in this study showed the approaching the rectangular-shaped as shown in Fig. 4. It indicates that the carbon electrodes have a potential for CDI cell application. The greater the value of the electrode capacitance possessed the greater the ability of the electrode to reduce salt because salt ions more easily attracted and retained on the electrode surface. The greater the value of the electrode capacitance possessed the greater the ability of the electrode to reduce salt because salt ions more easily attracted and retained on the electrode surface [5]

The EIS test was used to determine the kinetics of electrochemical parameters related to electrical quantities such as capacitance, resistance, and inductance. Electrical impedance is the measure of the opposition that a circuit presents to a current when a voltage is applied. Based on the CV measurement, the carbon electrodes which is synthesized with 5 cycles using freezing-thawing method could potentially be used as an electrode for CDI cell, so that the EIS measurement only is done for it. EIS measurement was carried out for its ability to distinguish resistance from capacitance. Hereinafter, the specific capacitance can be calculated using equation as follows [8]

$$C = \left|\frac{1}{\omega Z''}\right| \tag{3}$$

Where ω is the angular frequency of alternating current (AC) signal and Z'' is the imaginary impedance data. EIS curve for the carbon electrode is shown in Fig 5. It can be analyzed that the largest specific capacitance using equation 2.

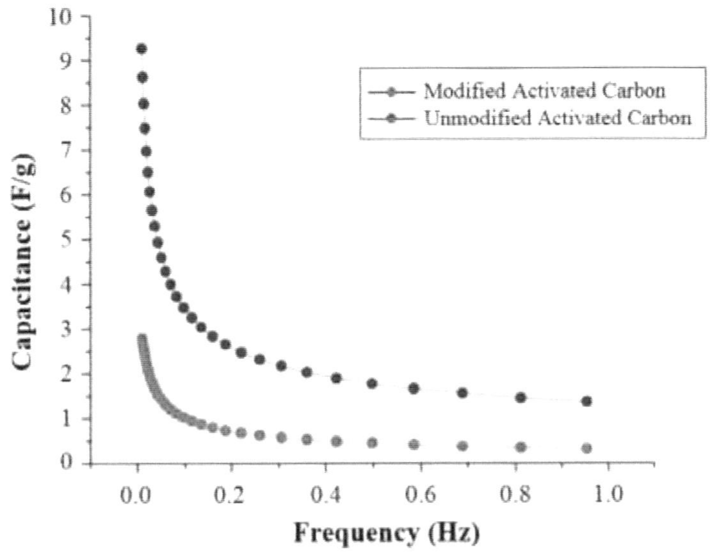

Figure 6. Specific capacitance to frequency chart on modified and unmodified carbon.

In addition to determining the nature of capacitive and resistive electrodes, EIS test can determine the value of the electrode capacitance. From Figure 5, it can be seen that the greatest value of specific capacitance measured at 9.25 F/g at 0:01 Hz frequency for modified electrode. Meanwhile, the electrodes without modification only at 2.78 F/g. The capacitance value is obtained using the equation 2. Specific capacitance value decreases with increasing frequency and constant at high frequencies; this is because the low-frequency current is applied, the electron transfer is not disturbed by the frequency [7]

3.4 Desalination performance of capacitive deionization

To evaluate the salt removal performance of the carbon electrodes, 596 μS/cm NaCl solution was pumped to the CDI cell with flow rate at 25 mL/min. The two pairs of carbon electrodes were constructed to CDI unit cell with each pairs consisted of two parallel electrodes separated by a spacer. The salt removal experiments consisted of two steps namely is desalination (adsorption) and regeneration (desorption). Desalination process was applied a cell potential of 2 V for 1 hour and 4000 sample data.

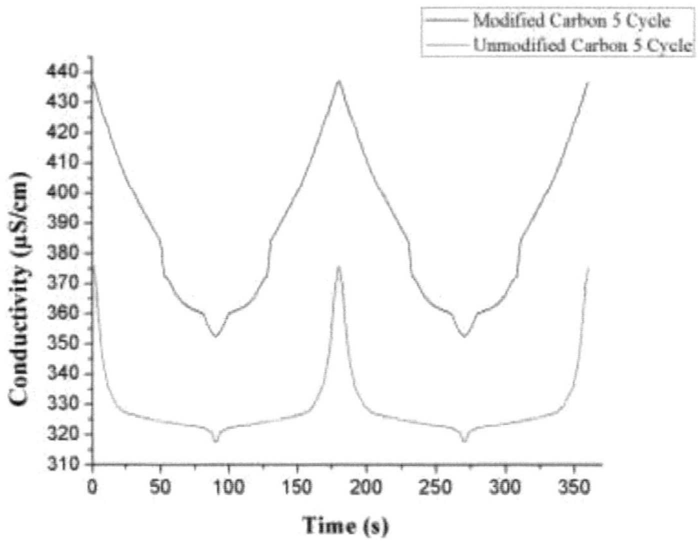

Figure 7. Graphic adsorption and desorption of processes in the CDI system for modified and unmodified carbon 5 cycle.

Fig. 7 shows the desalination process and regeneration process for CDI cell with modified carbon electrode using freezing thawing 5 cycles. After that, the salt-removal percentage can be calculated using equation as follows [9]:

$$\% \Delta Salt = \frac{\sigma_f - \sigma_p \, x \, 100\%}{\sigma_f} \tag{4}$$

where σ_f is the initial conductivity ($\mu S/cm$) and σ_p is the final conductivity ($\mu S/cm$). Finally, It can be deduced from figure that the greatest value of salt-removal percentage was achieved at 55.70 % when the solution flowed to the CDI cell with the modified carbon prepared with freezing-thawing for 5 cycles. The salt removal value of unmodified carbon 5 cycle was achieved at 24.80 %. It shows that modified carbon more efficience than unmodified carbon 5 cycle.

4. Conclusion

A desalination system with stack-type CDI method has been successfully fabricated. Desalination process is performed at 0.0025 M NaCl solution. The result showed that the salt removal percentage of the carbon electrodes with modified activated carbon has greater than the carbon electrodes without modified activated carbon, with reduction level at 55.7 and 24.8 %. It showed that the most of NaCl ions adsorbed by carbon electrodes. Characterization and optimization of the results of the CDI system acquired good results so that the system is highly prospective to be developed into a desalination system to produce fresh water.

5. References

[1] Y. Oren 2008 *Capacitive deionization (CDI) for desalination and water treatment — past, present and future (a review)* Desalination, **228**(1–3) 10–29

[2] C.-H. Hou, J.-F. Huang, H.-R. Lin, and B.-Y. Wang *2012 Preparation of activated carbon sheet electrode assisted electrosorption process* J. Taiwan Inst. Chem. Eng **43**(3) 473–479

[3] M. A. Anderson, A. L. Cudero, and J. Palma 2012 *Capacitive deionization as an electrochemical means of saving energy and delivering clean water. Comparison to present desalination practices: Will it compete?* Electrochimica Acta **55** (12) 3845–3856

[4] B. Jia and L. Zou, *Graphene nanosheets reduced by a multi-step process as high-performance electrode material for capacitive deionisation* 2012 Carbon **50** (6) 2315–2321

[5] Nadakatti S, Tendulkar M, Kadam M. 2011 *Desalination* **268** 182–188

[6] N. E. Vrana 2009 *Use of Poly Vinyl Alcohol (PVA) Cryogelation for Tissue Engineering: Composites, Scaffold Formation and Cell Encapsulation.* (School of Mechanical and Manufacturing Engineering - Dublin CitY University, Ireland).
[7] Li Haibo 2010 *Separ. Purif. Technol.* **75** 8–14
[8] Park B-H, Kim Y-J, Park J-S, Choi J 2011 *J Ind Eng Chem.* **17** 717–722
[8] Endarko, I. Fatimah and I.P. Sari 2016 *Measurements of the salt-removal of NaCl, KCl and MgCl using a carbon electrode prepared with freezing thawing method in capacitive deionization* by AIP Conference Proceedings **1725**, 020015
[9] Endarko and I.P. Sari 2016 *Enhanced salt-removal percentage in capacitive deionization with addition of ionexchange membrane using carbon electrode synthesized with freezing thawing method.* by AIP Conference Proceedings **1725** 6

IOP Conf. Series: Materials Science and Engineering **214** (2017) 012025 doi:10.1088/1757-899X/214/1/012025

The Effect of Thermo-mechanical Treatment of Substrate Preparation and Carburizing Temperature on The Morphology and Hardness of Carburizing on Low Carbon Steel

Raden Dadan Ramdan[1], Andra Adetia[1] and Rochim Suratman[1]

[1]Faculty of Mechanical and Aerospace Engineering, Institut Teknologi Bandung, Jl. Ganesa no 10 Bandung, West Java, Indonesia, 40132

Email: dadan@material.itb.ac.id

Abstract. Low carbon steel has a lot of applications in daily life because of its excellent properties. Among its excellent properties are good ductility, toughness, formability, and weldability. For the case of application that requires good wear resistant, carbon steel is not sufficient. For this case it is required to increase the hardness of carbon steel, such as by carburizing method. In this method one factor that affects the carburizing layer is metal substrate condition, i.e grain size. Another factor is temperature that could have significant role on the carburized layer characteristics. Therefore the present works focus in the correlation between degree of reduction of cold rolling and carburizing temperature on the carburized layer characteristic. Thermo-mechanical treatment was applied to the specimen with reduction of 0%, 30%, and 80% by mean of rolling before carburizing process. Carburizing processes were carried out at temperature of 850°C, 900°C, and 950°C. Examinations on carburized specimens were conducted by metallography and micro hardness test. The result showed that the specimen of 80% reduction giving the thickest layer and the hardest surface layer. In addition, at temperature of 900°C, the hardest layer was formed with the hardness at 1003 HV. However, the thickest layer was formed at the temperature of 950°C.

1. Introduction

Carbon steel is the most widely applied metals for various applications due to its excellent mechanical properties that combine high strength and sufficient ductility and the relatively good manufacturing ability. However hard coating are normally applied on the steel substrate in order to increase wear resistance characteristic of this metal [1-5] for application that requires excellence of this characteristic. Among of this method is carburizing process, which can be applied in various method of processes [6-10].

Carburizing is a method of surface hardening, a method to improve the mechanical properties of metal surfaces. This method is also considered as thermo-chemical treatment since combining of chemical treatment (by diffusing element on the metal surface) and heat treatment processes. This method is widely applied due to relatively low cost of processes involving relatively cheap raw material precursor and the easiness of application as compared with other surface hardening processes.

Based on the above description, carburizing play important role in industries and therefore various attempts have been performed by researchers in order to optimize this method. In the present work, substrate preparation and carburizing temperature have been selected as variable parameters to be optimized. Substrate preparation has been performed by thermo-mechanical treatment, which is combination between rolling and heat treatment processes. Different degree of rolling reduction is expected to give different microstructure characteristic and thereby giving different diffusion characteristic during carburizing process on the steel substrate. It has been well understood that after forming process, such as rolling, high density of dislocation (a type of metal defect) occurred after this process. This high density dislocation of metals are susceptible for recrystallization that produce finer grain structure, if after forming process is followed with heat treatment at recrystallization temperature of the metals. Different degree of forming process might create different density of dislocation and thereby creating different degree of recrystallization during the following heat treatment that results in different grain size of microstructure of metal. Since during carburizing, carbon enter the metal substrate by diffusion process, and diffusion is predicted occur at the faster rate at the grain boundary, then it is predicted different grain size of the metal substrate might produce different carburizing layer structure. Few researches have been conducted on the topic of the effect of thermo-mechanical treatment on the carburized layer structure, and most of them are working with commercial carburizing resources. The present research utilizing coconut shell charcoal as carbon resources, and therefore the type of carburizing conducted in the present research is pack carburizing process. the On the other hand, different carburizing temperature is also expected to give different diffusion characteristic as well as substrate characteristics during carburizing processes.

2. Experimental method

Metal substrate used in the present research is low carbon steel which is then prepared by thermo-mechanical treatment (TMT) in order to obtain substrate sample with different grain size before carburizing process. TMT processes were performed by combining cold rolling process and heat treatment. There are three type of samples that will be evaluated regarding TMT process, which are sample without TMT, sample with degree of cold rolling reduction of 30% and 80%. After cold rolling process each samples are heated at 625°C for one hour in order to create recrystallization on the cold rolled samples.

After metal substrate preparation, carburizing processes were performed by single quenching method and utilizing solid pack-carburizing method. By single quenching method, two heat treatment cycles have been performed. The first cycle is putting the sample in the container contains solid pack carburizing powder agent and heated at austenization temperature in order to diffuse carbon on the steel substrate and followed with air cooling process. The second cycle is heating this steel sample up to austenization temperature and follow with quenching by immersing the sample in the water. For the first cycle, variations of heating temperature were performed at 850, 900 and 950°C. After carburizing process characterization on the carburized samples were performed. Two main characterizations were performed including micro-vicker hardness test on the cross-section sample and microstructure characterization by using optical microscope on this section. Beside hardness and microstructure data obtained from these two characterizations, it can also be obtained the case depth of carburizing which is related with the kinetic of the process.

3. Data and Analysis

Figure 1 shows microstructure of steel sample before treatment, showing ferrite and pearlite structure, characteristic of low carbon steel structure. On the other hand, Figure 2 shows microstructure of steel samples after given thermo-mechanical treatment. Almost insignificant differences with original sample observed for the samples given 30% reduction of rolling and subsequent heat treatment process. However significant microstructure changing observed for the sample given 80% reduction of rolling and the following heat treatment treatment (Figure 2(c) and (d)). After this degree of rolling, it can be seen highly elongated structure, a characteristic of cold formed structured. This structure is heavily deformed and contained much higher density of dislocation as compared with untreated

IOP Conf. Series: Materials Science and Engineering **214** (2017) 012025 doi:10.1088/1757-899X/214/1/012025

sample. This condition induces in the massive recystallization during the following heat treatment and results in the fine grain structure as is shown in the Figure 2(d).

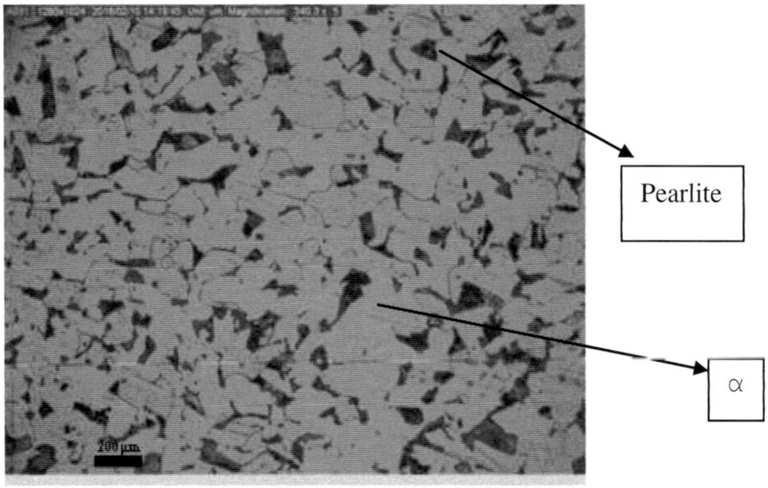

Figure 1. Original steel sample microstructure before treatment.

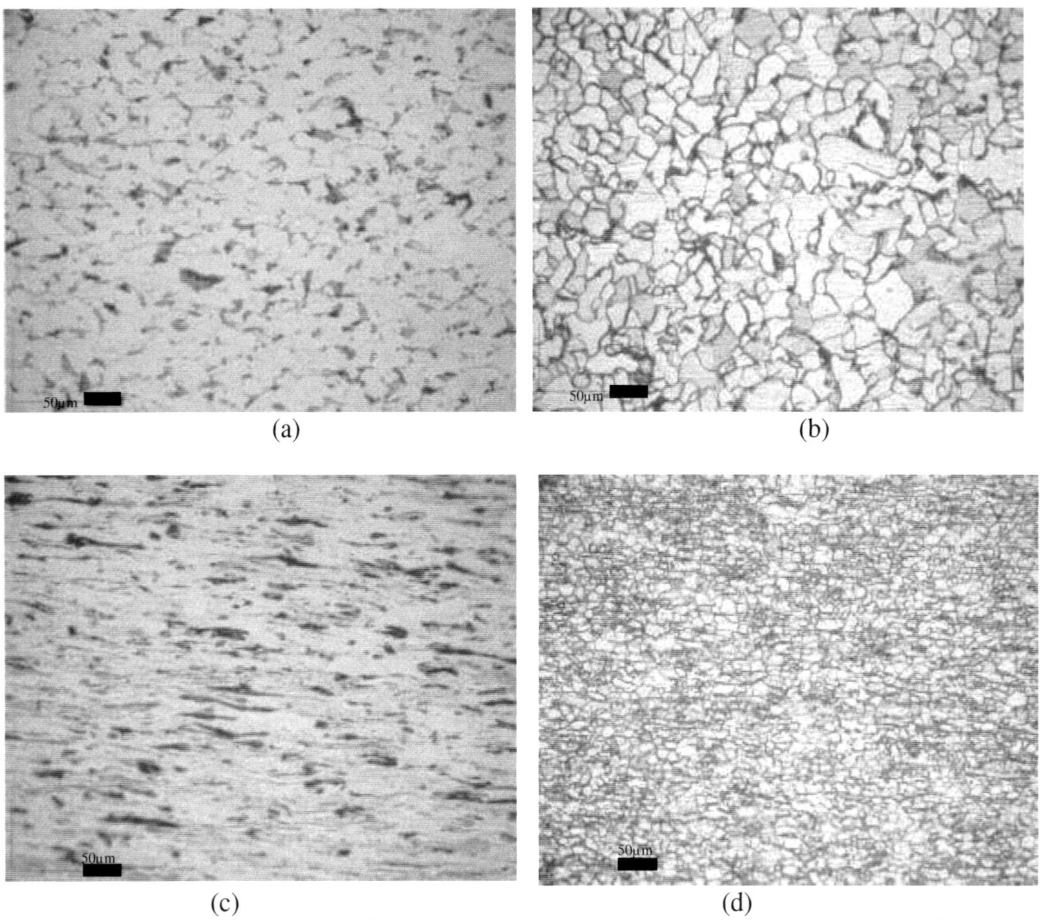

Figure 2. Microstructure of (a) Cold rolled sample at 30% reduction, (b) 30% cold rolled sampled plus annealing, c) Cold rolled sample at 80% reduction, (d) 80% cold rolled sampled plus annealing.

Figure 3 shows cross-section microstructure of sample without thermo-chemical treatment preparation, after carburizing process. Point a,b and c in the figures are points where hardness test performed on the samples, and points a is the outermost points of the carburized layers in each figure. Carburized layer is considered as the layer with minimum hardness of 550 HV, and therefore sample that carburized at 900°C shows the thickest layers. In addition from Table 1, it can be seen that this sample also having the highest hardness at the outermost point (poin a) of measurement.
On the other hand, Figure 4 shows cross-section microstructure of sample prepared with 30% cold rolling and heat-treatment process. Refer to Table 2, it was obtained that sample carburized at 950°C having hardness above 550HV at point c, and therefore have highest thickness of case depth. However, the highest hardness at the outermost layer position is shown by the sample carburized at 900°C.

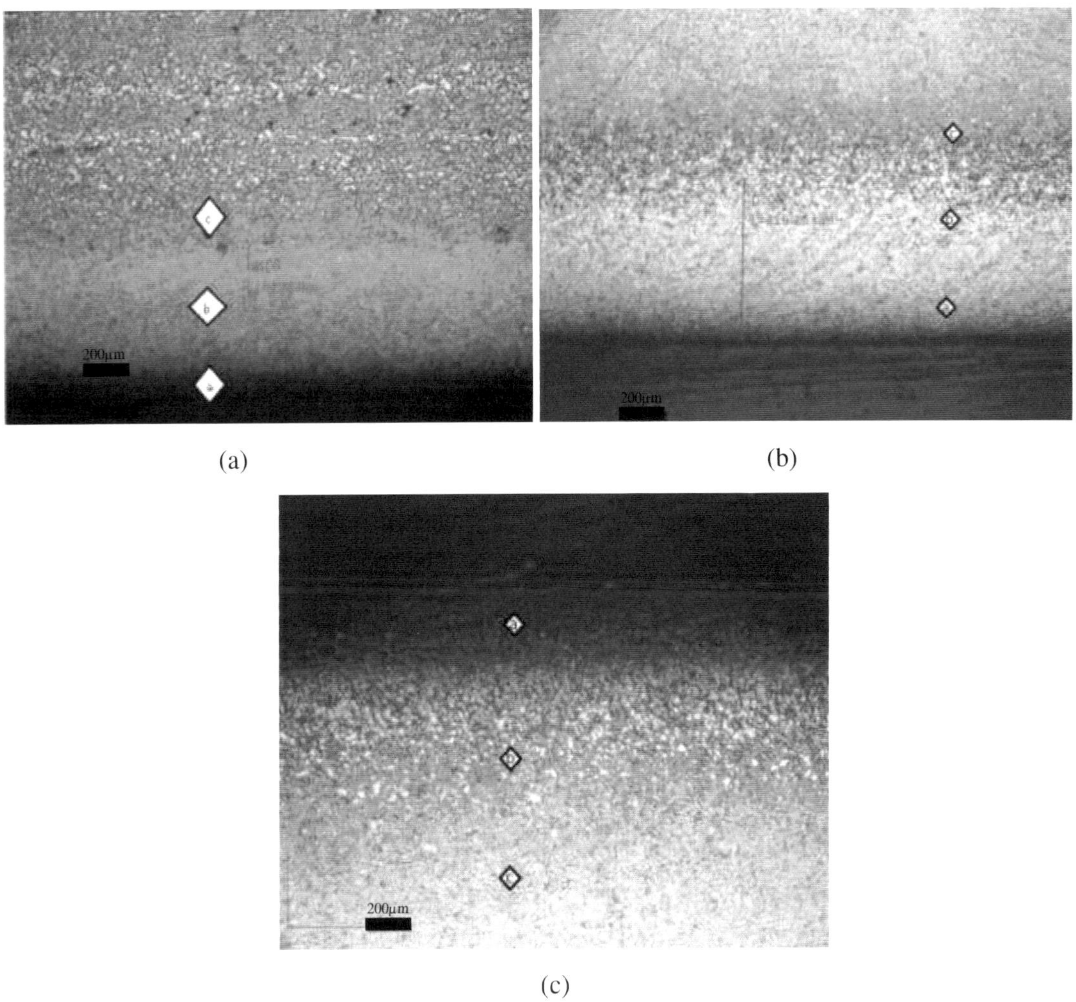

Figure 3. Microstructure of sample prepared without treatment after carburizing at (a) 850°C, (b) 900°C, (c) 950°C.

Table 1. Hardness of carburized samples, samples prepared without thermo-mechanical treatment

Samples	**Hardness at point a (HV)**	**Hardness at point b (HV)**	**Hardness at point c (HV)**
Carburized at 850°C	641	566	296
Carburized at 900°C	847	706	407
Carburized at 950°C	660	557	282

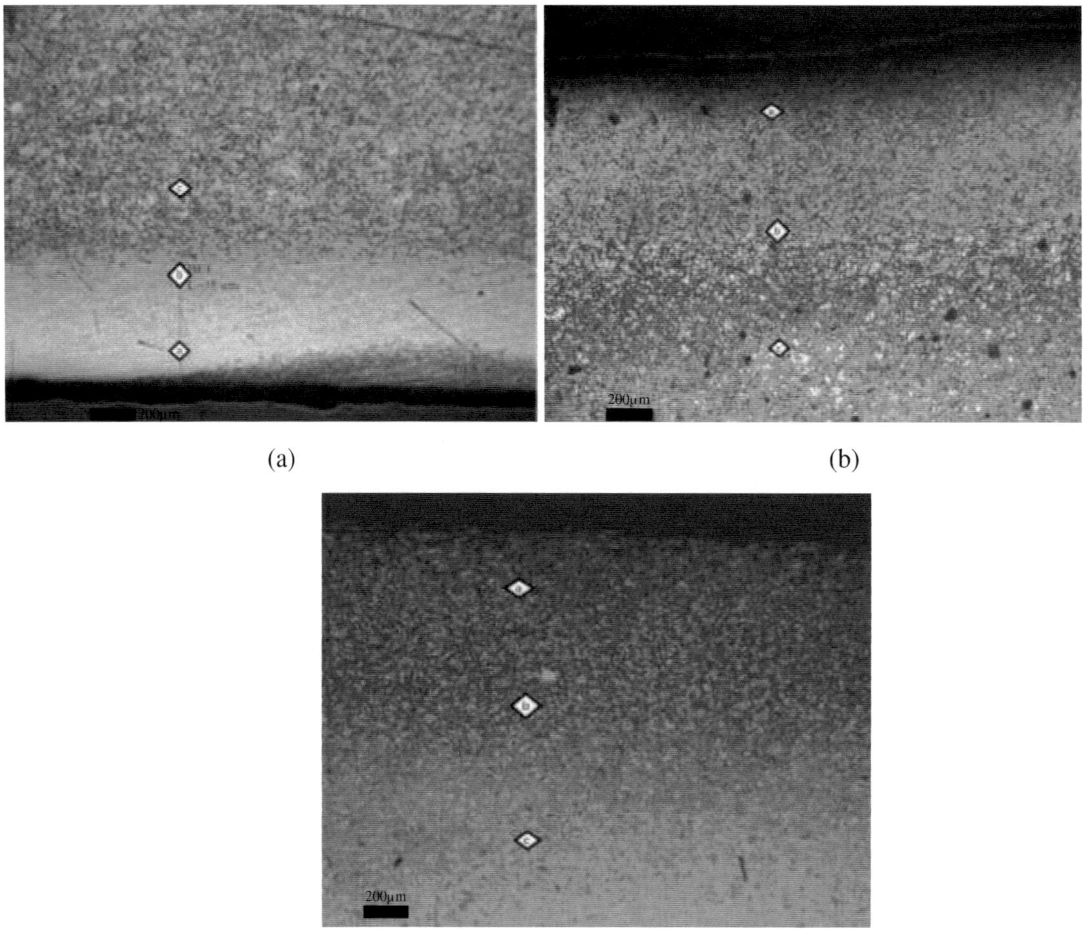

(a) (b)

(c)

Figure 4. Microstructure of sample prepared with 30% cold rolling+heat treatment, after carburizing at (a) 850°C, (b) 900°C, (c) 950°C.

Table 2. Hardness of carburized samples, samples prepared with 30% cold rolling+heat treatment

Samples	**Hardness at point a (HV)**	**Hardness at point b (HV)**	**Hardness at point c (HV)**
Carburized at 850°C	782	566	327
Carburized at 900°C	914	762	517
Carburized at 950°C	724	710	608

In addition Figure 5 shows cross-section microstructure of sample prepared with 80% cold rolling and heat-treatment process. Hardness data at the points of measurements shown in this figure are tabulated in the Table 3. Interesting facts are observed here that both samples carburized at 900°C and 950°C are having hardness above 550HV along their cross-section points, which means the case depth is along the cross-section of both samples. However, sample carburized at 900°C still showing highest hardness at its outermost side.

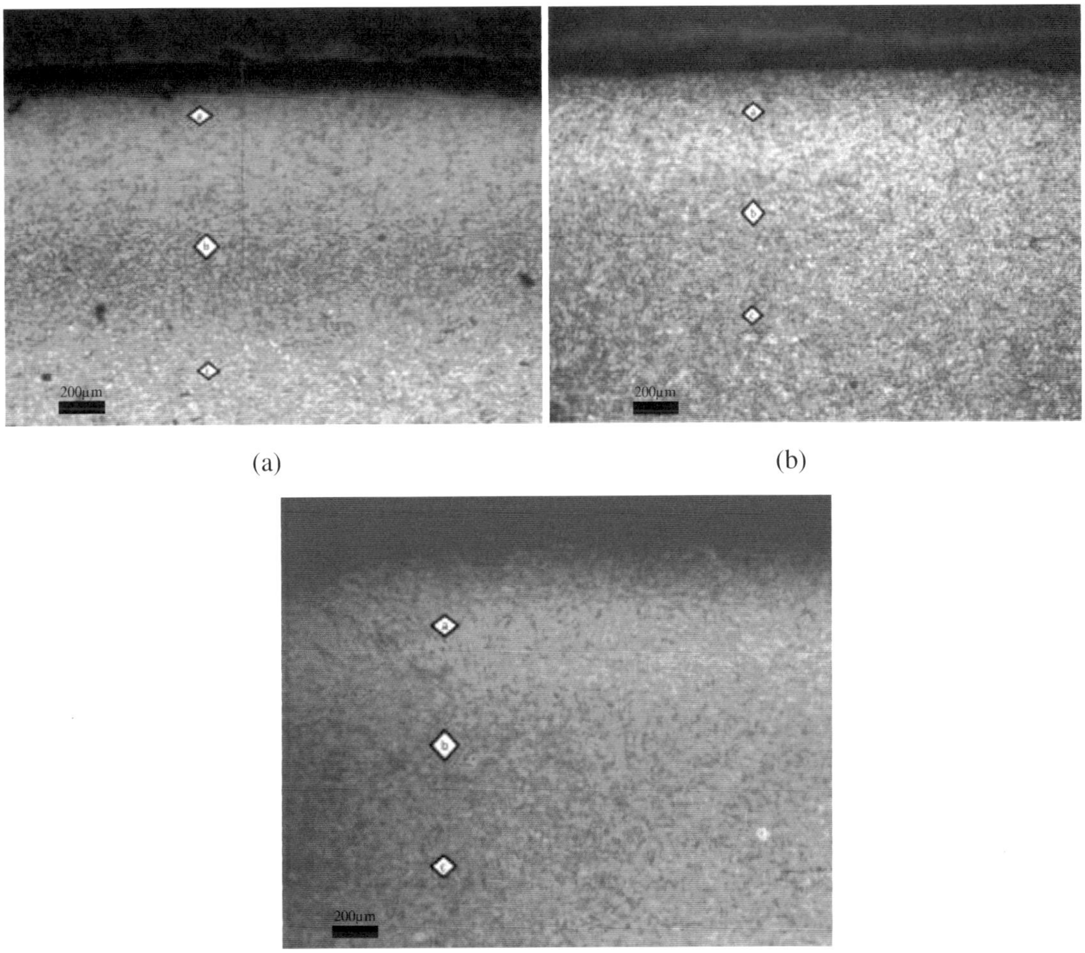

(a) (b)

(c)

Figure 5. Microstructure of sample prepared with 80% cold rolling+heat treatment, after carburizing at (a) 850°C, (b) 900°C, (c) 950°C.

Table 3. Hardness of carburized samples, samples prepared with 80% cold rolling+heat treatment

Samples	**Hardness at point a (HV)**	**Hardness at point b (HV)**	**Hardness at point c (HV)**
Carburized at 850°C	805	724	316
Carburized at 900°C	1003	825	587
Carburized at 950°C	813	782	636

Table 4 summarized the effect of degree of rolling on the hardness of carburized layer at each carburizing temperature. It can be seen from this table that sample with degree of rolling at 80% show highest hardness at each carburizing temperature. This condition is considered related with the microstructure condition of these samples which having finest grain structure as is shown in the Figure 2. Finer grain structure provide more massive diffusion path during carburizing process and therefore results in the more compact carburized layers than the coarser grain structure.

From this research, surface hardness of the steel substrate obtained after the carburizing process due to the formation of martensite structure at the surface as shown in the Figure 3,4 and 5. The steel substrate which having low carbon content is normally cannot be hardened by heat treatment since transformation to martensite structure is difficult for the low carbon steel. During carburizing process, carbon is diffused into the surface of the steel, and therefore increases carbon content at the surface. This condition change the hardenability of the steel surface, and transformation to martensite structure become possible by heat treatment process. In addition, it was also obtained that carburizing temperature at 900°C applied in the present research shows highest hardness at each degree of rolling condition. This condition suggests that there is optimum temperature to obtain highest hardness of carburized layer. High temperature is required in the carburizing process in order to provide diffusion energy for carbon into the steel surface. However, high temperature also accelerates other mechanism that imparts the softening of steel, such as grain growth process. Therefore it is important to determine and select the optimum temperature, in order to obtain the best character of steel substrate after the carburizing process

Table 4. The effect of degree of rolling on the hardness of carburized layer at each carburizing temperature

Carburizing temperature (°C)	**Hardness at degree of rolling (HV)** 0%	30%	80%
850	641	782	805
900	847	914	1003
950	660	724	813

4. Summaries

The present research focuses on the effect of thermo-mechanical treatment as substrate preparation and carburizing temperature on the carburized layer. Carburized layer with the thickness of hundreds micrometer have been obtained after the process. It was observed that highest hardness obtained for the samples cold rolled at 80% degree of reduction. On the other hand, temperature of 900°C gives the optimum carburizing temperature (regarding hardness properties) from the range temperature applied in the present research.

5. References

[1] Bahri A, Kaçar E, Akkaya S.S, K. Elleuch K and M. Ürgen M 2016 *Surf.Coat.Tech* **304** 560
[2] Momeni S, Tillmann T 2016 *Wear* **368-369** 53
[3] Morita T, Inoue K, Ding X, Usui Y and Ikenaga M, 2016 *Mater. Sci. and Eng. A* **661** 105
[4] Vettivel S.C., Jegan R, Vignesh J and Suresh S 2017 *Surf and Interf.* **6** 1
[5] Sigolo E, Soyama J, Zepon G, Kiminami C S, Botta W J and Bolfarini C 2016 *Surf. And Coat. Tech.* **302** 255
[6] Bepari MMA 2017 *Ref. Module in Mater. Sci. and Eng, from Comprehensive Mater. Finishing* **2** 71
[7] Adachi S and Ueda N 2014 *Surf. And Coat. Tech.* **259** 44
[8] Edenhofer B, Joritz D, Rink M and Voges K 2015 *Thermochem. Surf. Eng. of Steel* 485
[9] De Rege F 1998, *Metal Powder Report* **53** 41

[10] Bataev IA, Golkovskii MG, Losinskaya AA, Bataev AA, Popelyukh AI, Hassel T and Golovin DD 2014 *Appl. Surf. Sci.* **322** 6

Acknowledgments

Authors wishing to acknowledge support from the Faculty of Mechanical and Aerospace Engineering Institut Teknologi Bandung Indonesia for the support of the present research.

IOP Conf. Series: Materials Science and Engineering **214** (2017) 012026 doi:10.1088/1757-899X/214/1/012026

Release Profile and Inhibition Test of The Nanoparticles *A. Paniculata* Extract as Inhibitor of α-Glucosidase in The Process of Carbohydrates Breakdown Into Glucose Diabetes Mellitus

Farisa Imansari[1], Muhammad Sahlan[2,4] and Rita Arbianti[3,1*]

[1]Chemical Engineering Department, University of Indonesia, Kampus UI, Depok, Indonesia ,16424

[2]Chemical Department, Institute of Technology Bandung, 10 Ganesha, Bandung, Indonesia, 40132

[3]Chemical Department, University of Indonesia, Kampus UI, Depok, Indonesia, 16424

[4]Chemical Engineering Department, TUAT Japan

Correspondent Author: arbianti@che.ui.ac.id

Abstract. *Andrographis paniculata* (*A.paniculata*) contain the main active substances Andrographolide which helps lower glucose levels in diabetics by inhibiting the enzyme α-glucosidase. The ability of the extract *A.paniculata* in lowering glucose levels will increase with the technique encapsulation with a coating of composition Chitosan-STPP as a drug delivery to the target organ. This study aimed to get an overview of *A.paniculata* release profile of nanoparticles in a synthetic fluid media with various concentrations of coating and inhibition testing nasty shard extract in inhibiting the enzyme α-glucosidase. This research resulted in nanoparticles by coating efficiency and loading capacity of chitosan greatest variation of 2% and 1% STPP 60% and 46.29%. chitosan greatest variation of 2% and 1% STPP 60% and 46.29%. The ability of *A.paniculata* extracts as α-glucosidase enzyme inhibitors has been demonstrated in this study, the percent inhibition of 33.17%.

1. Introduction

Diabetes mellitus (DM) is a disease that cannot be cured but the quality of life of patients can be maintained with a good metabolic control. Dm is a major cause of morbidity and mortality in the world. Mortality in DM increased two times higher due to the complications of diabetes which include cardiovascular disease, retinopathy, nephropathy, and neuropathy DM. Indonesia is a country with the number of people with diabetes mellitus fourth largest in the world. That number is expected to increase from 8.4 million in 2000 to around 21.3 million by 2030 in future [1]. The DM generally divided into two types namely, DM type 1 or insulin-dependent diabetes mellitus (TYPE 1) and type 2 DM or noninsulin-dependent diabetes mellitus (NIDDM). Type 1 diabetes is a chronic metabolic disease or disorder with multiple etiologies characterized by high blood sugar levels followed by impaired metabolism of carbohydrates, lipids, and proteins as a result of insufficiency of insulin function. Insufficiency of insulin function may be caused by impaired or deficient production of insulin by the β Langerhans cells of the pancreas that caused reaction autoimmune [1]. Type 2 diabetes occurs due to interference with the binding of glucose by the production of insulin receptor but still within normal limits so that patients are not dependent on insulin. In development, business pharmacological treatment for diabetics is already widely used, such as therapy with oral hypoglycemic drugs such as sulfonylureas and acarbose, insulin therapy that is useful to stabilize blood sugar concentration, or a combination of both. Therapy insulin is one the requirement for patients dm type 1.Antidiabetic drugs used often have a side effects like therapy insulin in DM type 1 [2]. A side effect is a decrease in blood glucose levels below normal and obese.

Herbal plant extract of *sambiloto* (*Andrographis paniculata*) contain the main active substances i.e. andrographolide responsible for pharmacological activity and efficacious as an antidiabetic [3]. Andrographolide can stimulate the release of insulin and inhibiting the absorption of glucose by α-glucosidase and α-amylase [4]. Sambiloto has less soluble in water, limiting the distribution of the bio, unstable in alkaline and acid on digestion which is extreme, and have very short biological half-life (2 hours). The application of nanotechnology can help the problems of *A. paniculata* in delivering the active compound into the body. Advantages of the application of nanotechnology include increasing the surface area, increasing solubility, improving bioavailability in oral administration, drug protection from degradation, and maintain release in the long term. Polymers that facilitates the preparation of nanoparticles can be selected in the form of water-soluble polymers, one of which is chitosan. Chitosan has the ideal properties, i.e., biocompatible, biodegradable, non-toxic, and not expensive [5]. Use of Chitosan as an encapsulator has been widely used previously, such as the coating of vitamin C [6], the coating of tea catechins [7], and recent studies shows that absorption of chitosan nanoparticles can improve insulin and has the ability to interact with the intestinal mucosa [8]. Meanwhile, chitosan as encapsulator has disadvantage that the mechanical properties of chitosan fragile. The addition of a crosslinking agent sodium tripolyphosphat (STPP) can improve mechanical properties chitosan, so when drug passing a very acidic condition, the matrix of chitosan formed is not degraded until the active substances that released reaching the target. Profile release from *A. paniculata* extract which has been nanoencapsulated with the chitosan coating is important for review in order to know the performance of the drug in the body to reduce blood glucose levels for patients with diabetes, as well as α-glucosidase inhibitory activity of nanoencapsulation extracts *A. paniculata* by chitosan in vitro.

2. Equipment and Materials

2.1 Equipment.
Sonicator from *Elmasonic*, *Blender*, *Beaker glasss*, Spectrophotometry, digital incubator from *Boekel* , *rotary evaporator* from *IKA*, *magnetic stirer*, *Freeze dryer* from *Eyela Word*, *syringe*, *centrifuge* from Hanil MF 50, FESEM by *Zeiss*.

2.2 Material.
Andrographis paniculata, Ethanol 70%, Aquade, NaOH, HCl, Chitosan (*low*), STPP, Tween 80, Acetat acid, enzim α-*glucosidase* from sigma aldrich, *p*-Nitrophenyl α-D-glucopyranoside from sigma aldrich, KH_2PO_4, dan KCl.

2.3 Simplisia Preparation.
At this stage carried out preparation of plant leaves from sambiloto simplicia (*Andrographis paniculata*). Preparation of simplicia consists of sorting the leaves dried with rods downsizing, and sieving to the size of the resulting homogeneous ± 1.15 mm

2.4 A.paniculata Extraction.
Extraction leaves A. panniculata is done using ethanol 70% by sonication method. Simplicia that have been prepared and weighed as much as 50 grams of ethanol, and then mixed with 70% with a ratio of 1:20 (w / v) in the glass beaker and covered with plastic wrap and aluminum foil. Extraction with a sonicator lasted for 60 minutes with the operating conditions on the frequency of 40 kHz and a

temperature of 600C. The filtrate is obtained subsequently separated using a vacuum rotary evaporator until it formed a thick extract.

2.5 Phytochemicals Qualitative Test.
Qualitative test aims to identify the presence of phytochemicals andrographolide class. Silica gel 60 F254 plates prepared with a size 2 × 75 cm with a mobile phase (eluent) used chloroform: methanol (9: 1). Then the eluent was taken by 4 ml and poured in chamber which is then allowed to stand for 30 minutes. Next, as much as 5μL extract spotted on spotted point that is within 1 cm of the bottom line using a capillary tube. After 30 minutes, the plate was spotted eluted by inserting a plate into the chamber and the elution process that occurs until the eluent reaches the upper limit. Then the silica plate is taken to stop the elution process. The stains formed were examined under UV light at a wavelength of 254 nm to measure the value of Rf.

2.6 Nanoparticles Process.
Making nanoparticles *A.paniculata* leaf extract in this study using ionic gelation method with a coating composition concentration ratio of chitosan and STPP. This study will be conducted for the release profile variation of chitosan-STPP (1%: 1%; 1% to 1.5% and 2%: 1%). Making starts with 1% acetic acid as the solvent chitosan. Chitosan is dissolved into a solution of acetic acid 1% (w / v) with the desired ratio (1% and 2%) using a magnetic stirrer. Furthermore, *A. paniculata* extract as much as 0.15 grams of prepared and then diluted with an ethanol 70% about 5 drops. *A. paniculata* leaf extract mixed into a solution of chitosan. Preparing STPP solution in aquades to the desired concentration (1% and 1.5%), dripping a solution of chitosan extract using a syringe into a solution of STPP followed by magnetic stirer. The process of size reduction is done by sonicator at a frequency of 20 kHz for 60 minutes. Centrifugation carried out on the mixed solution to separate the nanoparticles (sediment) from the supernatant with a speed of 10,000 rpm for 15 minutes. Nanoparticles have been obtained used as solids (powders) using freeze drying.

2.7 The Analytical Determination of Nanoparticles.
Analytical determination nanoparticles were obtained by FESEM.

2.8 Test In-Vitro Release Profile.
Test in vitro release profile of nanoparticles *A.paniculata* done using synthetic fluid media. Media synthetic fluids used were simulated gastric fluid (pH 1.2) and Simulated Intestinal Fluid (pH 7.4). The testing process starts by mixing 0.1 g matrix of any variation with 50 ml of synthetic fluid pH 1.2 and save in an incubator with a temperature of 37^{0}C. Sampling was done by taking 10 ml solution from a beaker glass with syringes and moved into the tube centrifugation, centrifugation done at a speed of 2000 rpm for 10 minutes. A total of 4 ml samples were taken from the centrifuge tube and inserted into the bottle as a sample vial, then the liquid in the tube centrifugation returned into a beaker glass and put in 4 ml of synthetic fluid new media to keep synthetic fluid volume constant. The succession of buffer is done on hour-3 with pH 7.4 and additional enzymes α-amilase 0,003% (w/w). The data taken until the hour to the 7 th. Sample absorbance measured using UV spectrophotometry with wavelength 208 nm.

3. Result and Discussion

3.1 A. paniculata Extraction.
The value of the average yield obtained on extracting the leaf *A.paniculata* is 15.96% where percent *yield* has values that are almost the same with the research done by [9] using the extraction method and the same solvent, of 15.88%.The results of different research done by [10], where the use of solvent methanol produces *yield* value is higher than the solvent ethanol of 39,8%.This is due to the solvent with low viscosity, more easily to channeled. Methanol has a lower viscosity values (0,597 mPa.s) compared with ethanol merely 1,200 mPa.s. The viscosity is reduced (cohesion are also reduced) resulting in decreased surface tension. The surface tension decreases can help solvent get into

the solid matrix, thereby increasing the speed of the reaction. This causes the percentage of yield of methanol.

3.2 Phytochemicals A.paniculata results.
Based on the results of these observation, Rf value obtained on the same andrographolide standards such as the Rf values obtained on the leaf extract of *A.paniculata* 0.61.This may indicated that in the samples of *A.paniculata* extract contains a bioactive compound andrographolide. In this case the selection of solvent affects the existence of active compounds in the extract. This indicates that the use of ethanol solvents to extract andrographolide active compounds contained in leaf *A.paniculata* can produce the extraction process of active compounds target properly.

3.3 A.paniculata Nanoparticles.
Tables 1 shows the data encapsulation efficiency and loading capacity gained from the leaf extract *A.paniculata* nanoparticles

Table 1. The results of the encapsulation efficiency and loading capacity leaf extract of *A.paniculata* nanoparticles

Variation	Encapsulation efficiency	Loading Capacity
Chitosan 1% : STPP 1%	55.89%	30.62%
Chitosan 1% : STPP 1,5%	28.79%	11.09%
Chitosan 2% : STPP 1%	60%	46.29%

Based on table 1 of the greatest efficiency values obtained on the variations of Chitosan: STPP (2:1) by 60%, where the higher the concentration of Chitosan then the percentage encapsulation efficiency also higher. This is due to the larger number of nanoparticles of chitosan, then the amount of the coated andrographolide also increasingly due to the surface area of the particles increases so that efficiency may be even greater. While the higher concentration of TPP then the less andrographolide that absorbed because the function of the TPP as a substance that strengthens cross-matrix binding nanoparticles. Nanoparticle become more compact and dense due to the concentration of TPP the larger, so that the rate of diffusion out ingredients (andrographolide) declining, which is in accordance with the theory that the more TPP will further strengthen the ties of Ionic revolt on nanoparticle so that the drugs will be difficult to remove the back [11].The largest percentage of loading capacity in line with the largest percentage encapsulation efficiency. This is due to the more TPP used the more ions tripolyphosphat that interact with NH_3 which cause the bond between the two ions stronger so that the structure of the solid-matrix more compact because mass ratio increased to chitosan : TPP, which lead to an increase in the amount of nanoparticle formed, resulting in increased efficiency encapsulation and loading capacity in nanoparticle. The results of this study are consistent conducted by [12].

3.4 The Analytical Determination of Nanoparticles Using FESEM.
The results of measurement with zoom FESEM 1,000 – 10,000 times can be seen in Figure 1.

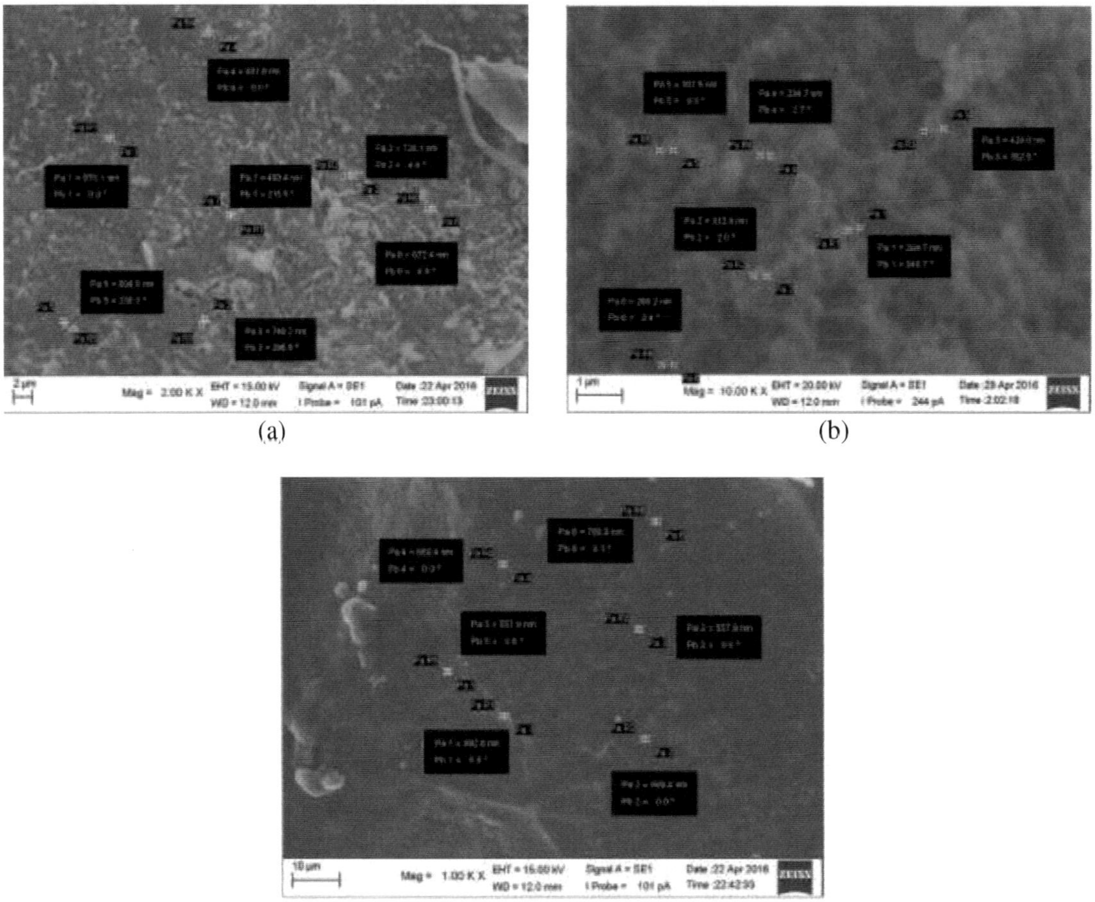

(a) (b)

(c)

Figure 1. FESEM test results (a) variation 1 (Chitosan1%: TPP1%) (b) the variation 2 (Chitosan1%: TPP 1.5%) and (c) the variation 3 (Chitosan 2% : TPP 1%).

Table 2. The Diameter of the particle size on the variation 1, variation 2, variation 3

Variations	Diameter of Particles (nm)					Min	Max	Average	Deviation Standard
	1	2	3	4	5				
1	837.6	604	728.1	480.4	672.4	480.4	837.6	664.5	133.81
2	301.5	234.7	264.7	284.7	268.2	234.7	301.5	270.76	24.91
3	669.4	557.9	557.4	892.6	669.4	557.4	892.6	669.3	136.74

From the figure 2 can be known that the size of the smallest particles found on the second variation with a diameter 234.7 nm, while the size of the largest particles found on the third variation with a diameter 892.6 nm. The distribution of the particle size approaches the uniform is occupied by the second variation as seen from the above data with the value of the standard deviation variation 2 variation is much smaller than the other, which indicates the diameter of the particle size distribution variation of 2 is homogeneous .It can be concluded that the size of the particles obtained from this research have reached the size of the nano, but to be an effective drug delivery required particle size <100 nm. The form is obtained from the variation of 1 to 3 variations show that less spheric form and the size of its particles relatively large. According to Wahyono [11] the addition number of number of Chitosan with a fixed number of TPP led to coagulation (agglomeration) which causes the shape of nanoparticles is not spheric. The particle size is not uniform is also expected because of *A. paniculata* leaf extracts not only entered into the matrix of chitosan nanoparticles but also attached to the nanoparticle surface.

3.5 α-Glucosidase Inhibition.

Extraction of *A. paniculata* leaf positively can inhibit α-glucosidase enzymes are shown in Figure 2

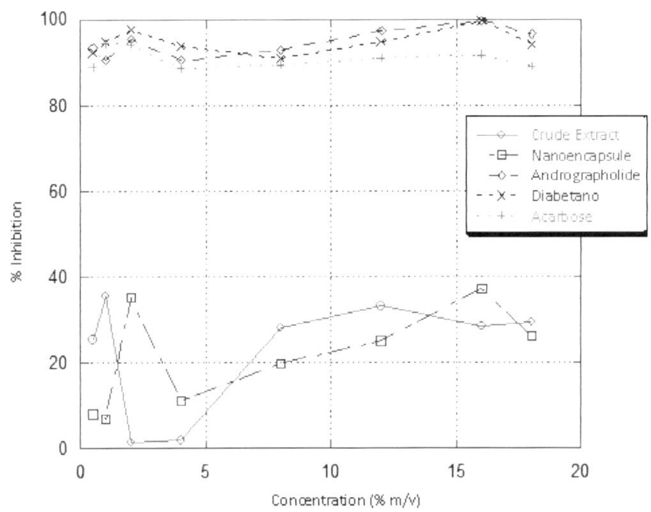

Figure 2. Comparison of inhibition test all samples.

The figure above shows that the standard andropholide have percent highest enzyme inhibition while extracting the leaf *A. paniculata* have percent enzyme inhibition low against the enzymes α-glucosidase. The percentage increase enzyme inhibition andrographolide occurs along with the addition of the concentration of andropholide, where the maximum absorbance peak is at a concentration of 16% with the percent inhibition of 99.56% and at a concentration of 18% inhibition andrographolide power has decreased. This means that the concentration of andrographolide 16 percent is the maximum concentration to inhibit troubleshooting enzymatic reactions pati or carbohydrates into glucose in the small intestine.

Percent inhibition crude extract of leaf *A. paniculata* is much smaller than the comparison substance andrographolide under the same conditions, where the percent inhibition highest crude extract of *A. paniculata* leaf is 33.17% which in konsentrasi12%.The percent difference significant inhibition is the possibility of other chemical compounds. Such compounds may be no synergistic in inhibiting the enzyme α-glucosidase. In nanoencapsulation *A. paniculata* leaf extract, the highest percent inhibition obtained, the same as the percent inhibition of crude extract of 33.17% but this occurred at a concentration of 16%.Percent enzyme inhibition is still smaller than the substance of comparison such as acarbose and diabetano. This is due to the fact that there are still other chemical compounds in the leaf extract *A. paniculata* nanoencapsulation. In addition, the percent inhibition nanoencapsulation little *A. paniculata* leaf extract also caused quite a small andrographolide content in the sample, of 2.5 mg / ml, whereas according to the research that has been done Subramanian et al, [4] required 62.5 mg / ml andrographolide to reach a maximum of 89% percent inhibition. Inhibition of the action of the enzyme can effectively reduce complex carbohydrate digestion and absorption, thereby reducing post-prandial glucose levels in diabetics.

Nanoencapsulation leaf *A. paniculata* extract also inhibit the enzymes α-amilase pancreas working in hydrolyzing polysaccharide in caused by small intestine. Nanoencapsulation leaf *A.paniculata* extract as inhibitor enzymes α-glucosidase also function in inhibits the expression of the glucose transporter SGLT1 and GLUT-4 translocation mRNA is increased by 3 times in diabetic patients who are permanently in the membrane apical resulting in an enzyme α-amylase and α- glucosidase also increased so that the taking of glucose in the intestine increased 3 times faster than the normal. This transporter functions to move glucose from one compartment to another compartment to pass through biological membranes so that glucose can be absorbed by the body [13].

Standard Andrographolide, diabetano, and acarbosa have the same concentration in inhibit the activity of the enzymes α-glucosidase of 16%. It is needed 3 times to extract the leaves *A.paniculata* to inhibit the enzymes α-glucosidase in accordance with drugs acarbose and diabetano.

3.6 Release Profile. These test results are shown in Figure 3.

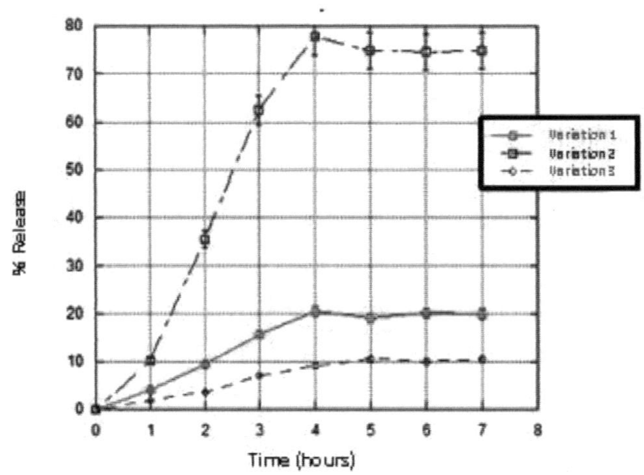

Figure 3. Percent cumulative release variation 1, variation 2 and variation 3, where the clock to 1- 3 indicates the condition of the stomach and the clock to 4 - 7 shows the small intestine.

Based on Figure 3 it can be seen that the variation 2 produces the largest cumulative release percent ie 74.83% at 7, followed by the first variation of 19.92% and 10.45% of the variation 3 at the same hour. The result of this release is inversely related to the results of encapsulation efficiency and loading capacity that produces optimum value in three variations. This shows that the concentration ratio of chitosan : STPP on third variation of 2%: 1% (g/ml) resulted in a strong cross linking between them and the characteristics of a strong matrix so that difficult broken and make the release of the drug lasts longer. The cumulative amount of drug release produced variations on the hour 3 to 7 is worth the least. This is in line with the results of research [14] which stated that the increased concentration of chitosan make the secretion of drugs requires a longer time.

The release profile of nanoencapsulation extract of *A. paniculata* leaf describes the events burst release on the condition of gastric and intestines sustained released on conditions. Events burst release is due to the dissociation of the drug on the surface of chitosan nanoparticles that cause diffusion of the drug entrapped in nanoparticles contained in the release of high level experience. Characteristics of chitosan unstable under acidic conditions led to the electrostatic repulsion between molecules on the surface of chitosan nanoparticles that produces the drug adsorbed diffusion, thereby making the drug release rate that is higher than the matrix [15] and [13]. Events burst release indicates that some drugs (in this study is an extract of *A. paniculata* leaf) entrapped on the surface of the nanoparticles [16].

Events sustained release can be caused by the nature of chitosan as the main material of the matrix has not dissolved at pH under 6.0, but have durability at pH above the value, and hydrogen bonds between the molecules chitosan make nanoparticles stronger that cause soft diffused into the molecules become more difficult and then followed by the secretion of the drug lasted longer [15].

4. Conclusion

The highest percent inhibition α-glucosidase obtained diabetano herbal medicine by 99.631% on the concentration of 16 percent and percent enzyme inhibition lowest obtained nanoencapsulation *A.paniculata* extract of 33.17 on should concentrate on 16%. To match the ability of inhibition diabetano it takes 3 times the dose nanoencapsulation of *A. paniculata* extract that can inhibit the enzyme α-glucosidase optimally. The application of nanotechnology is proven to help the problems of *Andrographis paniculata* is limited to the distribution of bio in delivering the active compound into the body by maintaining the release in the long term. This research proves the nanoencapsulation characteristics of chitosan 2% and STPP 1% is able to withstand burst release and bring *A. paniculata* extract to the target organ small intestine.

5. Reference

[1] WHO Department of Noncommunicable Disease Surveillance Geneva. 1999.Definition, Diagnosis and Classification of Diabetes Mellitus and its Complications. *Report of a WHO Consultation Part 1: Diagnosis and Classification of Diabetes Mellitus.*

[2] Dennis, M. L., Scott, C. K., Funk, R., & Foss, M. A. 2005. The duration and correlates of addiction and treatment careers. *Journal of Substance Abuse Treatment*, **28**(1), S51-S62.

[3] Widyawati, Tri. 2007. *Aspek Farmakologi Sambiloto (Andrographis paniculata Nees).* Universitas Sumatera Utara, Medan.

[4] Subramanian, R., Asmawi, M.Z., Sadikun, A. 2008. In vitro alpha-glucosidase and alpha-amylase enzyme inhibitory effects of Andrographis paniculata extract and Andrografolid, *Act.J.Biochim.Pol.*, **55**(2): 391-398.

[5] Tiyaboonchai, W. 2003. Chitosan nanoparticles: Apromising system for drug delivery. *NaresuanUniversity Journal,* **11**(3): 51–66.

[6] Desai, K. G. H. and Park, H. J., 2005 Encapsulation of vitamin C in tripolyphosphate crosslinked chitosan microspheres by spray drying. *J. Microencapsulation*, **22**, 179-192.

[7] Hu, B., Pan, C., Sun, Y., Hou, Z., Ye, H., Hu, B. And Zeng, X. X., 2008.Optimization of fabrication parameters to produce chitosan tripolyphosphate nanoparticles for delivery of tea catechins. *J. Agric. Food Chem.*, **56**, 7451-7458.

[8] Prego C., Torres D., Alonso M.J., 2005 The potential of chitosan for the oral administration of peptides, *Expert Opin. Drug Deliv*, **2**, 843-854.

[9] Yonanda, Rona Jutama. 2011. Formulasi Ekstrak Sambiloto (*Andrographis paniculata*) dan Brotowali (*Tinospora crispa*) sebagai Inhibitor α- Glucosidase dan Analisis Sidik Jari Menggunakan Teknik Kromatografi. Institut Pertanian Bogor, Bogor, 9-10.

[10] Mathew,Arhewoh., Falodun, Abiodun., Okhamafe, Augustine O.,Bao, Yang., Sheng, Qui. 2011. Ultrasonic Assisted Extraction and radical scavenging activity of some selected medicinal plants. *Journal of Pharmacy Research*, **4**(2), 408-410.

[11] Wahyono, Dwi. 2010. Ciri Nanoparticle Chitosan dan Pengaruhnya pada Ukuran Partikel dan Efisiensi Penyalutan Ketroprofen. IPB, Bogor.

[12] Hou, Dongzhi., Gui, Ruyi., Hu, Sheng., Huang, Yi., Feng, Zuyong., Ping, Qineng. 2015. Preparation and Characterization of Novel Drug-InsertedMontmorillonite Chitosan Carriers for Ocular Drug Delivery. *Scientific Research Publishing*, **84**:73-74.

[13] Dounighi, Mohammadpour, et al. (2012). Preparation and In Vitro Characterization of Chitosan Nanoparticles Containing *Mesobuthus eupeus* Scorpion Venom as an Antigen Delivery System. *The Journal of Venomous Animals and Toxins Including Tropical Diseases*, **18**(1) 44-52.

[14] Nagarajan, E., *et.al.* (2015). Development and Evaluation of Chitosan Based Polymeric Nanoparticles of an Antiulcer Drug Lansoprazole. *Journal of Applied Pharmaceutical Science,* **5** (04), pp. 020-025.

[15] Dhanasingh, Selvakumar & Nallaperumal, Shunmuga Kumar. (2010). Chitosan/Casein Microparticles: Preparation, Characterization and Drug Release Studies. *World Academy of Science, Engineering and Technology*, **4**, 08-23.

[16] Nesalin, J. Adlin Jino & Smith, A. Anton. (2012). Preparation and Evaluation of Chitosan Nanoparticles Containing Zidovudine. *Asian Journal of Pharmaceutical Sciences* 7 (1): 80-84.

IOP Conf. Series: Materials Science and Engineering **214** (2017) 012027 doi:10.1088/1757-899X/214/1/012027

Phase-Pure of $BiFeO_3$ Ceramic Based on Citric Acid – Assisted Gel by Sintering Time Variation

Dwita Suastiyanti[1] and Ismojo[2]

[1]Mechanical Engineering Department – Institut Teknologi Indonesia (ITI)
Puspiptek Raya Street, Serpong, Tangerang Selatan, Banten, Indonesia
[2]Automotive Engineering Department – Institut Teknologi Indonesia (ITI)
Puspiptek Raya Street, Serpong, Tangerang Selatan, Banten, Indonesia

Email: dwita.suastiyanti@iti.ac.id

Abstract. Bismuth ferrite powder ($BiFeO_3$/BFO) with high purity was synthesized by sol-gel process. It was used $Bi_5O(OH)_9(NO_3)_4$ and $Fe(NO_3)_3.9H_2O$ as main compound sources. Citric acid ($C_6H_8O_7$) was used as fuell. As multiferroic material, BFO promises important technological applications in several devices like data strorage, spinotronics, sensor, actuator devices etc. This research would know the optimum process condition of sol-gel process to produce BFO powder by varying of sintering time. The novelty of this research is how to produce BFO in single phase by simple method. It was used calcination condition at 160°C for 4 hours and sintering condition at 600°C with varying of sintering time of 4, 6 and 8 hours. Thermogravimetric Analysis/Differential Thermal Analysis (TGA/DTA), X Ray Diffraction (XRD) and Scanning Electron Microscope (SEM) were used to characterize the powder. Loss of mass and heat flow were seen at TGA/DTA test at 160°C approximately (used as reference of calcination temperature). BFO powder sintered at for 8 hours has no secondary phase, meanwhile for another sintering time (4 and 6 hours) it has Bi_2O_3 as secondary phase. It is also show at SEM observation result that powder with sintering time of 8 hours has finer grain than of 4 and 6 hours sintering at the same temperature. The grains of BFO powder has heterogenous in size, shape and still agglomerated.

1. Introduction

Multiferroics are a class of materials with coupled electric, magnetic and structural order parameters that yield simultaneous effects of ferroelectricity, ferromagnetism and ferroelasticity in the same material[1]. These materials offer a wide opportunity for potential applications in information storage such as spintronic devices and sensors where both electric and magnetic polarizations can be coupled, giving enough opportunity for manipulating devices. Quality of the devices and sensors depends on purity of BFO. $BiFeO_3$ (BFO) especially synthesized through solid solution exhibits weak ferromagnetic properties at room temperature due to residual moment from a canted spin structure[2], which is believed to be generated from the impurity phase related to Fe phase separation . In addition, a major problem generally encountered in transport measurements is the high leakage current due to the low resistivity of the sample, which creates difficulties in the observation of the ferroelectric hysteresis loops. The high leakage current in bulk is attributed to defects and nonstoichiometry, which is mainly due to the difficulty in preparing pure BFO phase based on the kinetic and thermodynamic properties of the Bi_2O_3–Fe_2O_3 system

[3]. Therefore, this material becomes unsuitable for applications. There are several reports on production of BFO dispersed in $BaTiO_3$ matrix to restrict the second phase, leakage current and low resistivity in the sample. Recently single phase BFO ceramic samples have been synthesized by rapid liquid phase sintering technique[4]. These ceramic samples show relatively high resistivity and saturated ferroelectric hysteresis loops at room temperature and could be useful for potential applications. The characteristic is due to purity of BFO which is produced by rapid liquid phase sintering method. Wet chemical methods are a promising route to prepare fine and homogeneous powder. Various wet chemical methods such as hydrothermal [5], co-precipitation, combustion synthesis [6], molten-salt method [7], thermal decomposition, and sol–gel process [8,9] have been developed and designed to prepare pure BFO nanopowder. Recently, acid-assisted gel strategy has been proved to be an effective way to synthesize metastable BFO nanopowder. Pure BFO powder ca be directly synthesized through the acetic acid-assisted or the tartaric acid-assisted sol–gel method. However, BFO powder synthesized by the organic acid-assisted sol–gel method maybe has relatively low purity resulting from the easy formation of bismutite phase during calcining. Therefore, mineral acid should be considered as anadjuvant to prepare BFO nanopowder. In the present paper, it is reported the synthesis of a pure BFO by sol-gel method using citric acid as a fuel. The method is more simple, energy saving, cost effective and requiring lower process temperature than other methods. Another advantages of using sol-gel method include: reagents required are simple compound, produces nanoparticles, no special equipment is needed, the elements of dopants could be easily incorporated into the final product, there is little possibility of agglomeration of particles, and uniform grain shape [10].

2. Methods

It was used basic compound pro analysis Merck product with a purity of 99.99% $Bi_5O(OH)_9(NO_3)_4$ (bismuth nitrate), $Fe(NO_3)_3.9H_2O$, HNO_3 (Fe nitrate), H_2O and citric acid $C_6H_8O_7$ as a fuel. The basic compounds dissolved in aquabidestilate which was then heated on a hot plate at 80-90°C to form a gel (approximately for 4 to 5 hours). The gel formed was then heated in a furnace at a temperature of 160°C for 4 hours (it was obtained from TGA/DTA test result). The goal was to evaporate the water and the elements C, N and H. The powder obtained was then carried out by heating the sintering process in the furnace at temperature of 600° C for 4,6 and 8 hours respectively (it was also obtained from TGA/DTA test result). The flowchart of the synthesis process was shown in Figure-1.

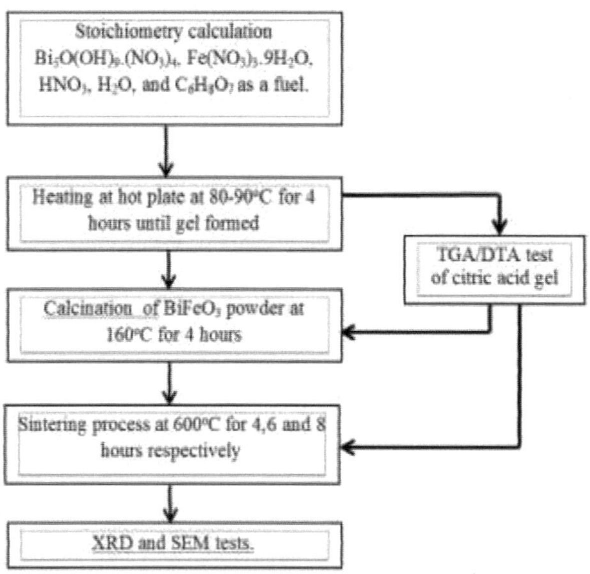

Figure 1. Flowchart of BFO Synthesis by Sol Gel Method Using Citric Acid Gel.

X-Ray Diffraction (XRD) test was performed using an XRD PW 1835 Phillips type with diffraction angle of 20°-100° and using CuKα radiation. The test was used to confirm the establishment phase. The characterization using Thermogravimetric Analysis (TGA) / Differential Thermal Analysis (DTA) aimed to observe changes in mass and heat of samples (still in gel form) to the increase in temperature, using a TGA / DTA Thermal Balance Research type LINSEIS L81-Series I / L81- STA (TGA-DTA). The test was used to know the reference of calcination and sintering temperatures. The characterization using Scanning Electron Microscope (SEM) was used to know the morphology of powder grain.

3. Results and Discussion

To determine the temperatures of sintering, it has been already done TGA / DTA test showed in Figure 2.

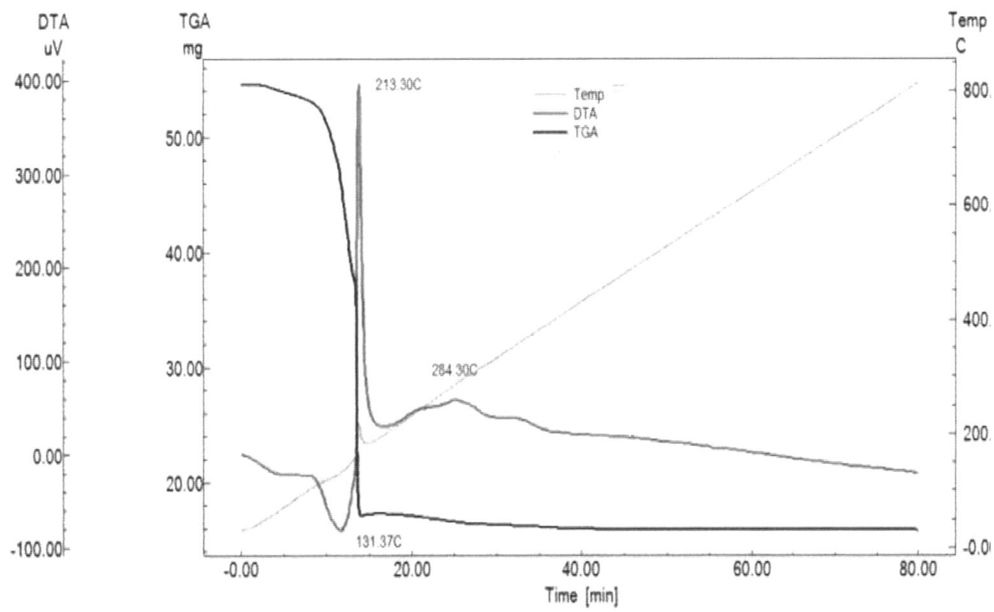

Figure 2. TGA/DTA Test of Citric Acid - $BiFeO_3$ Gel.

The result shows that phase transition occurs at range of 131.37-213.30^{O}C approximately due to weight reduction (blue line) and increasing of energy (red line) at the temperature. The temperature is used as reference to calcination temperature (it is used 160°C from the range as the calcination temperature for this research). The second transition occurs at range of 284.30-785°C approximately, and the temperature is used as reference to sintering temperature (it is used 600°C from the range as the sintering temperature for this research). To confirm the formation of BFO phase, it is performed XRD test for all samples and the results are shown in Figure 3.

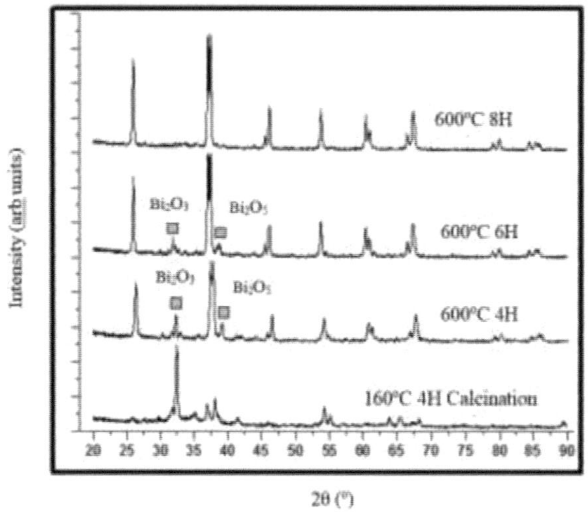

Figure 3. X-Ray Diffraction Pattern of BFO Powder.

Figure 3 shows that after calcination process at 160oC for 4 hours the pattern has not had yet peaks of BFO. So then powder sintering process is carried out. Time of sintering that is applied varied at 600oC for 4,6 and 8 hours respectively. Sintering for 4 dan 6 hours produce powder having impurity phase, Bi2O3 and Bi2O5. Meanwhile sintering for 8 hours produces phase-pure of BFO ceramic, no impurity phase, Bi2O3 and Bi2O5. Producing phase-pure of BFO ceramic due to longer time of sintering, 8 hours that causes perfect of crystallization and forming of BiFeO3. It is still Bi2O3 not yet oxidized to BiFeO3 when powder is sintered for 4 and 6 hours. The morphologies of BFO powder are shown in Figure 4.,5 and 6.

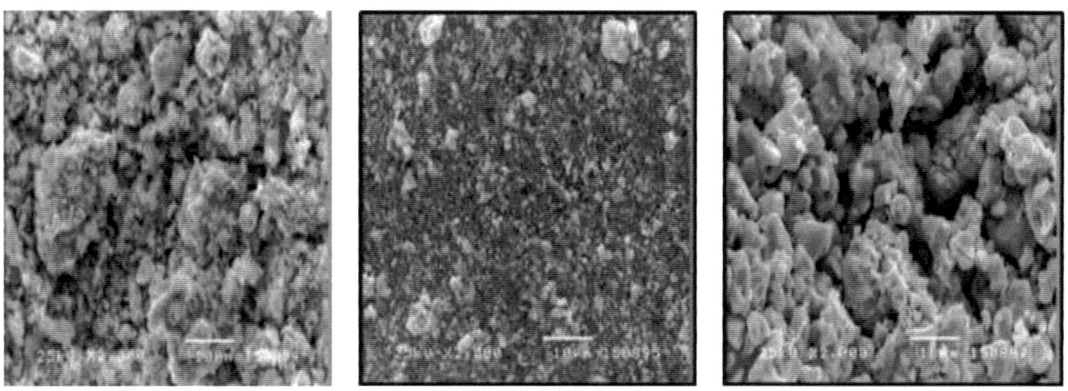

(a) 4 Hours Sintering (b) 6 Hours Sintering (c) 8 Hours Sintering

Figure 4. Scanning Electron Microscopy Observation of BFO Powder for Sintering at 600oC , 4, 6 and 8 Hours.

Figure 4 shows that all process has not yet produced homogen grain and still agglomerated. Meanwhile powder which is sintered for 8 hours has the finest grain at magnification 2000 x. Grain shape of all samples is different between powder of 4, 6 and 8 hours sintering. Powder of 8 hours sintering has agglomerated level lower than other samples.

4. Conclusion
The method using sintering temperature of 600°C for 8 hours and calcination temperature of 160°C for 4 hours could produce phase-pure of $BiFeO_3$ ceramic, no secondary phase of Bi_2O_3 and Bi_2O_5. It also has the finest grain among other sintering treatment for 4 and 6 hours. Grain produced by all process is still agglomerated and heterogenous grain structure.

5. References
[1] Salje.E.K.H, 1990, Cambridge University Press, Cambridge
[2] Pradhan A.K., Kai Zhang, Hunter D, Dadson J.B, Loutts G.B 2005 Journal of Applied Physics.
[3] Shami MY, Awan MS, Anis-Ur-Rahman M 2011 *J.Alloy Compd* **509** 10139-44
[4] Bea H, Gajek M, Bibes M, Barthelemy A 2008 *J.Physics : Condens Matter* **20** 434221
[5] Di LJ, Yang H, Xian T, Li RS, Feng YC, Feng WJ 2014 Ceram Int, **40** 4575-8
[6] Yang J, Li X, Zhou J, Tang Y, Zhang Y, Li Y 2011 *J.Alloy Compd*, **509** 9271-7
[7] Liu Y, Qian Q, Yi Z, Zhang I, Min F, Zhang M 2013 *Ceram Int* **39** 8513-6
[8] Dwita S, Marlin W 2016 *ARPN Journal Engineering and Applied Sciences* **11**(2) 901- 905
[9] Dwita S, Marlin W 2016 *ARPN Journal Engineering and Applied Sciences* **11**(2) 891- 895
[10] Dwita S, Bambang S, Muhammad H 2014 *Applied Mechanics and Materials* **493** 634-939

Acknowledgment
This research is supported financially by Fundamental Research Grant 2016 through letter of decree No. 0299/E3/2016, January 27th, 2016, Ministry for Research and Technology DIKTI, the Republic of Indonesia.

IOP Conf. Series: Materials Science and Engineering **214** (2017) 012028 doi:10.1088/1757-899X/214/1/012028

Effect of Calcination Temperatures on Ratio of Atomic Weight of Al/O in Sol-Gel Method for Synthesis γ Al_2O_3 as a Buffer Catalyst

Dwita Suastiyanti[1], Sri Handayani[2], Maykel T.E. Manawan[3]
[1]Mechanical Engineering Department, Institut Teknologi Indonesia (ITI)
Puspiptek Raya Street, Serpong, South Tangerang, Banten, INDONESIA
[2]Chemical Engineering Department, Institut Teknologi Indonesia (ITI)
Puspiptek Raya Street, Serpong South Tangerang, Banten, INDONESIA
[3]Politeknik Negeri Jakarta (PNJ)
Depok, West Java, INDONESIA

Email: dwita.suastiyanti@iti.ac.id

Abstract. The aim of this research was to determine the effect of calcination temperatures on the ratio of atomic weight of Al/O that could affect the formation of γ-Al_2O_3 phase. The novelty of this research is how to produce γ-Al_2O_3 in single phase and nanosize by simple method (sol-gel method).The calcination process was performed at a temperature which varied of 190°C, 275°C and 320°C for 4 hours respectively and sinter process carried out at a temperature of 420°C for 6 hours. Calcination process at temperature of 320°C for 4 hours produces powder with the ratio of atomic weight of Al/O in accordance with the ratio of atomic weight of Al/O in Al_2O_3 compound, 0.6667 (2/3). This ratio is as expected for a compound according to formula of Al_2O_3. This condition also produces alumina powder with the smallest particle size on the nanometer scale of 84.5 nm. SEM test results show that the grain is still heterogenous in size and shape. The results also show that the grain is still agglomerated.

1. Introduction

Catalyst of monometallic, Pd supported by γ Al_2O_3, are used in a variety of applications including selective hydrogenation, catalytic combustion, and vinyl acetate synthesis. Commercial catalyst of Pd/γ-Al_2O_3 typically are prepared by aqueous impregnation using inorganic salt precursors. The most common technique, incipient wetness impregnation, often results in non-uniform Pd distributions, poor Pd dispersions, and unwanted modification of the surface chemistry of the support[1]. In recent years, increasing attention has been focused on the development of nano-sized alumina powders for advanced engineering materials [2,3]. Conventional processes for synthesizing ceramic of nano powder involve mechanical synthesis, vapor phase reaction, precipitation, and combustion and sol–gel methods [4]. Catalytic dehydration of methanol over solid-acid catalysts offers a potential process for dimethyl ether synthesis. Several catalysts having activity and selectivity for the catalytic conversion of methanol to DME are known, the so called acidic dehydration catalysts [5,6,7]. Commercially γ-Al_2O_3 is used to a large extent for this reaction at a temperature range of 270 to 380°C. The sonochemical synthesis of nanophase materials has the advantage that various classes of materials can be generated simply by changing the reaction medium. So it is worthwhile to overview the different applications, where cavitations can be used efficient. Preparation of metal and metal oxide nanoparticles immobilized on various materials is one of the key researches in nanoscience and nanotechnology, because an excellent synergy and Bifunctional effect would be expected [8]. It is well known that the alternative method for generating stabilized metal nanoparticles involves synthesizing them in or on nanoporous supports, which help define particle size and serve to immobilize the resulting particles [9]. Compound of γ-Al_2O_3 is most known transition alumina as a catalyst support compound, where the compound of γ-Al_2O_3 is stable at high temperature, physically stable, strong and malleable in the manufacturing process. However, commercially γ-Al_2O_3 is not available in single phase and still relatively expensive. The aim of this research is to produce buffer catalyst of γ-Al_2O_3 in single phase, nanoparticle size and the weight ratio of atomic Al/O = 2/3 (0.6667). It is produced by method which is easier and more economical. One way to improve the performance of the catalyst is made in size to nanoparticles of the buffer. Use of the catalyst nanoparticles have been studied by several researchers using ZSM-5 catalyst in cracking LDPE (Low Density Poly Ethilen). The results showed that the nanoparticles ZSM-5 had higher catalytic activity than the microparticles. Therefore, in this research the size of the compound of γ-Al_2O_3 was produced in the size of the nanoparticles. The method used in this research was sol-gel because this method has many advantages such as the process takes place at low temperatures, the process is relatively easy, could be applied in all circumstances (versatile), producing products with high purity and homogeneity if the parameter is varied. In addition, the most impressive of the sol-gel process is relatively cheap, and products in the form of silica xerogel produced non-toxic [10] .Alumina have had a relatively hard physical properties, relatively stable at high temperatures, low electrical conductivity, melting point high, large pore structure, and has a surface area in the range of 100-200 m^2 / g. With these characteristics, causes the alumina is often used in industries, such as absorbents, abrasives, catalysts and catalyst support. In the active form, alumina has a polar surface that is able to adsorb polar compounds. In research conducted by Rahmanpour [11], γ-Al_2O_3 synthesized from $Al(NO_3)_3.9H_2O$ (0.26M), NH_4OH (3.2%), and deionized water at pH of 7.5-8.5 with a temperature of 310 to 340 ° C for 15 hours and had crystal size of 1-2 nm.

2. Methods

Aluminum nitrate {$Al(NO_3)_3$-$9H_2O$, 99.5%}(Merck), Ammonia {NH4OH, 32%} (Merck) and deionized water were used as starting chemicals. A transparent gel-like precursor containing Al cations is precipitated at pH ~7.5–8.5 when ammonia (3.2%) and Al nitrate salt solutions (0.26 M) are mixed together in 400 ml deionized water. The solution was mixing under ultrasonic vibration and maintained at a temperature of 70°C for 2 h. The following chemical reactions occurred during preparation:

$$Al(NO_3)_3 + 3\ NH_4OH \longrightarrow Al(OH)_3 \downarrow + 3NH_4NO_3$$

$$Al(OH)_3 \longrightarrow AlOOH \downarrow + H_2O$$

$$AlOOH \longrightarrow Al_2O_3 + H_2O$$

Sol-gel process was preceeded by the formation of a gel by heating the solution on a hot plate at a temperature of 70-80°C until the gel was formed (approximately 4-5 hours). Gel was then tested by TGA / DTA to determine the temperatures of the phase transformation. The temperatures were used as references of the calcination and sintering processes. The parameters varied were the calcination temperatures of 190°C, 275°C and 320°C respectively for 4 hours. The sintering process was done after the calcination process at a temperature of 420°C for 6 hours. Characterization of the resulting powder was done by X-Ray Diffraction / XRD (Phillips type) test to confirm the formation of phases and determine the ratio of atomic weight Al/O. Observation by Scanning Electron Microscope/SEM (type JEOL/EO, JEM-1400 Version 1.0) was to determine the morphology of the grains. To determine the particle size, it was used Particle Size Analyzer/PSA type of Beckman Coulter DelsaTM Nano by using a solution of dispersing Ethyl Alcohol left for 4 days then the particle / powder was broken again by using ultrasonic. Tests using TGA / DTA (Thermal Gravimetry Analyzer / Differential Thermal Analysis) performed on gel (before the calcination process) was to determine the temperatures of the phase transition that could be observed through the powder mass reduction and a decrease in energy when the gel was heated at temperatures of up to 1000°C. This research was conducted by the flow chart as shown in Figure 1.

Figure 1. Equipments of γ Al_2O_3 Synthesis.

3. Results and Discussion

To find out the calcination and sintering temperatures, they were tested by the TGA/DTA that the results are shown in Figure 2.

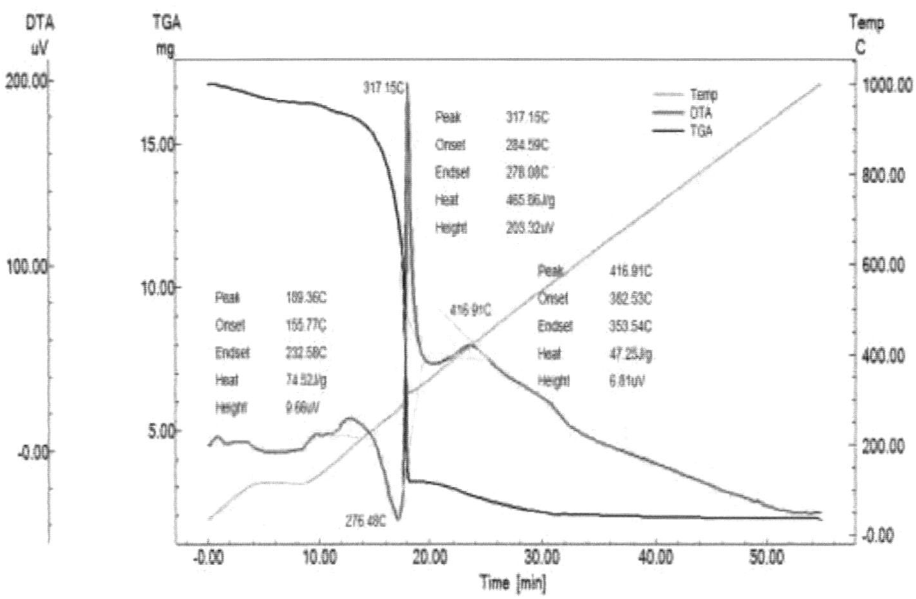

Figure 2. TGA/DTA Test of citric acid gel based on γ Al_2O_3.

Figure 2 shows that the loss of mass and energy reductions occur simultaneously in a temperatures range of 189.36°C-317.15°C. In the temperatures range, H_2O and other elements evaporated which is derived from precursor used. The calcination process is performed at temperature in the range that is at 190°C, 275°C and 320°C for 4 hours. The next energy changes on the diagram TGA / DTA occurs at temperatures of 416.91°C. At this temperature it begins to form alumina phase and it happens crystallization of alumina. So Sinter process is carried out at 420°C. The XRD results are shown in Figure 3, 4 and 5.

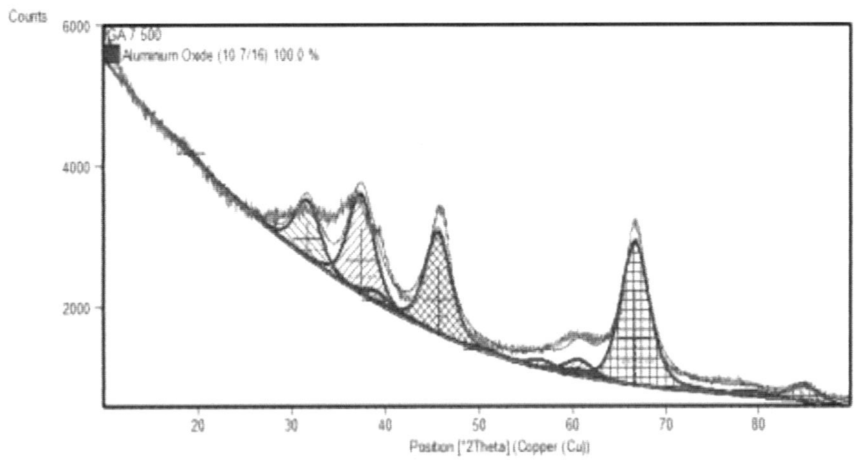

Crystallite Size	Ref. Code	Score	Compound Name	Displacement [°2Th.]	Scale Factor	Chemical Formula
35 A	98-002-8260	69	Aluminium Oxide (9.6666/ 12.9999 =0.7436)	0.000	0.482	Al9.6666 O12.9999

Figure 3. XRD Result and Crystal Pattern List of Alumina Calcined at 190°C for 4 Hours.

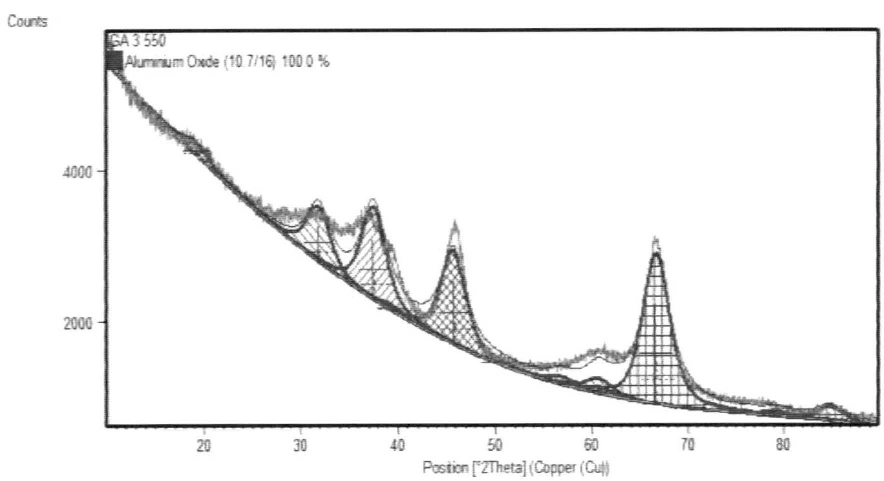

Crystallite Size	Ref. Code	Score	Compound Name	Displacement [°2Th.]	Scale Factor	Chemical Formula
31.4 A	98-002-8260	58	Aluminium Oxide (10.6666/ 16.9999= 0.6275)	0.000	0.470	Al10.6666 O16.9999

Figure 4. XRD Result and Crystal Pattern List of Alumina Calcined at 275°C for 4 Hours.

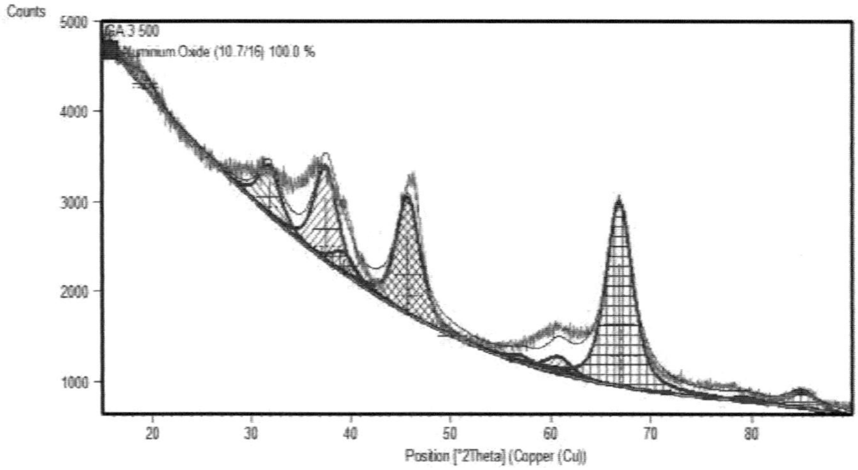

Crystallite Size	Ref. Code	Score	Compound Name	Displacement [°2Th.]	Scale Factor	Chemical Formula
28.7 A	98-002-8260	69	Aluminium Oxide (10.6666/ 15.9999 =0.6667)	0.000	0.881	Al10.6666 O15.9999

Figure 5. XRD Result and Crystal Pattern List of Alumina Calcined at 320°C for 4 Hours.

Figure 3, 4 and 5 show that in fact the whole powder calcined at varied temperature having γ-phase 100% Al_2O_3. There is no impurity phase contained by powder that is calcined at temprature of 190°C, 275°C and 320°C for 4 hours respectively. It could be seen from the pattern shape in which all peak are swept by one color (blue). However ratio of atomic weight Al/O for powder calcined at 190°C and 275°C are not same with ratio of atomic weight of Al_2O_3 (2/3). Powder calcined at 320°C for 4 hours has an ratio of atomic weight of Al/O equal to the ratio of atomic weight of Al/O on Al_2O_3 powder (2/3). The lower temperature of calcined causes unexpected formation of Al_2O_3, due to insufficient of heat energy provided to achieve a balance of Al_2O_3 chemical compounds. From the results of this refinement could also be known that the γ Al_2O_3 powder calcined at 190°C, 275°C and 320°C for 4 hours has crystallite size ranged 2.87-3.50 nm. This leads to the formation of particles in nano size anyway (<100 nm). Crystal Pattern list of powder for all calcined temperatures shown in Table 1.

IOP Conf. Series: Materials Science and Engineering **214** (2017) 012028 doi:10.1088/1757-899X/214/1/012028

Table 1. Crystal Pattern List For All Samples

NO	Treatment (Sintering at 420°C for 6 Hours)	Chemical Formula	Al/O Ratio	Crystallite Size (nm)
1	Calcined at 190°C for 4 Hours	**Al**9.6666**O**12.9999	0.7436	3.50
2	Calcined at 275°C for 4 Hours	**Al**10.6666**O**16.9999	0.6275	3.14
3	Calcined at 320°C for 4 Hours	**Al**10.6666**O**15.9999	0.6667 (2/3)	2.87

To know the chemical formulas shown in Table 1, it is used the result of XRD test which has been refined by HighScorePlus (HSP) software.To determine particle size of powder, the PSA test results are shown in Figure 6. The PSA test used a solution of ethyl alcohol to dissolve the powder tested, then the crushing process is carried out by ultrasonic. Crushing process is done to break down the particles of the powder to prevent agglomeration.

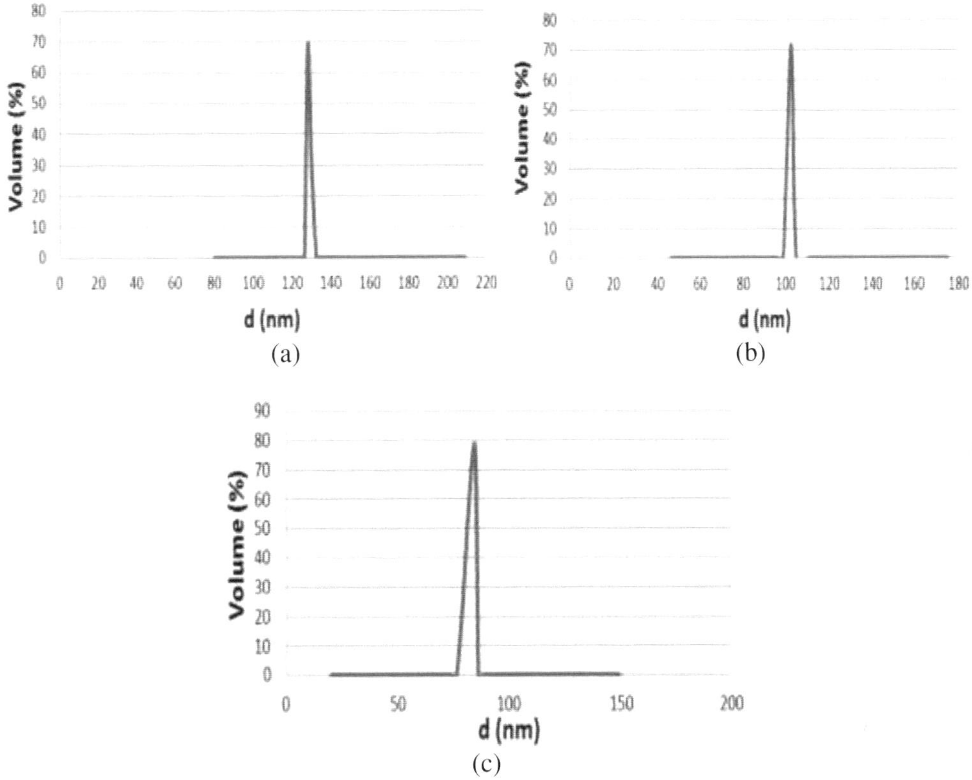

Figure 6. Particle Size Distribution (a) Calcined at 190°C (b) Calcined at 275°C (c) Calcined at 320°C.

Figure 6 shows that the powder has particle in nanosize (<100 nm) except powders calcined at 190°C for 4 hours. Powder with ratio of atomic weight of Al/O = 2/3 has the finest particle (85 nm). The results of the measurements of the particles showed that the sol-gel process has not yet reached the optimum conditions, Expected condition is a powder with a particle size of 20-30 nm when referring to crystal size of between 2.87 - 3,5 nm. To determine the morphology of the grain is observed by Scanning Electron Microscope (SEM), the result is shown in Figure 7.

Figure 7. SEM Observation of Powder Calcined at 320°C for 4 Hours.

Figure 7 shows that the powder is still in agglomeration and has not been homogenous in size and shape. Nevertheless, it could be still used as supporting material for catalyst because it would be wrapped up with other elements (for examples : Pd, Pt etc). The most important thing is purity of γ Al_2O_3 powder.

4. Conclusion

The condition of sol-gel process which is at the calcination temperatures of 190°C, 275°C and 320°C for 4 hours and sintering temperature of 420°C for 6 hours produce γ Al_2O_3 powder in single phase without impurities phase. But this condition produces powder of γ Al_2O_3 in differences of atomic weight of Al/O. Powder having ratio of atomic weight of Al/O = 2/3 (which is expected as according to Al_2O_3) is powder which is calcined at 320°C for 4 hours. The powder having nanoparticles is powder which is calcined at 275°C and 320°C (98 nm and 85 nm respectively). The morphology of the grain is still heterogenous in size, shape and still agglomerated. Although powder has particle in nanosize but it is not as expected particle size is about 20-30 nm when referring to the crystal size is about of 2.87-3.5 nm.

5. References

[1] Jason Kelly M, Jaehoon Kim, George W, Roberts, Henry Lamb H 2006 *North Carolina State University, Raleigh NC* 27695 (USA)
[2] Takeguchi T, Yanagisawa K, Inui T, and Inoue M 2000 *Appl. Catal. A: General* **192**
[3] Brown D.M, Bhatt B.L, and HsiungT.H 1991 *Catal*. Today **8**, 279
[4] Omid Rahmanpour, Ahmad Shariati, and Mohammad Reza Khosravi Nikou 2012 *International Journal of Chemical Engineering and Applications* **3** (2)
[5] Ohno Y, Inoue N, Ogawa T, Ono M, Shikada T, and Hayashi H, 2001, *NKK Technical Review*,
[6] Singh L. P., Agarwal, S. K., Bhattacharyya, S. K., Sharma, U., Ahalawat, S 2011 *Nanomater Nanotechnol* **1**(1) 44-51
[7] Wang A.W, Weigel S, and Muraro G, *Air Products and Chemicals Inc* 2002
[8] Kim J.H, Park M.J, Kim S.J, Joo O.S, and Jung K.D 2004 *Appl. Catal. A: General*. 26437–41.
[9] Gedanken A 2001 *Elsevier Science Ltd* 9450-9456
[10] Okitsu K, Mizukoshi Y, Yamamoto T.A, Maeda Y, and Nagata Y 2007 *Materials Letters* **61** 3429-3431
[12] Rahman O 2012 *International Journal of Chemical Engineering and Applications* **3** (2)

Acknowledgment

This research is supported financially by INSINas-RistekDikti Research Grant 2016 through letter of decree No. 266/SP2H/LT/DRPM/III/2016, Ministry for Research and Technology DIKTI, the Republic of Indonesia.

IOP Conf. Series: Materials Science and Engineering **214** (2017) 012029 doi:10.1088/1757-899X/214/1/012029

Preparation of Ni–YSZ Cermet through Reduction of NiO–YSZ Ceramic for SOFC Anode

P.S.N.Baity, B. Budiana and S. Suasmoro*
Dept. of Physics, Institut Teknologi Sepuluh Nopember Surabaya 60111,Indonesia

* Corresponding author: suasm@its.ac.id

Abstract. Research on the synthesis of Nickel–Yttria Stabilized Zirconia (Ni–YSZ) cermet for Solid Oxide Fuel Cell (SOFC) anode has been performed. The preparation was carried out through the reduction process of the Nickel Oxide–Yttria stabilized Zirconia (NiO–YSZ) ceramic. NiO and YSZ were prepared separately, the NiO powder was prepared by calcination of $CH_4Ni_3O_7.xH_2O$ at 500°C for 3 hours, while YSZ powder was prepared by calcination of 7mol% Y_2O_3 and 93mol% ZrO_2 mixture at 1350°C for an hour. The NiO–YSZ ceramic preparation was carried outby mixing of YSZ and NiO powder with natural white starch by weight ratio NiO: YSZ: natural white starch = 4:6:1 followed by sintering at 1200°C for 4 hours. The completion of reduction process of NiO–YSZ ceramic was performed at 1000°C in flowing Argon (Ar) containing 10% Hydrogen (H_2) up to 4 hours. The characterisations include thermogravimetric analysis (TGA), XRD, SEM–EDX and Impedance Analyzer meter. The synthesised Ni–YSZ cermet at composition 33wt% Ni and 67wt% YSZ, shows relative density 70% and electrical conductivity 10^{-2} S/cm at 700°C, it qualifies as anode for SOFC.

1. Introduction

One of the energy conversions can be achieved by solid oxide ceramic fuel cell (SOFC)[1], this conversion has advantages such as high efficiency and low pollutant emissions[2]. Anode is one of the part in SOFC which the main function is to promote the electrochemical oxidation of fuels[3]. Anode material requires stability during process, sufficient porosity (20–40%) and high electrical conductivity[3]. Merceroz, et al[4], prepared $La_{0.05}Ba_{0.95}Ti_{0.9875}O_3$ (LBTss), $La_{0.05}Ba_{0.95}Ti_{0.9375}Ce_{0.05}O_3$ (LBTCss), and $La_{0.05}Ba_{0.95}Ti_{0.9375}Mn_{0.05}O_3$ (LBTMss) as anode materials for SOFC using sol-gel method. Powder calcinations in air at 700°C for 5 hours and followed by sintering of pellets at 1400°C for 24 hours in air. The sintered pellet were heat treated at 900°C for 48 hours inAr/H_2(98%/2%) atmosphere. The electrical conductivity for LBTCss at 840°C respectively were σ=1.94.10^{-1} S cm^{-1} in Ar/H_2(98%/2%) and 9.9.10^{-3} S cm^{-1} in air, for LBTM were σ=2.2.10^{-2} S cm^{-1} at 780 °C in Ar/H_2 (98%/2%) and 6.9.10^{-4} S cm^{-1} at 800°C in air, while LBTMss were σ=4.2.10^{-3} S cm^{-1} at 825 °C inAr/H_2(98%/2%) and 3.9.10^{-4} S cm^{-1} at 800°C in air. While Zhiping He, et al[5]synthesized $SrMo_{0.94}Fe_{0.06}O_{3-\delta}$-$Gd_{0.2}Ce_{0.8}O_{19}$ through complexing combustion process, the electrical conductivity achieved 2676 S cm^{-1}. Besides that, TaherehTalebi, et al[6]prepared Ni–YSZ (50wt%:50wt%) as SOFC anode, it was obtained through reduction process of NiO–YSZ at 850°C for 5 hoursin flowing Ar:H_2 (93%:7%) atmosphere, the electrical conductivity was 1380–1710 S cm^{-1}. However, this leads to the instability of microstructure and low mechanical strength of anodes. In this study to improve the mechanical properties of Ni–YSZ anode for SOFC, the analysis was focused on lower the NiO content was reduced to 33%, however electrical conductivity must still in the range of the necessity for SOFC anode.

2. Experiment

The Ni–YSZ cermet synthesis was performed in three steps: first the formation of NiO compounds (Nickel Oxide), second YSZ (Yttria stabilized Zirconia) synthesis and finally NiO–YSZ synthesis. Termograviteric analysis was carried out to analyse the decomposition of $CH_4Ni_3O_7.xH_2O$ as source of NiO powder. The YSZ was performed by calcinations of 7% mol Y_2O_3 and 93 mol% ZrO_2 at 1350 °C for an hour the mixing process id carried out by a planetary milling. NiO and YSZ were then mixed with natural white starch addition for pore performer by the composition 4:6:1.The mixed powder was pressed with disc, then sintered at 1200°C for 4 hours. Afterward, NiO–YSZ was reduced by varying temperature and holding time in flowing gas of 10% H_2/90% Arat tubular furnace to perform Ni–YSZ. Variations of temperature were 850°C, 900°C, 1000°C for 3 hours and for completing the reduction the holding time was prolonged to 4 hours. Phase identification was measured byX-ray diffraction of Cu Kαradiation (X-ray Difractometer X Philips), the microstructure of samples were performed by the Scanning Electron Microscopy (SEM) and Energy Dispersive X-ray (EDX) (EVO@ MA10) and electrical properties by Solartron impedance analyzer 1260.

3. Results and Discussion

Termograviteric analysis of $CH_4Ni_3O_7.xH_2O$ powder with initial mass 4.38 mg is depicted in Figure 1, it can be seen there are three weight loss phenomena. Firstly, sample was reduced by 0.76% that is 0.03 mg at 40–60°C indicating the evaporation of moisture form $CH_4Ni_3O_7.xH_2O$ powder. Secondly, sample reduce by 14.02% that is0.61 mg at 60–220°C, indicating the release of hydrate. The third, the sample is reduced by 22.87%that is 1 mg at 220–483°C, indicating the release of carbon compounds. To determine the equation of $CH_4Ni_3O_7.xH_2O$ decomposition into NiO, the mass reduction percentage in experimental and theoretical then calculated.

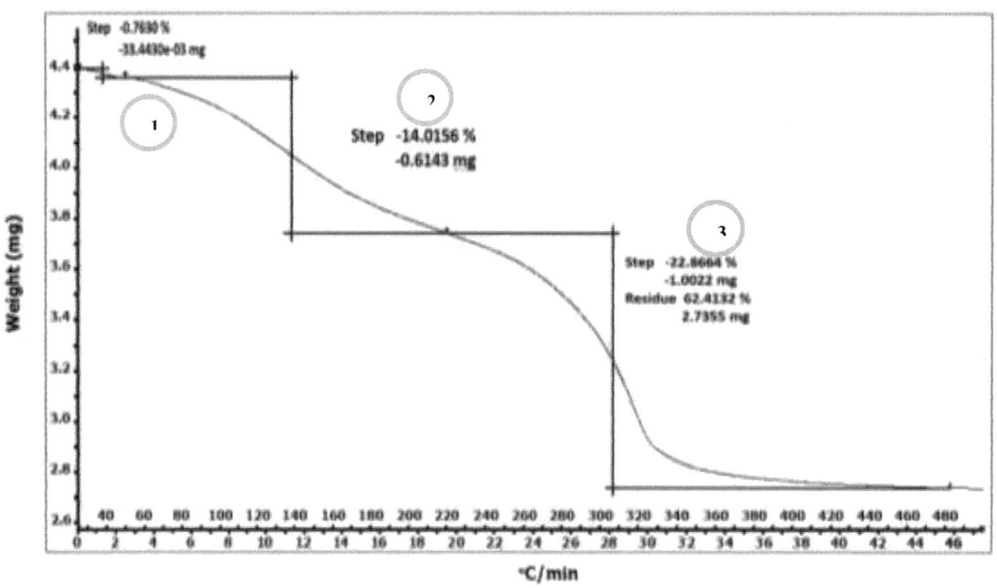

Figure 1. Termograviteric analysis on$CH_4Ni_3O_7.xH_2O$ powder.

Table 1. The mass reduction percentage of $CH_4Ni_3O_7.xH_2O$ decomposition into NiO in experimental and theoretical

Stage	Experimental	Teoritic	Expected
1–2	0.76 %	-	moisture
2–3	14 %	14%	xH_2O
3–4	22.87 %	26%	$2H_2O_{(g)} + CO_{2(g)}$

The difference of the mass percentage reduction in experimental and theoretical shows a relatively small, so it can be deduced that the decomposition reaction in $CH_4Ni_3O_7.xH_2O$ into NiO is as follows:

$$CH_4Ni_3O_7.xH_2O_{(s)} \rightarrow 3NiO_{(s)} + xH_2O_{(g)}$$

$$CH_4Ni_3O_{7(s)} \rightarrow 3NiO_{(s)} + 2H_2O_{(g)} + CO_{2(g)}$$

Figure 2 shows that XRD analysis on $CH_4Ni_3O_7.xH_2O$ before and after calcinations. It shows that an amorphous phase of uncalcined, indicated by the absence of peaks. After $CH_4Ni_3O_7.xH_2O$ through a calcination at 500°C for 3 hours, $CH_4Ni_3O_7.xH_2O$ underwent the phase change from amorphous into crystalline NiO phase this is indicated by the formation of diffraction peaks at 37.24°, 43.3° and 62.92°. Furthermore there is no impurity peaks was detected.

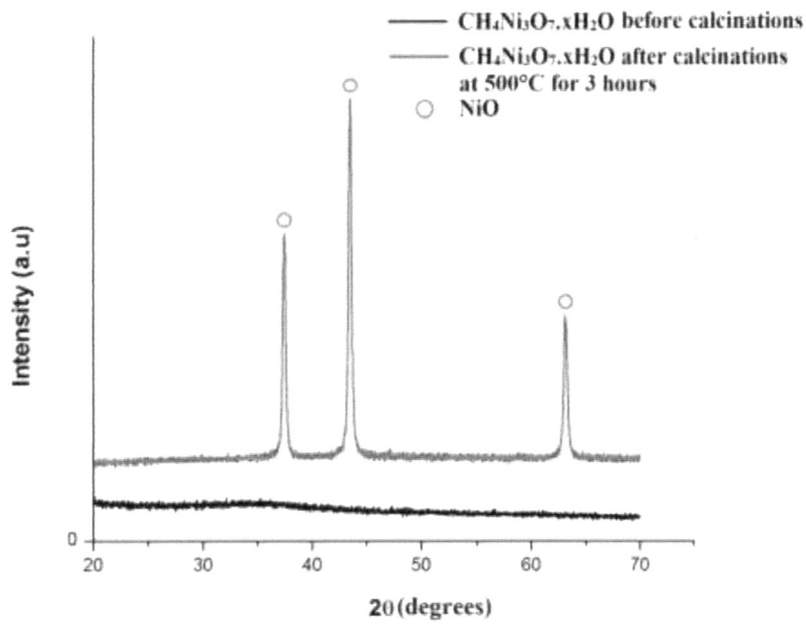

Figure 2. $CH_4Ni_3O_7.xH_2O$ before and after the calcinations.

The calcinations of mixed powder 7% mol Y_2O_3 and 93 mol% ZrO_2 at 1350°C for an hour (figure 3) shows three phases :cubic YSZ, monoclinic ZrO_2, cubic Y_2O_3. Weight percentage of cubic YSZ = 46.77%, monoclinic ZrO_2 = 50.70%, cubic Y_2O_3 = 2.53%. The calcinations process dissolve Y^{3+} into the structure of ZrO_2 create point defect Y'_{Zr} and $V_O^{\cdot\cdot}$ for netrality.

$$Y_2O_3 \xrightarrow{ZrO_2} 2Y'_{Zr} + 3O_O^x + V_O^{\cdot\cdot}$$

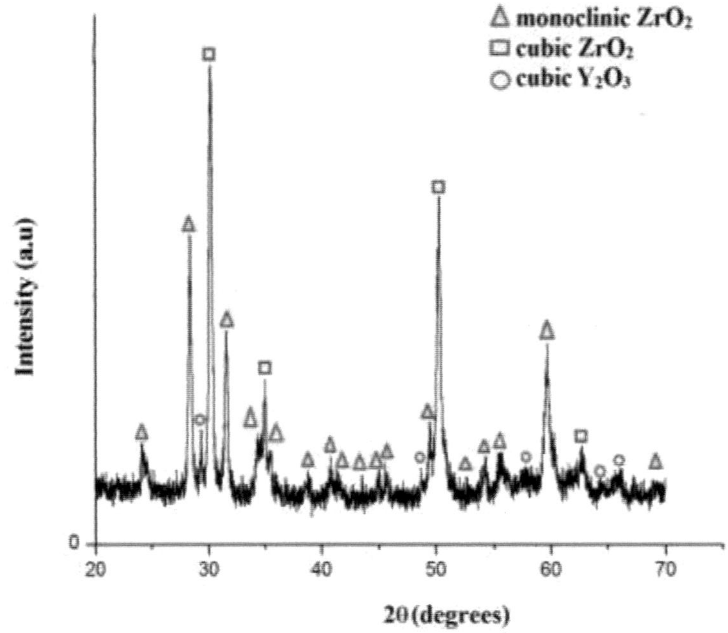

Figure 3. XRD characterization on YSZ.

XRD pattern of NiO–YSZ sintered at 1200°C for 4 hours, figure 4 shows that three phases are identified: cubic YSZ, monoclinic ZrO_2, cubic NiO. Further, using Rieveld refinement analysis the percentage of the phases are cubic YSZ = 26.13wt%, monoclinic ZrO_2 = 31.51wt%, cubic NiO = 42.36wt%. By this treatment the sample density attain 6.897 gram/cm^3 equal to 70% of theoretical. To prepare Ni–YSZ cermet electrode, the reduction process then carried out.

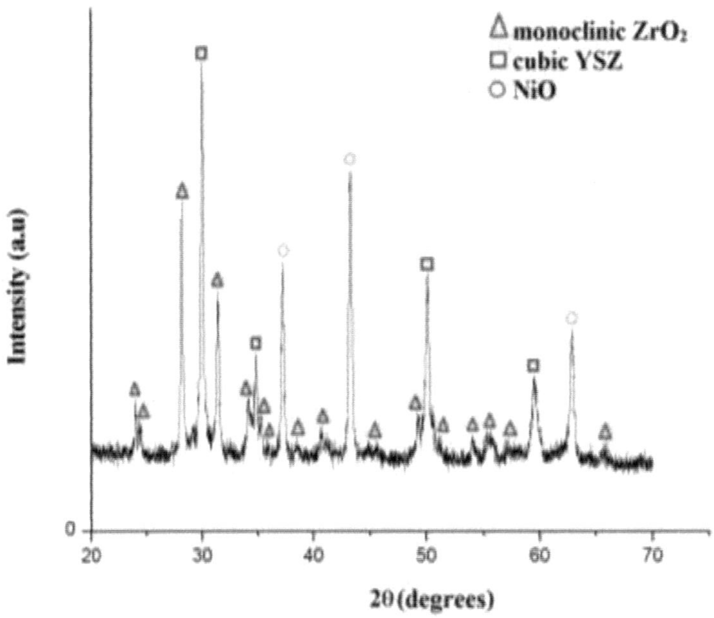

Figure 4. XRD characterization on NiO–YSZ.

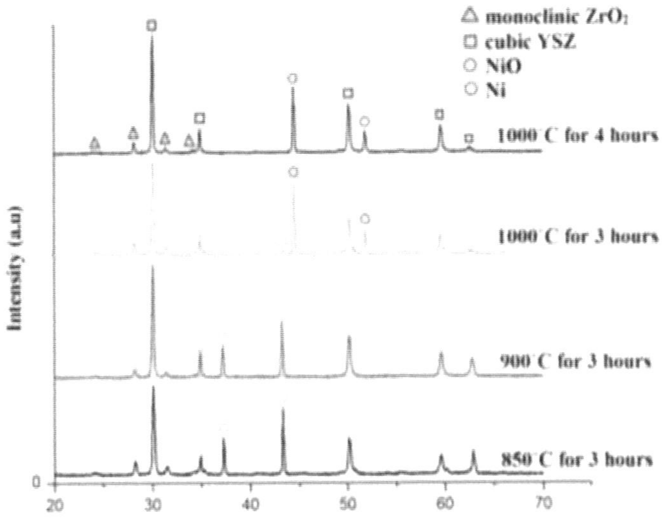

Figure 5. NiO–YSZ–natural natural white starch (4:6:1) with variations in temperature and holding time reduction : 850°C, 900°C, 1000°C for 3 hours and 1000°C for 4 hours in flowing 10% H2/90% Ar.

The reactions of reduction process NiO become Ni include the reaction as follows :

$$NiO–YSZ_{(s)} + H_2 \rightarrow Ni–YSZ_{(s)} + H_2O_{(g)}$$

Based on the XRD pattern, figure 5 indicating that the reduction temperature T=850°C for 3 hours and T=900°C and for 3 hours, NiO has not been reduced to Ni, this is indicated by absence of Ni peaks. However when reduction temperature was increased to 1000°C for 3 hours, NiO has been reduced to Ni, as seen in Figure 5, the Ni peaks were detected, this mean that reduction process was take place, even though there is a little peak of NiO were observed. When reduction time prolonged to 4 hours, the reduction process of NiO was complete. To quantify the phase percent after reduction process at 1000°C the Rietveld refinement analysis were performed. Figure 6 shows quantitative analysis on Ni–YSZ after reduction at 1000°C for 4 hours.

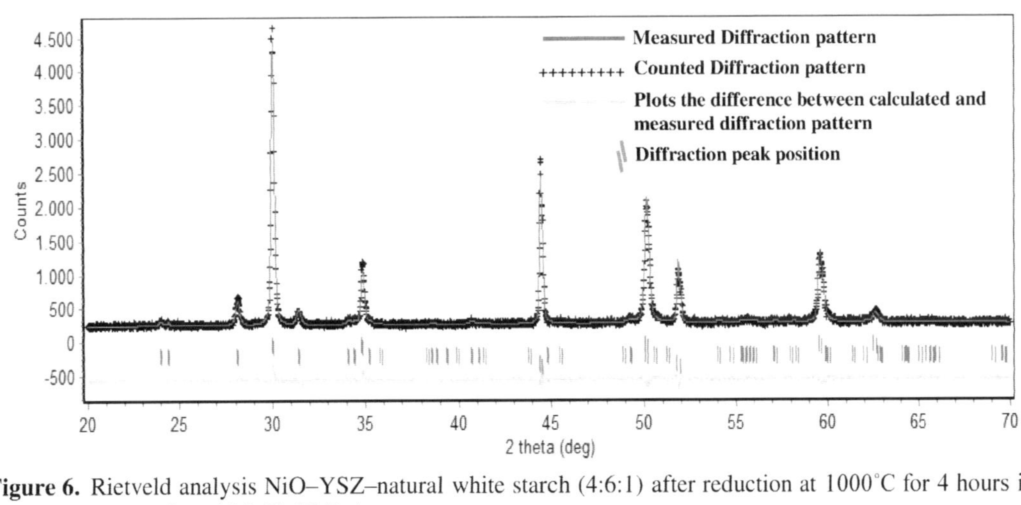

Figure 6. Rietveld analysis NiO–YSZ–natural white starch (4:6:1) after reduction at 1000°C for 4 hours in the gas flow 10% H_2/90% Ar.

Table 2. Relative weight fraction of phases percent before and after reduction process at 1000°C by rietveld refinement

Reduction Process	Relative weight fraction(%)			
	Cubic YSZ (%)	Monoclinic YSZ (%)	Cubic Ni (%)	Cubic NiO (%)
Without reduction	26.13 (3)	31.51 (4)	–	42.36 (5)
1000°C 3 hours	46.98 (4)	18.39 (4)	31.22 (4)	3.40 (3)
1000°C 4 hours	52.99 (4)	13.90 (4)	33.11 (3)	–

Figure 7a shows SEM image of sintered sample, it is seen that sample contained high porosity. Based on the Archimedes method the density 6.897 gram/cm^3 correspond to 30% porosity. It is compare to 27% of porosity determined using Image-J software of SEM figure8. This percentage of porosity is suitable for Ni–YSZ for SOFC anode.

Figure 7. SEM–EDX results on Ni–YSZ.

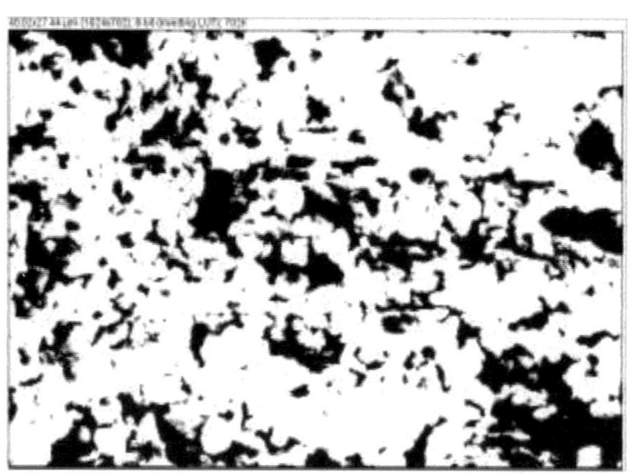

Figure 8. Analyse percentage of porosity Ni – YSZ use Image-J software of SEM, total area = 1098.15, porosity calculated area = 297.43 um and porosity percentage area =27,09 %.

Conductivity measurement of Ni–YSZ at 700°C for sample described above in 90%Ar/10%H_2 was $\sigma = 10^{-2}$ S/cm. This value smaller then obtained by TaherehTalebi, however still in the range of value indicated by Marceros et al between 10^{-4} S/cm–10^{-1} S/cm [4]. EDX analysis of Ni Kα in figure 7b show poor connecting network of Ni metal and it implies to small charge carrying cross section, this is the reason of the under value Ni–YSZ conductivity of this specific sample (33%Ni). Therefore, to increase the conductivity should be increased the percentage of NiO.

4. Conclusion

Ni–YSZ cermet was obtained through reduction process on flowing 10% $H_2$and 90% Argon of sintered NiO–YSZ. The optimum reduction process achieve at T=1000°C for 4 hours. The sample porosity attend 30% while conductivity $\sigma=10^{-2}$ S/cm by this condition Ni–YSZ is suitable anode however need to increase the conductivity.

5. Reference

[1] Pollet BG, Staffell I, Shang JL 2012 Current status of hybrid, battery and fuel cell electric vehicles: From electrochemistry to market prospects *Electrochimica Acta* **84**:235–249

[2] Horri BA, Selomulya C, Wang H 2012 Electrochemical characteristics and performance of anode-supported SOFCs fabricated using carbon microspheres as a pore-former. *Int J Hydrog Energy* **37** (24) 19045–54

[3] Fergus J, Hui R, Li X, Wilkinson DP, Zhang J 2008 *Solid oxide fuel cells: materials properties and performance*. CRC press

[4] Périllat-Merceroz C, Roussel P, Capoen E, Rosini S, Gélin P, Vannier R-N, et al 2015 Synthesis and properties of La0.05Ba0.95Ti1−xMyO3 (M=Mn, Ce) as anode materials for solid oxide fuel cells. *Solid State Ion*. **283** 21–9

[5] He Z, Huang X, Xia L, Yu Y, Yu J 2016 Electrical conductivity and electrochemical performance of SrMo0.94Fe0.06O3−δ–Gd0.2Ce0.8O1.9 composite for SOFC anode. *J Alloys Compd* **660** 108–14

[6] Talebi T, Sarrafi MH, Haji M, Raissi B, Maghsoudipour A 2010 Investigation on microstructures of NiO–YSZ composite and Ni–YSZ cermet for SOFCs. *Int J Hydrog Energy* **35**(17) 9440–7

IOP Conf. Series: Materials Science and Engineering **214** (2017) 012030 doi:10.1088/1757-899X/214/1/012030

Potential Application of Silica Mineral from Dieng Mountain in Agriculture Sector to Control the Release Rate of Fertilizer Elements

Solihin[1,*], Anggoro Tri Mursito[1], Eki N. Dida[1], Bagus D. Erlangga[1], Widodo[1]

[1]Indonesian Institute of Science Research Center for Geotechnology, Jl. Sangkuriang, Bandung 40135 Indonesia

[*]Email: solihin@lipi.go.id

Abstract. Silica mineral, which comes along with geothermal fluid in Dieng, is a product of erosion, decomposition and dissolution of silicon oxide based mineral, which is followed by precipitation to form silica mineral. This silica cell structure is non crystalline, and it contains 85,60 % silicon oxide, 6.49 volatile elements, and also other oxide elements. Among the direct potential application of this silica is as raw material in slow release fertilizer. Silica in compacted slow release fertilizer is able control the release rate of fertilizer elements. Two type of slow release fertilizer has been made by using silica as the matrix in these slow release fertilizer. The first type is the mixing of ordinary solid fertilizer with Dieng silica, whereas the second one is the mixing of disposal leach water with Dieng silica. The release test shows that both of these modified fertilizers have slow release fertilizer characteristic. The release rate of fertilizer elements (magnesium, potassium, ammonium, and phosphate) can be significantly reduced. The addition of kaolin in the first type of slow release fertilizer makes the release rate of fertilizer elements can be more slowed down. Meanwhile in the second type of slow release fertilizer, the release rate is determined by ratio of silica/hydrogel. The lowest release rate is achieved by sample that has highest ratio of silica/hydrogel.

1. Introduction

Silicon oxide is among the abundant compound in world in the form of silica or silicate mineral [1]. Silicate or silica usually can be found in nature as gangue mineral in some ores such as iron, nickel, gold and bauxite ore [2]. Silica also can be found as disposal mineral in hydrothermal power plant [3[. Silica cell structure is various, but the most reactive silica is the amorphous one [4]. One of the places where non crystalline silica can be found is Dieng Mountain. This non crystalline silica is actually a solid part of geothermal suspension fluid that has been sucked toward the surface of the earth during geothermal power plant activity. Non crystalline silica is formed through erosion and decomposition of crystalline silica along with other mineral, which is followed by dissolution and precipitation of SiO_4^{2+} ion to form non crystalline silica [5]. Up to now, in geothermal power plant, this silica is labeled as by product mineral and it is still treated as disposal.

Silica or silicate can be directly used in some simple but important application after being treated through certain simple physical treatment [6.7.8]. Silica can be used as raw material in slow release fertilizer synthesis through certain physical method [9]. Slow release fertilizer is a new kind of fertilizer, in which the release rate of the elements in fertilizer can be controlled [6,7]. Silica in slow release fertilizer can be used to control the release rate of fertilizer elements. The application of slow release fertilizer can increase fertilizer efficiency, while at the same time decrease fertilizer cost. From the environmental point of view, the application of slow release fertilizer can also reduce the ground water contamination that is caused by excessive leaching of fertilizer elements [7,9]. The aim of this research is to characterize silica from Geothermal Power Plant Site in Dieng Mountain and also to characterize the slow release fertilizer made by using this silica mineral

2. Experimental Procedure

Silica mineral used in this research was obtained from hydrothermal power plant site located at Dieng Mountain, Bondowoso, Central Java, Indonesia. This silica is actually the mineral that comes along with geothermal fluid in Dieng. The silica was washed with water to remove the salt prior to drying for 24 hours. The chemical analysis and XRD evaluation of this disposal was conducted through wet analysis and XRD, respectively.

The fertilizer material used in this experiment was commercial NPK fertilizer that is commonly sold in Bandung area, Indonesia. Another source of fertilizer element was disposal leach water that was collected in the Central Disposal Processing Unit located in West Bandung, Indonesia. The elemental composition of NPK and disposal leach water was identified and measured through Atomic Absorption Spectrometry (AAS) and gravimetry analysis.

There are two type of slow release fertilizer we have made. The first type of fertilizer was made by using ordinary solid fertilizer and silica from Dieng. We prepared two formulas by using these raw materials. The first sample (namely A1) consists of silica and ordinary commercial fertilizer. The second sample (namely A2) consists of silica, ordinary commercial fertilizer, and kaolin. The detail composition of both formulas is shown in Table 1. The mixed raw material is pressed and heated at 10 kg/cm2 and 80 oC, respectively, for 24 hours.

Table 1. Composition of slow release fertilizer synthesized by using commercial fertilizer, silica from Dieng, and kaolin

Sample Code	Weight of Silica from Dieng (gram)	Weight of Commercial Fertilizer (gram)	Weight of Kaolin (gram)
A1	130	15	-
A2	90	15	20

Another type of slow release fertilizer was made by using silica and disposal leach water, as source of fertilizer nutrients. The disposal leach water was obtained from Garbage Disposal Unit in West Bandung District, West Java, Indonesia. The disposal leach water was mixed with hydrogel (commercial hydrogel) and silica from Dieng. The mixing was shaped in the form of tablets and air dried for 24 hours. The total weight of mixed silica-hydrogel and the volume of disposal leach water was fixed at 200 gram and 400 ml, respectively. Table 2 shows 5 (five) sample with different weight ratio of silica and hydrogel.

Table 2. The weight variation of silica and hydrogel in the samples

Sample Code	Weight of Silika (gram)	Weight of Hidrogel (gram)
B1	190	10
B2	180	20
B3	170	30
B4	160	40
B5	150	50

The release rate of the elements from the fertilizer was determined through the measurement of elements concentration in the beaker glass that was previously filled with sample and de-ionized water. For a certain range of time, 2 ml of water was taken, diluted and analyzed through Atomic Absorption Spectrometry (AAS) and Spectrophotometer.

3. Result and Discussion

The chemical composition of washed and dried geothermal power plant mineral is shown in Table 3. The table shows that the main compound in the mineral is silicon dioxide (SiO_2). The XRD pattern (Figure 1) shows that the crystal structure of this silica is non crystalline structure. Non crystalline silica can be commonly found at geothermal site. The composition of NPK fertilizer used in our experiment is shown in Table 4. The fertilizer elements in this fertilizer are potassium, magnesium, calcium, phosphorous, and volatile compounds (in which ammonium related compound exist). In our research, the elements that were evaluated are potassium, magnesium, and nitrogen

Table 3. Composition of geothermal power plant mineral

Components	% Weight
Silicone dioxide (SiO_2)	85.60
Alumunium trioxide (Al_2O_3)	0.04
Iron trioxide (Fe_2O_3)	0.21
Manganase oxide (MnO)	0.05
Magnesium oxide (MgO)	0.03
Calcium oxide (CaO)	0.04
Potassium oxide (K_2O)	1.86
Sodium oxide (Na_2O)	0.05
Phosphoric (P_2O_5)	0.32
Moisture content (H_2O^-)	2.25
Volatile content (H_2O^+)	6.49
LOI (Ignition Loss)	11.59

Table 4. Composition of NPK Fertilizer

Components	% Weight
Silicone dioxide (SiO_2)	0.57
Alumunium trioxide (Al_2O_3)	0.64
Iron trioxide (Fe_2O_3)	1.02
Manganase oxide (MnO)	0.02
Magnesium oxide (MgO)	0.15
Calcium oxide (CaO)	3.38
Potassium oxide (K_2O)	18.10
Sodium oxide (Na_2O)	0.77
Phosphoric (P_2O_5)	20.51
Moisture content (H_2O^-)	0.70
Volatile content (H_2O^+)	49.8
LOI (Ignition Loss)	4.4

The XRD analysis to mixed, pressed and heated sample shows that, although it consists of crystalline NPK elements, these samples are amorphous (the figure is not shown). Based on this result, it seems that after being pressed and heated, silica network in the sample was polymerized. The polymerization of silica takes usually takes place when non crystalline silica is pressed and heated

[10]. The polymerization of silica traps the fertilizer elements; hence, the release of fertilizer elements is retarded.

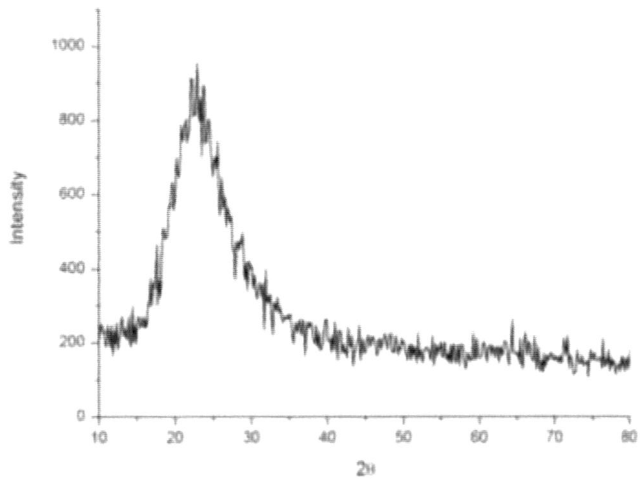

Figure 1. XRD pattern of solid part (silica mineral) of hydrothermal suspension.

The result of the release test of magnesium, potassium, sodium, and ammonium ions from modified fertilizer is shown in Figure 2, Figure 3, Figure 4, and Figure 5, respectively. The release test graph actually represents the release behavior of each element from particular slow release fertilizer.

In the first part of our work, there are two sample of slow release fertilizer we have made, namely A1 and A2. Both of these samples used silica mineral as one of the raw material. The difference is that in Formula A2, some part of silica was replaced by kaolin. Figure 2 shows that magnesium release behavior from both samples seems to follow the behavior of common slow release fertilizer. The release of elements from slow release fertilizer usually follows a polynomial curve. After 2 hours of release test, magnesium concentration is only 80 ppm, but it is eventually increases as test time is increased until it finally reaches a quite stable region after 3 hours. Although both of formulas show similar slow release behavior of magnesium, the quantity of magnesium that is released at each stable region is very different. The stable region of sample A1 is at around 250 ppm, whereas for that of A2 is at around 150 ppm.

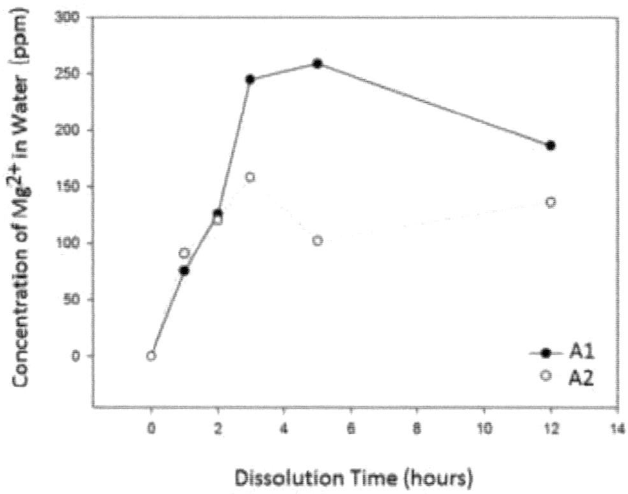

Figure 2. Release profile of magnesium (Mg) from slow release fertilizer.

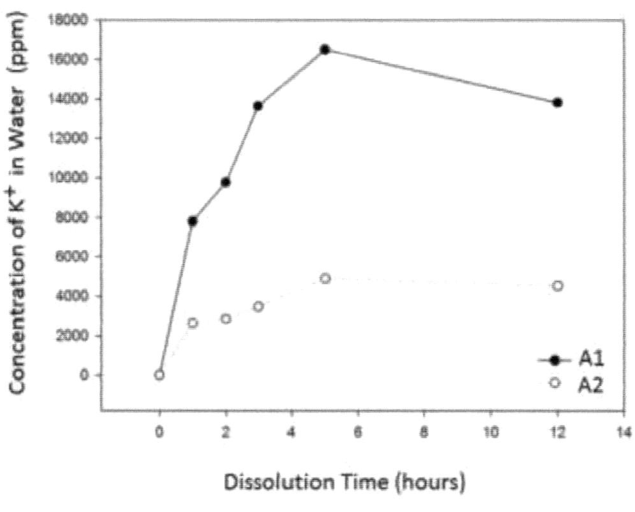

Figure 3. Release profile of potassium (K) from slow release fertilizer.

Figure 3 shows the release profile of potassium in water media. The release behavior of potassium is slightly different from that of magnesium. The quantity of potassium being released from slow release fertilizer is larger than that of magnesium. After one hour, the concentration of potassium released from Formula A1 and Formula A2 media reaches 2100 ppm and 8000 ppm, respectively. Both formulas seem to follow the slow release behavior of common slow release fertilizer. The difference in the quantity of magnesium and potassium being released in water media might be caused by difference of the nature of magnesium and potassium containing mineral. Magnesium and potassium containing mineral in ordinary fertilizer is in the form of magnesium and potassium oxide. Magnesium oxide is less soluble in water than potassium [7]. Potassium oxide readily reacts with water to form potassium hydroxide, but in contrast, only small amount of magnesium oxide dissolves in water.

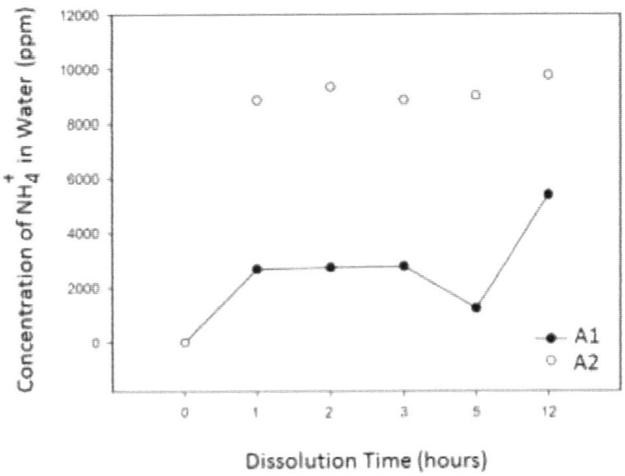

Figure 4. Release profile of Ammonium (NH_4^+) from slow release fertilizer.

Figure 4 shows the release profile of nitrogen, which is represented by .ammonium ion. The ammonium ion released from sample A1 and A2 reach stable reagion in 1 hour at 9000 ppm and 2500 ppm, respectively. It is quite different with those of potassium and magnesium. Unlike ammonium ion, the stable region of magnesium and potassium is reached in longer time, 3 hours for magnesium and 4 hours for potassium. The difference in the time needed to reach stable region might be caused by the difference in solubility value. The time neded to reach stable region for magnesium, potassium, and ammonium is in accordance with the solubility profile of these components. Magnesium oxide is the least soluble component compared to potassium oxide and ammonium hydroxide. Its solubility at 30 ^{o}C is only 0.0086 gram/100 ml. Potassium oxide, when it meets with water, decomposes and reacts into potassium hydroxide. This reaction takes time before the product of reaction dissolves in water. On the other hand, ammonium hydroxide dissolves readily in water. Thus, the solubility sequence of these three elements are ammonium hydroxide, potassium oxide, and magnesium oxide.

Another source that contains fertilizer elements is disposal leach water. The composition of disposal leach water is shown in Table 5. The major component in this water are potassium, phosphate, sodium, nitrogen, magnesium, iron and calcium. These component are actually the common components in ordinary fertilizer. In the second part of our work, we have mixed, pressed and dried samples consisting of disposal leach water, silica from Dieng and hydrogel. The mixing, pressing and drying of these samples results in the formation of slow release fertilizer. The result of release test of this fertilizer is presented in Figure 5, and Figure 6.

Table 5. Elemental composition of disposal leach water obtained from Bandung Barat, West Java, Indonesia

Element/Component	Concentration (ppm)
Ca	3.17
K	2959.55
Mg	11.9
Fe	9.58
Mn	0.45
Na	1475.6
N	9440
PO_4	765.33
pH	8.4

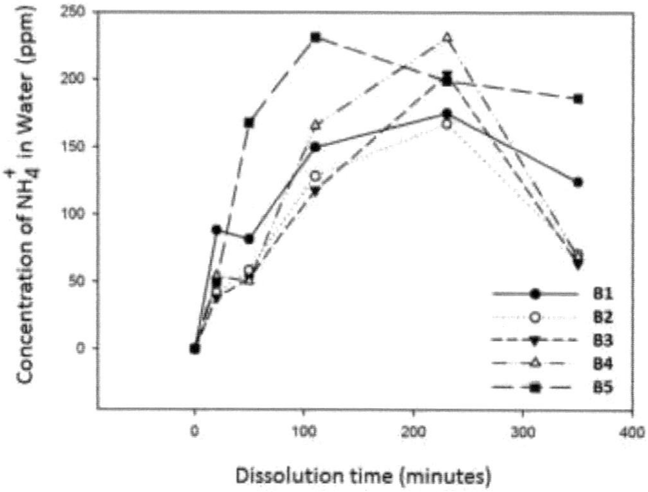

Figure 5. The release profile of ammonium ion from second type fertilizer.

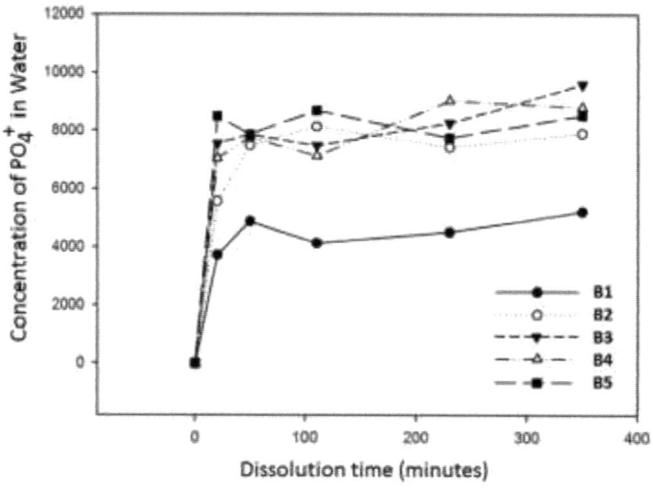

Figure 6. The release profile of phosphate ion from second type fertilizer.

Figure 5 shows the release profile of ammonium from the slow release fertilizer that was synthesized by using disposal leach water and Silica from Dieng. All the samples shows similar behaviour, in which the ammonium is released at relatively slow speed. After 100 minutes, all samples shows the achievement of stable region, but the concentration of ammonium at stable region is quite different. The ascending sequence of ammonium concentration at stable region in each sample are B3, B2, B1, B4, and B5, with the weight ratio of silica/hydrogen are 6, 9, 19, 4 and 3, respectively. It seems that the highest weight ratio of silica/hydrogel (lowest amount of hydrogel) gives highest concentration of ammonium. Hydrogel absorb disposal leach water during slow fertilizer synthesis and release it during release test in water. The total mass fraction of liquid/hydrogel determines the amount of disposal leach water in each hydrogel particle. Low fraction of hydrogel leads to the high amount of disposal leach water in each particle. The higher the amount of disposal leach water in each particle the easier the liquid is released from hydrogel, which leads to the highest ammonium concentration at lowest ratio of silica/hydrogel. The difference in silca/hydrogel ratio seems to affect the concentration of other elements such as phosphate and potassium as shown in Figure 6. The highest amount of phosphate in water is achieved by sample that has lowest ratio of silica/hydrogel.

4. Summary

Silica mineral from geothermal site in Dieng, which is the product of erosion, decomposition, dissolution of silicon oxide based mineral, can be used directly as slow release fertilizer. Two type of fertilizer has been successfully synthesized. The first type of slow release fertilizer consists of silica mineral and ordinary commercial fertilizer; whereas the second type consists of silica mineral and disposal leach water. Both type of synthesized slow release fertilizer shows slow release characteristic. The release rate of magnesium, potassium, ammonium and phosphate can be decreased. The release rate of magnesium, potassium, and ammonium in first type of synthesized slow release fertilizer is determined by the addition of kaolin. The release rate of fertilizer elements in the sample with kaolin is slower than the sample without kaolin. Meanwhile, the release rate of ammonium and phosphate in the second type of synthesized slow release fertilizer is determined by ratio of silica/hydrogel. Sample with highest ratio of silica/hydrogel gives slowest release rate of fertilizer element

Reference

[1] Railsback L B 2016 Some Fundamentals of Mineralogy and Geochemistry, Department of Geology, University of Georgia, Athens, Georgia USA
[2] Djamaluddin H, Thamrin M 2012 Prosiding Hasil penelitian fakultas Teknik Universitas Hasanudin, Makasar, Indonesia
[3] Riyanto N, Sumardi P, Perdana I 2012 *Jurnal Rekayasa Proses* **6** 1
[4] Fournier R O, Rowe J J, 1977, *American Mineralogi* **62** 1052-1056
[5] Yuliyanti A., Nurohman H, Al Kautsar A, Sudarsono, Indarto S, Bakti H, Permana H, Gaffar E Z, 2015, *Seminar Pemaparan Geotekologi LIPI*
[6] Yuan W, Solihin, Zhang Q, Kano J, Saito F 2014 *Powder Technology* **260** 22–26
[7] Solihin, Zhang Q, Tongamp W, Saito F 2010 *Ind. Eng. Chem. Res* **49** 2213–2216
[8] Agustinus E T S., Sembiring H, Listiyowati L N, 2010, *Riset Geologi dan Pertambangan* **20** (2) 81-93
[9] Solihin, Zhang Q, Tongamp W, Saito F 2011 *Powder Technology* **212** 354–358
[10] Zou H, Wu S, Shen J 2008 *Chem. Rev* **108** 3893–3957

IOP Conf. Series: Materials Science and Engineering **214** (2017) 012031 doi:10.1088/1757-899X/214/1/012031

Corona Glow Discharge Plasma Treatment for Hidrophylicity Improvement of Polyester and Cotton Fabrics

A I Susan[1,3], M Widodo[2,3],and M Nur*[1,3]

[1]Department of Physics, Faculty of Science and Mathematics, Diponegoro University, Semarang, 50275, Indonesia
[2]Department of Textile Chemistry, Polytechnic of Textile Technology, Bandung, Indonesia
[3]Center for Plasma Research, Diponegoro University, Semarang, Indonesia

*Email: m.nur@undip.ac.id

Abstract. The effects of irradiation by a corona glow discharge plasma on hidrophylicity properties of polyester and cotton fabrics were investigated. We used a corona glow discharge plasma reactor with multiple points to plane electrodes, which was generated by a high voltage DC. Factors that affect the hidrophylicity properties were identified and evaluated as functions of irradiation parameters, which include duration of treatment, distance between electrodes, and bias voltage. It was readily observed from SEM examinations that plasma changed the surface morphology of both polyester and cotton fibers, giving result to an increased roughness to both of them. Results also showed that the hidrophylicityof polyester and cotton fabrics improved by the treatment, which is proportional to the time of treatment and voltage, but inversely proportional to the distance between electrodes. Time of treatment that provided the optimum enhancement of hidrophylicity for cotton is 15 minutes which improved the wetting time from 8.16 seconds to 1.26 seconds. For polyester, it took 15 minutes of irradiation time to improve the wetting time from 7340 seconds to 2905 seconds. The optimum distance between electrodes for both fabrics in this study was found to be 2 cm. Further analysis showed that the improved hidrophylicity properties is due to the creation of surface radicals by free radicals in the plasma leading to the formation of new water-attracting functional groups on the fiber surface.

1. Introduction

In textile, plasma treatment is recognized as an alternative technology to replace the chemical refining process and preparation of refinement. This technology offers a number of advantages greater than conventional chemical processes. Surface modification using plasma does not require the use of water and chemical compounds, so it is considered as an inexpensive and eco-friendly technology [1]. Plasma treatment is recognized to be able give a number of changes in the physical and chemical properties on the surface of the textile fabric. Plasma treatment has been reported could improve water absorption on the fiber surface, mainly by oxidation and etching. Plasma treatment has also been used for several applications such as anti shrink, increasing the rub resistance properties and ability to absorb the dye in the wool fibers, and is able to provide improved mechanical properties significantly both through radio frequency and low frequency which includes the shear strength, flexural strength and tensile strength [2-5].The properties changes are caused by the several factors, including: changes in surface morphology and/or the formation of the active chemical functional groups on the substrate fabric due to plasma treatment. Another possibility is the appearance of free radicals on the substrate by plasma treatment, then causing changes in the physical properties of fabric.

The process that takes place in generation of plasma treatment on substrate is that when the power increased, the plasma gas will get more energy for ionization and can be more easily ionized. On the other hand, the speed of electrons under strong electric field will increase, causing an increase in kinetic energy of electrons. These two factors will increase the action of plasma on fiber's surface. The action will cause the surface roughness and the formation of polar oxygen functional groups, which contribute to give hydrophilic properties and improve adhesion on polyester fabrics [6-8].

This study examines the wetting properties changes on polyester and cotton fabrics, respectively selected to represent synthetic and natural fibers, due to the plasma treatment. The type of plasma used in this study is corona glow discharge plasma which generated by multi point to plane electrode configuration. The properties changes observed, as indication of the influence of plasma treatment, were associated with morphology changes and the appearance of free radicals on the surface of treated fabrics.

2. Methods

2.1. *Characteristic of electric current – voltage*

The research begins by performs characteristic of electric current as a function of voltage on the series of multi-point to plane reactor. The characteristics procedure was performed on some variation of the distance between the two electrodes. While in irradiation process, the treated fabric sample was placed on the surface of plane electrode.

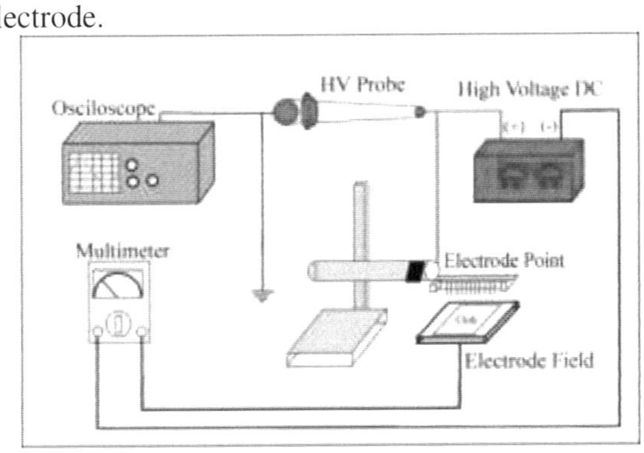

Figure 1. The schematic of the multi-point to plane reactor series.

To generate plasma, the point electrode is given positive polarity. Electric current in the circuit of plasma reactor is measured using a multimeter (Sanwa Japan YX360TRF), while the value of the voltage generated by the high voltage (HV) DC power supply was measured using an oscilloscope connected to probe high voltage (HV Probe DC max Voltage DC 40 kV, AC 28 kV EC code number 1010, EnG1010, Made in Taiwan). There are three parameters were varied in this study, ie: voltage, distance between electrodes, and time of treatment. The distances between electrodes are 1 cm, 2 cm, 3 cm, 4 cm and 5 cm, while the times of treatment are 0, 15, 60, and 90 minutes.

The assessments of physical-chemical properties on treated-fabric are consisting of wetting ability, capillary absorption, surface's morphologycal, and radical density tests. All of these tests results were then compared to the properties of control- and treated-fabrics.

2.2. *Ion mobility*

When the charge carrier particles of a material were in the electric field would move with the average velocity which called as drifting velocity or drift velocity. The magnitude of the drift velocity is proportional to the magnitude of the electric field and the value of mobility, in accordance to Eq. (1)

$$I_s = \frac{2\mu\varepsilon_0 V^2}{d} \tag{1}$$

where I_s is an unipolar saturated current (ampere), V is voltage (volt), μ is unipolar ion mobility (m/Vs), ε_0 permitivy of vaccum (F/m) and d is space between electrode (m) (Sigmond,1982).

Variations of irradiation parameters, such as space between electrodes (d) and applied voltage (V), in addition will effect on the measured current (I) they will also affect the value of unipolar ion mobility (μ). The value of unipolar ion mobility can be calculated using Eq.(2).

$$\mu = \frac{I_s.d}{V^2.2\varepsilon_0} \tag{2}$$

But for the case of the multi-point electrode it required to modify the equation above by adding a factor of the number of points of electrode, which is considered as N.In addition, the presence of fabric samples in a series of corona plasma reactor can be considered as a dielectric material with a specific dielectric constant value, which can be considered as ε_1, so the equation above becomes:

$$\mu = \frac{I_s.d}{V^2.2\varepsilon_t N} \tag{3}$$

In the characteristic curves, the value of mobility can be calculated based on the value of a linear gradient of the curve so that the above equation becomes:

$$\mu = \frac{(\tan\alpha)^2 d}{2\varepsilon_t N} \tag{4}$$

where: $\tan\alpha = \frac{\sqrt{I_s}}{V^2}$, while the total dielectric constant was calculated for a series configuration by assuming that the air and sample were installed in series.

2.3. *Physical and chemical properties assessment*

Wetting ability is the rate of water spread on the fabric. The principle of this test is to observe a drop of water dropped from a certain height on the surface of a sample. Time for disappearance of water on the sample was measured and recorded as wetting time. This Test conducted under the atmospheric conditions. The fabric sample is placed 1 cm below the tip of the burette droplets.

Capillary absorption is the ability of the fabric to absorb water by capillary absorption. Capillary absorption rate was calculated based on the height of the water absorbed in the fabric h (cm), using the equation:

$$v = \frac{h}{t},$$

where v is capillary absorption rate, and t is absorption time of 2 minutes.

Morphology of the samples was observed using SEM (Scanning Electron Microscope) to observe changes on fabric's surface. While the density of surface radicals tested using DPPH solution. The wetting properties tests, including wetting ability and capillary absorption rate, have been conducted with immediately and delayed tests to see how the permanence properties obtained from plasma treatment.

3. Results and Discussion

3.1. *Characteristic of electric current – voltage*

The purpose of this characterization in general is to: get the area corona glow at some variations of distance between electrodes, and to analyze the charge carrier mobility which allegedly influence the changing of wetting properties on the surface of treated fabric. In addition to distance, the characterization of the current-voltage function is also carried out in a state by and without fabric sample between the electrodes. It aims to analyze the influence of the textile material presence to electric current at the same voltage, and to get the area range of glow discharge corona plasma when there is a fabric sample between the electrodes.

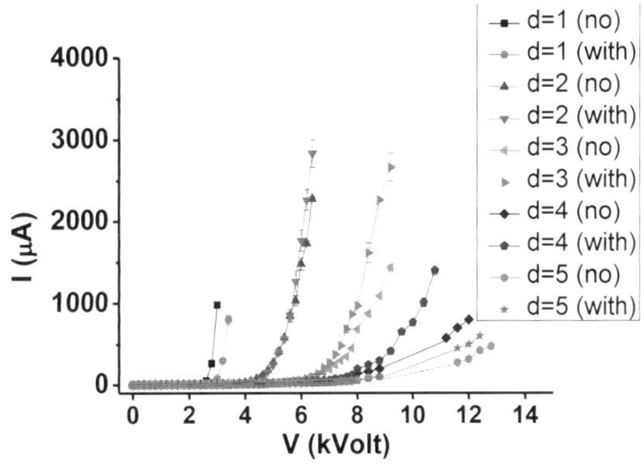

Figure 2. Figure 2. Characteristic of electric current as a function of voltage. The characteristics has been observed with variations of distance between electrodes *d,* with and without samples between electrodes.

In general it can be concluded that the presence of the fabric sample between electrodes has increased the value of electric current. It is occurs because the fabric has a certain thickness and appears to be a negatively charged when electrons flowing from the electrode field causing an effective reduction in distance between the positive electrode and the field electrode, and finally rising the electric current.

3.2. *Ion Mobility*

The calculation of ion mobility has been conducted based on the characteristic graphs of electric current as function of voltage. Figure 3 shows the ion mobility as a function of distance between electrodes in corona discharge region.

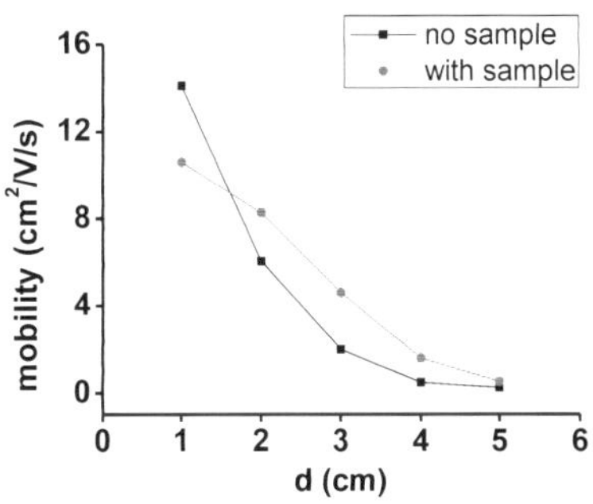

Figure 3. Ion mobility as a function of distance between electrodes *d.*

The results showed that the ion mobility values obtained from some variation of *d*, approaches the ion mobility under atmospheric pressure from the study by Langevin. In this study, ion mobility value is from 0.25 to 14.1 cm2/V.s. while mobility from Langevin's study is approximately ± 2 cm2/V.s for atmospheric pressure [9].

3.3. *Wetting properties*

The results show that the treatment with maximum glow provide the greater wetting time reduction than with minimum glow, both on cotton and polyester fabrics. Another result obtained is the distance between electrodes o f 2 cm provides maximum decrease compared to the distance of 1 and 4 cm. If it is associated with wetting time, the result has not the same tendency with ion mobility as a function of distance *d*, which is based on the graph it can be seen that the higher distance between the electrodes provide the smaller value of mobility,

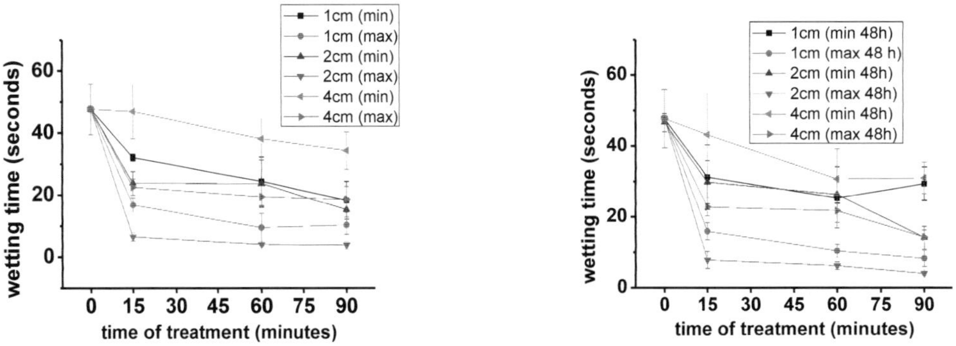

(a) (b)

Figure 4. time of absorption on cotton fabric (a) in immediately test, (b) delayed test, 48 hours after treatment.

Similar results were also obtained on the polyester fabric, the differences are on the order of wetting time. If the cotton fabric takes tens of seconds, the polyester fabric takes tens of minutes to hundreds of minutes (see Figure 5).

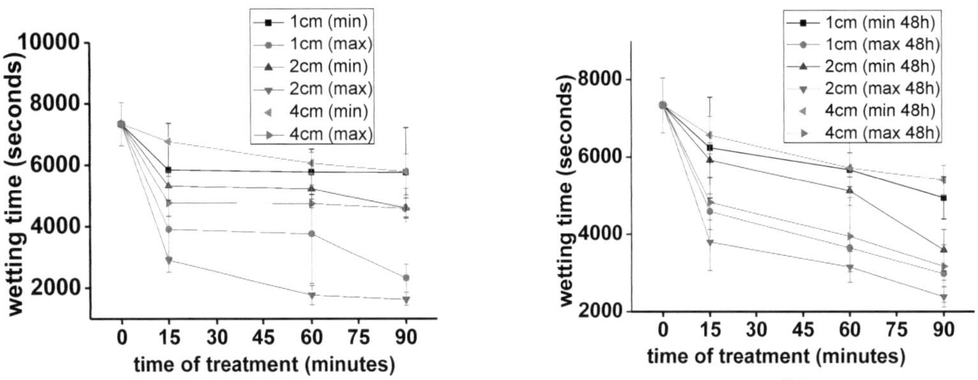

(a) (b)

Figure 5. time of absorption on polyester fabric (a) in immediately test, (b) delayed test, 48 hours after treatment.

Based on the observations using FTIR_ATR, Kan et al found that the polymeric hydroxyl groups on the cotton fabrics decreased after plasma treatment. It occurs due to the polymeric hydroxyl groups related with starch-based sizing material is removed by plasma treatment. This suggests that the plasma treatment can physically remove the starch-based sizing material and wax from the surface of the fibers and thus the wetting properties of the plasma-treated fabrics are sharply increased [10]. In this study, the effect of durability/permanency of plasma was observed only until the 2nd day, and not obtained any significant decline in wetting properties on polyester (Figure 5.b) and cotton (Figure 4.b) fabrics. These results are consistent with the results obtained by Cheng et al, that under 5 days after treatment does not occur enhancement of contact angle and even occurs a slightdecrease in the contact angle [11].

3.4. *Capillary absorption*

Figure 6 shows the change of capillary absorption rate on cotton fabrics. The capillary absorption rate increased with increasing time of treatment. In contrast to the results obtained in wetting ability, in immediately test, time of treatment of 15 minutes not sufficient to provides a significant change to capillary absorption (Figure 6.a). Then, it required longer time to provide sizeable change on capillary absorption rate. In accordance by the results obtained from wetting ability test, the capillary absorption test also showed that the highest capillary absorption rate obtained in distance between electrodes of 2 cm. Figure 6.b shows the changes of capillary absorption rate of 48 hours test after treatment. It's found that he highest enhancement in capillary absorption rate obtained on the 15 minutes treated fabric, while the capillary absorption rate on 60 minutes and 90 minutes treatment get very little enhancement and even tends to saturate.

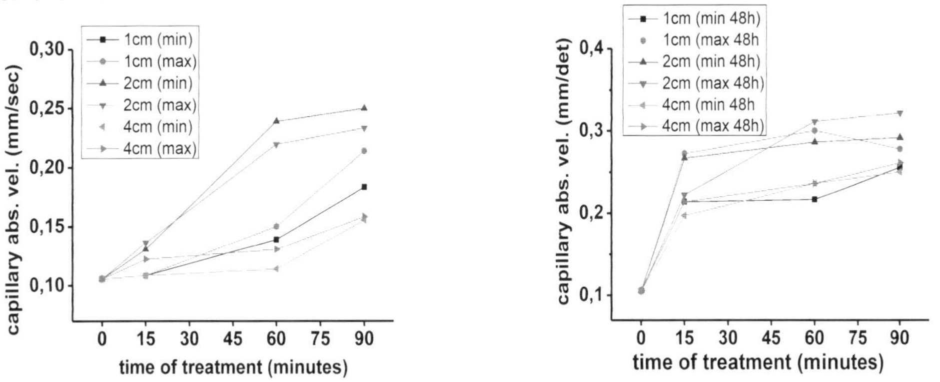

(a) **(b)**

Figure 6. Capillary absorption velocity on cotton fabric(a) in immediately test, (b) delayed test, 48 hours after treatment.

In polyester fabrics, capillary absorption rate is very low at under 0.1 mm/sec (see Figure 7.a.). Capillary absorption rate on the control fabric is 0 mm/sec. Both of immadiately and delayed test have provided the similar changes of capillary absoprtion rate. The maximum rate was about 0.03 and 0.04 mm/sec, but the highest rate was obtained at d=4cm in 90 minutes plasma treatment that was 0.11 mm/sec (Figure 7.a). Delayed test after 48 hours has relatively showed the similar rising as immediately test in capillary absorption test of cotton fabric (Figure 7.b).

Wang et al. confirmed that the increase in plasma discharge power will increase the height of the capillary absorbency of fabrics. Theoretically, as a surface modification technology, the forming of energetic and active plasma species was influenced by the discharge power. The Low discharge power typically causes the limited number of active plasma species. The larger discharge power allows for more effective plasma penetration into the fibers, tougher physical etching and plasma chemical reactions to be occur [12].

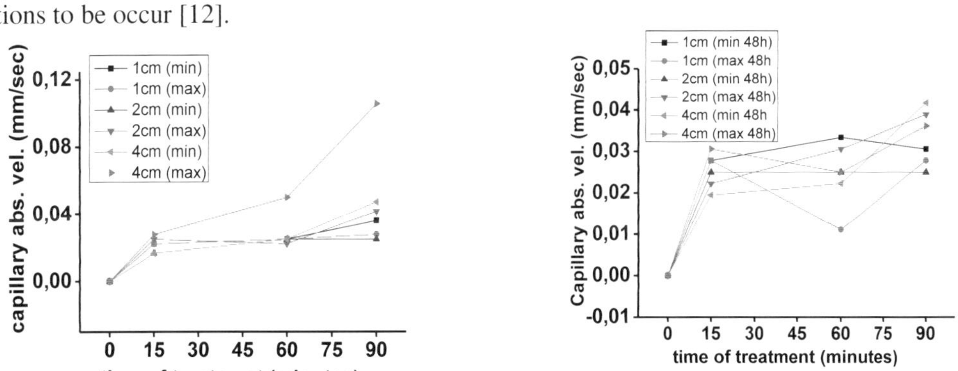

(a) **(b)**

Figure 7. Capillary absorption velocity on polyester fabric(a) in immediately test, (b) delayed test, 48 hours after treatment.

Similar results were also obtained by Ferrero were testing the wetting ability on synthetic fabrics due to treatment by RF plasma that generated at 0.64 mbar using capillary rise [13]. In a particular power, height of capillary absorption increased with increasing treatment time, where at a given time the height of absorption increased with the increasing of electric power but decreases with increasing aging time. These results are in accordance with the increase in the contact angle after the plasma treatment, due to the emergence of polar groups on the polymer surface or the chain breaking process.

It was reported that the effects of the surface tends to be unstable and and decreasing wettability as the increase of storage time [13].

3.5. *Radical Density*

In this study, the changes of surface radical density due to plasma treatment were only observed as function of time of treatment on cotton and polyester fabrics (Figure 8.a.) with d=2 cm, and as function of distance between electrodes in maximum glow and time of treatment of 15 minutes (Figure 8.b). Figure 8.a. showed that the surface radicals density increased with increasing plasma treatment time. While Figure 8.b showed that the surface radicals density ecreases with increasing distance between the electrodes. These results showed have the same tendency of ion mobility on corona glow discharge plasma. The higher distance between electrodes provides the smaller value of ion mobility (see Figure. 3). The radical activity was also affected by charge carrier mobility.

In his study, using atmospheric pressure jet plasma generated with radio frequency (RF), Widodo et.al. reported that surface radical density increased by increasing of electric power on kevlar fabric and PPTA film [14]. The similar results also obtained by Poncin-Epaillard and Wang, using microwave plasma and nitrogen gas, that surface radical density rised exponentially by the electric power. In addition, Poncin-Epaillard dan Wang also found that surface radical density decrease with increasing distance between electrodes and increase as function of time of treatment of 0 to 5 minutes and then decrease after over 5 minutes treatment [15].

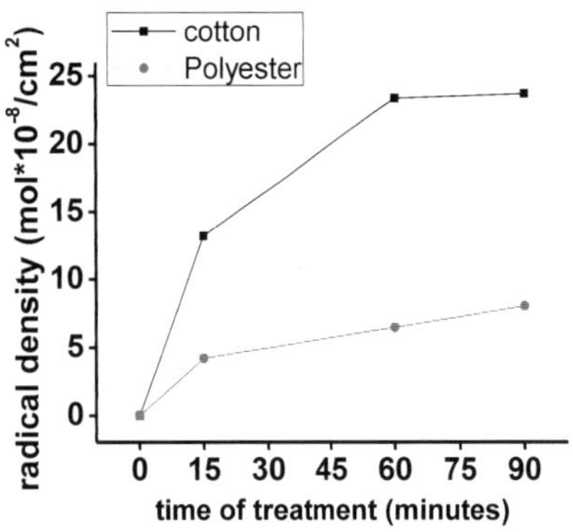

(a)

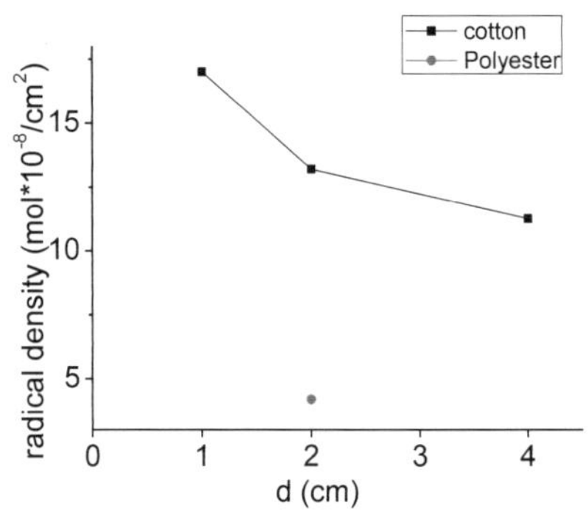

(b)

Figure 8. Radical density on fabric samples as function of **(a)** time of treatment with*d*=2 cm in maximum glow, **(b)** distance between electrodes*d* with time of treatment of 15 minutes in maximum glow.

3.6. *Morphological changes*

Interaction of active species in plasma with substrate may causing the particles addition to the substrate or particles release from the substrate. Morphological changes on fabric's surface was observed by comparing the morphology of control and treated fabrics. Figure 9 and 10 show the SEM image of cotton and polyester fabric, respectively. Based on the SEM image, it can be seen that control fibres, both cotton and polyester, have a smooth surface. While the fibers of treated fabrics have a rougher surface than control fabrics. The second thing can bee seen is that the surface roughness increase with increasing time of treatment. Then the third thing can be observed by SEM Image is that the polyester fabric seems to have rougher surface than cotton fabric (see Figure 9 and 10).

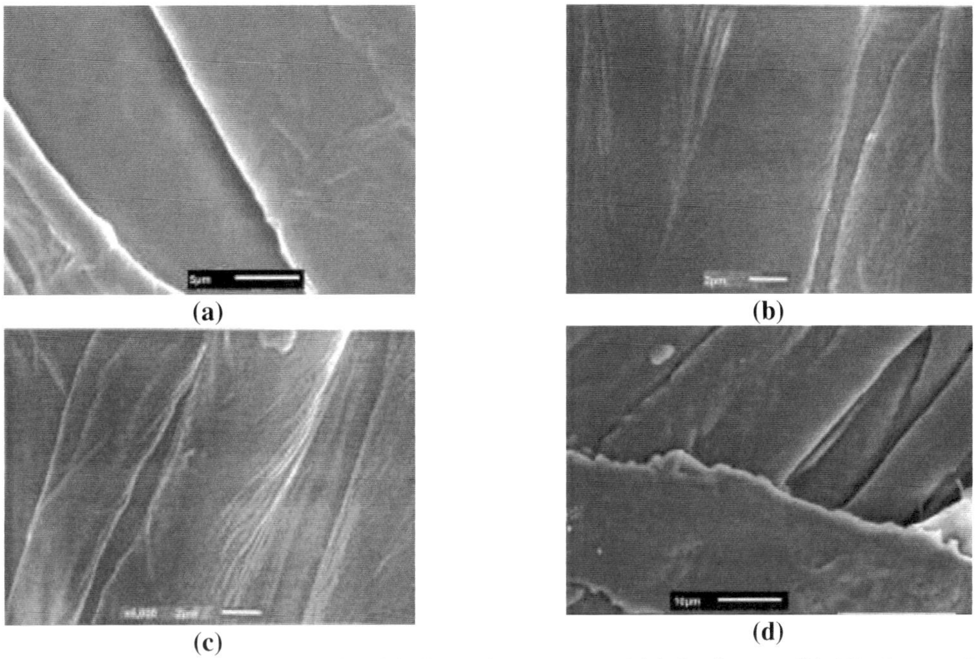

Figure 9. SEM image of cotton fabric with time of treatment of:(**a**) 0 minutes; (**b**) 15 minutes; (**c**) 60 minutes; (**d**) 90 minutes.

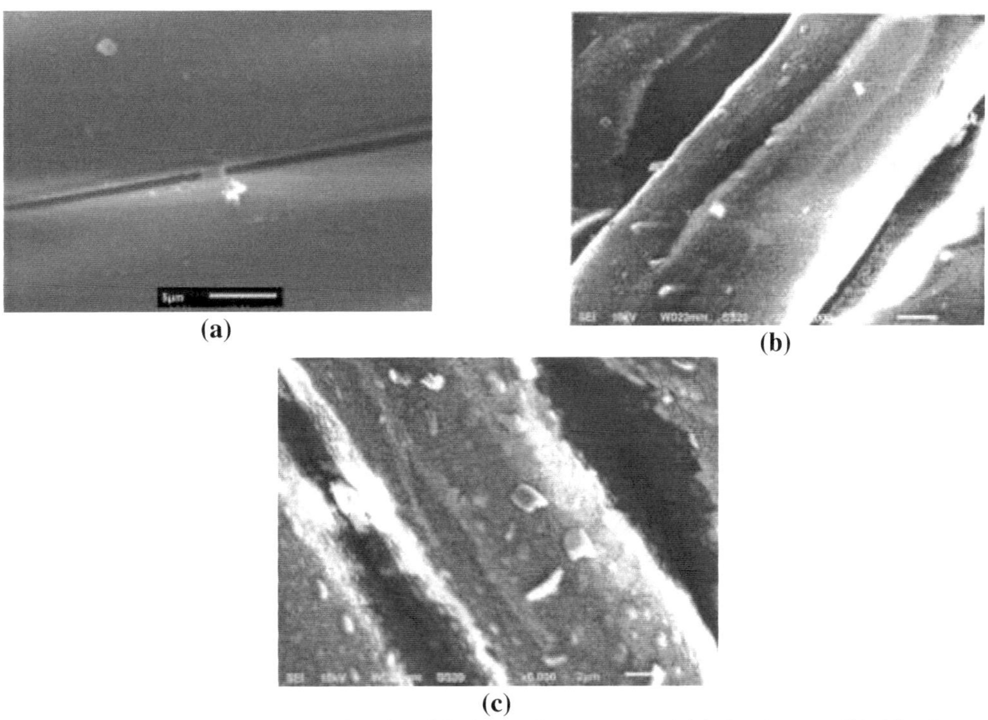

Figure 10. SEM image of polyester fabric with time of treatment of:(**a**) 0 minutes; (**b**) 15 minutes; (**c**) 60 minutes.

In a study on cotton fabric by using a non-thermal plasma and atmospheric air sources, Pandiyaraj and Selvarajan found that micro roughness on the surface of the treated fabric is higher than untreated fabric. In fact they also reported that there were fractures and holes in cotton fibers due to plasma treatment. These surface changes may provide a new pathway for water to enter the fibers and thus will increase hydrophilicity of the materials [16]. Microcracks found on the treated fiber surface indicates that the etching process occurs due to direct bombardment of the active species to the fiber surface [10, 17].

Plasma treatment has generated reactive groups and radicals on the treated-fabric's surface. In polyester fabrics, plasma treatment can oxidize the polyester surface by break the ester bonds and generate radicals. These radicals can react with gas produced and creates polar groups such as hydroxyl, carbonyl, and carboxyl group (or C-O, C=O, C=N, C≡N and N-C-O). Based on the carbon and oxygen compositions determined by XPS, Zhang et al. has reported the decrease in C composition level (from 81.77 to 71,45%) and the increase in O level (from 15.23 to 25.36) on the polyester surface after being treated by plasma, compared to before treated [18]. It expected that C-C bond on the polyester fiber surface has been broken by plasma treatment, then the C=O, C-OH and COOH generated by plasma to form the oxygen-containing polar groups [19]. The enhancement of hydrophylicity is caused by the increased polar groups formed on the fabric surface due to plasma treatment [20]. In addition to generates functional groups, an increase in surface roughness occurred due to the removal of material [21].

4. Conclusions

Corona glow discharge plasma treatment has caused the changes in wetting properties of polyester and cotton fabric. Some conclusions obtained from this study are:

Wetting properties increase with increasing time of treatment, both through wetting time and capillary absorption tests. Time of treatment of 15 minutes provides the optimum changes in wetting properties in this study.

The changes of wetting properties affected by the distance between electrodes, and the distance between electrodes of 2 cm provides the optimum enhancement of wetting properties, both for cotton and polyester fabrics.

One of the factors that caused changes in wetting properties is the morphological changes on fabric's surface. It's found that the enhancement of surface roughness provides enhancement of wetting properties, while the surface roughness increase with increasing time of treatment.

In addition, the presence of surface radical has also caused the changes of wetting properties due to plasma treatment. It was found that the surface radicals density increase with increasing time of treatment and then causing the enhancement of wetting properties. Beside it, it also found that the surface radical density decrease with increasing distance between electrodes.

The changes of wetting properties were caused by free radical reaction on the fabric's surface, the longer time of treatment give the greater radical density and causing the decrease of wetting time on fabrics.

5. References

[1] Carneiro N, Souto A P, Silva E, Marimba A, Tena B, Ferreira H and Magalhaes V 2001 *Color Technol.* **117** 298

[2] Mori M and Inagaki N 2006 Relationship between anti-felting properties and physicochemical properties of wool fibers treated with Ar-plasma *Textile Res J.* **79** 687–694

[3] Canal C, Erra P, Molina R and Bertra´n E 2007 Regulation of surface hydrophilicity of plasma treated wool fabrics *Textile Res J.* **77** 559–564

[4] Sun D and Stylios G K 2005 Investigating the plasma modification of natural fiber fabrics-the effect on fabric surface and mechanical properties *Textile Res J.* **75** 639–644

[5] Yoldas S and Mehmet S 2010 Effect of the low radio frequency oxygen plasma treatment of jute fiber on mechanical properties of jute fiber/poliester composite *Fiber dan Polymers* **8** 1159-1164

[6] Wong W, Chan K, Yeung K W and Lau K S 2000 Pulsed UV Laser and Low Temperature Plasma Modification on Poliester Microfibre Effect on the Dyeing Properties, *J. Textile Engineering* **46** (2) 32-41

[7] Liu Y C, Xiong Y and Lu D N 2006 Surface Characteristics and antistatic mechanism of plasma-treated acrylic fibers *Appl. Surface Science* **252** 2960-66

[8] Rauscher H 2010 *Plasma Technology and Hyperfunctional Surface* (Wiley-VCH Verloag GmbH & Co. Weinheim)

[9] Nur M, Bonifaci N and Denat A 1997 Non-Thermal Electron Mobility in High Density Gaseous Nitrogen and Argon in Divergent Electric Field In *XXIII International Conference on Phenomena in Ionized Gases* **4** 12–13

[10] Kan C W, Lam C F, Chan C K and Ng S P 2014 Using atmospheric pressure plasma treatment for treating grey cotton fabric *Carbohydrate Polymers* **102** 167– 173

[11] Cheng C, Liye Z and Zhan R J 2006 Surface Modification of Polymer Fibre by the New Atmospheric Pressure Cold Plasma Jet *Surface and Coatings Technology* **200** (24) 6659–65

[12] Wang L, Xiang Z Q, Bai Y L and Long J J 2013 A plasma adided process for greycotton fabric pretreatment *J. Cleaner Production* **54** 323–331

[13] Ferrero F 2003 Wettability Measurements on Plasma Treated Synthetic Fabrics by Capillary Rise Method *Polymer Testing* **22** (5) 571–78

[14] Widodo M, El-Shafei M and Hauser P J 2012 Surface Nanostructuring of Kevlar Fibers by Atmospheric Pressure Plasma-Induced Graft Polymerization for Multifunctional Protective Clothing *J. Polymer Science* **60** (16) 1165–72

[15] Poncin-Epaillard F, Chevet B and Brosse J 1994 Study of an Aramid Surface Reactivity: Modification with a Cold Plasma or an Electron Beam Followed by a Postgrafting Reaction, *J. Appl. Polymer Science* **52** 1047–61

[16] Pandiyaraj, Navaneetha K and Selvarajan V 2008 Non-Thermal Plasma Treatment for Hydrophilicity Improvement of Grey Cotton Fabrics *J. Materials Processing Technology* **199** (1) 130–39

[17] Kan C W and Yuen C W M 2012 Effect of atmospheric pressure plasma treatmenton the desizing and subsequent colour fading process of cotton denim fabric *Coloration Technology* **128**(5) 356–363

[18] Zhang C, Zhao M, Wang L, Qu L and Men Y 2017 Surface modification of polyester fabrics by atmospheric-pressure air/He plasma for color strength and adhesion enhancement *Appl. Surface Science* **400** 304–311

[19] Sanaee Z, Mohajerzadeh S, Zand K, Gard F S and Pajouhi H 2011 Minimizingpermeability of PET substrates using oxygen plasma treatment *Appl. Surf. Sci.* **257** 2218–2225

[20] Kan C W, Yuen C W M and Tsoi W Y 2011 Using atmospheric pressure plasmafor enhancing the deposition of printing paste on cotton fabric for digital ink-jetprinting *Cellulose* **18** (3) 827–839

[21] Molina J, Fernandez J, Fernandes M, Souto A P, Esteves M F, Bonastre J and Cases F 2015 Plasma Treatment of Polyester Fabrics to Increase the Adhesion of Reduced Graphene Oxide, *Synthetic Metals* **202** 110–22

Acknowledgments
Ade Ika Susan, S.Si would like to acknowledge to the Center For Plasma Research (CPR) Diponegoro University Indonesia which supported her all of the facilitiesfor this Research.
Conflict of Interest : The authors have no conflicts of interest to report in regard to this manuscript.

IOP Conf. Series: Materials Science and Engineering **214** (2017) 012032 doi:10.1088/1757-899X/214/1/012032

Synthesis and Characterization of ZSM-5 Catalyst at Different Temperatures

W Widayat[1,2,*] and A N Annisa[2]

[1]Laboratory of Advanced Material Center of Research and Service Diponegoro University

[2]Department of Chemical Engineering Diponegoro University

Jl. Prof. Soedarto SH, Tembalang, Semarang, 50275

[*]Email. widayat@live.undip.ac.id

Abstract. ZSM-5 is a synthetic zeolite which contains silica (Si) and alumina (Al) with the ratio of silica greater than the alumina. The catalyst is known as ZSM-5 because it has a pore diameter of 5 Å (angstroms) and it has more than five Si/Al ratios. The catalyst has a high production cost and a complex production process which is highly influenced by time and temperature. Based on that fact, the focus of this research is studying the production process of ZSM-5 and investigating the effect of time and temperature on the crystallinity and the morphology of ZSM-5. In this research, Latourette *et al.* method was used, followed by the calcination at different temperatures (500 °C, 600 °C, 700 °C, 800 °C) in 5 and 7 hours. Subsequently, ZSM-5 was characterized using XRD (X-ray Diffraction) and SEM (Scanning Electron Microscope). The XRD analysis shows that the peak of standard ZSM-5 synthesis is at 2 theta degree 23. Prior to the calcination, ZSM-5 has an average crystallinity of 37.7337%. However, after the calcination, the crystallinity of the catalyst increased as the temperature rises. At 500 °C, 600 °C, 700 °C , and 800 °C the crystallinity increases to 39.1959%, 44.0927%, 42.9425%, and 44.8806%, respectively. Meanwhile, calcination time affects the XRD peak intensities by increasing the ZSM-5 crystal cores. The longer the calcination time, the higher the peak intensity is.

1. Introduction

Cracking is a process that is widely used in research and industry, especially in hydrocarbon-related ones. However, the hydrocracking process traditionally requires high operating temperatures and can cause environmental problems. Therefore, the use of transition metals, silica, zeolite-modified zeolite, mesoporous mixed oxide catalyst, and metal organic frameworks, was developed in the catalytic cracking process [2]. The catalytic cracking process can break down complex hydrocarbons to simpler ones at low temperature and pressure [1]. Catalytic cracking is widely used to produce biofuel from vegetable oils [3].

A catalyst that is widely used in the catalytic cracking process is zeolite [1]. Zeolite is the best catalyst for oil cracking process because it has a high thermal stability, has microcrystalline structure, and able to concentrate the reactants that are contained in the pores [2]. In addition, the aluminosilicate framework of zeolite provides high cation-exchange capacity, high adsorption, and has hydration-dehydration property. However, the use of zeolite as catalysts is limited because it has narrow pores. Due to this fact, the reactant components which have a large molecular size will have difficulty in mass transfer process[4-5]. Furthermore, the zeolite is a non-metal mining product that has limited availability. These drawbacks encourage researchers to study about the synthesis of zeolite or ZSM-5.

Synthesis of ZSM-5 catalyst (Zeolites Socony Mobil-5) was first performed by Argauer and Landolt which is patented by Mobil Oil Corporation in 1972 [8]. ZSM-5 is a synthetic zeolite which contains silica (Si) and alumina (Al) with the ratio of silica greater than the alumina. The existence of aluminum and Si/Al ratio of ZSM-5 catalyst causes ZSM-5 to have acidic property [9]. This property makes the catalyst has the function as a catalyst support, adsorbent, etc. [10]. It has pore diameter range of 5 Å (angstroms) and it has more than five Si/Al ratio. ZSM-5 has a medium pore size (0.54 nm × 0:56 nm) and has two intersecting three-dimensional channels with one straight parallel channel and the other running parallel defined by sinusoidal 10-membered ring openings of 5.3 Å × 5.6 Å and 5.1 Å × 5.5 Å [2, 6-7].

ZSM-5 catalyst can be used to manufacture a variety of products through a catalytic cracking process. Types of products that can be produced by ZSM-5 catalyst are influenced by physical properties of the catalytic material and constituent of ZSM-5 catalyst. There are several parameters that can affect the properties of ZSM-5 catalysts. These parameters are the silica source, the alumina source, template, and crystallization conditions [6]. To understand these parameters, the characterization of the catalyst by means of morphology, crystallinity, and catalyst framework structure analysis is needed. Analysis of the catalyst morphology can be done by using a Scanning Electron Microscope (SEM), while the crystallinity analysis and the structure of the catalyst can be analyzed by using XRD.

In this research, synthesis of ZSM-5 catalyst with Latourette *et al.* method [11] at the various temperatures, calcination time, and catalyst supports was conducted. After that, the produced catalysts were analyzed by using SEM and XRD to understand their characteristics.

2. Methods

2.1. Materials

Materials that are used in this research are: Aluminum Sulfate ($Al_2(SO_4)_3.18H_2O$), Cobalt Nitrate ($Co(NO_3)_2.6H_2O$), Nickel Nitrate ($Ni(NO_3)_2.6H_2O$), Sulfuric Acid 98% (H_2SO_4), Sodium Silicate and Sodium Hydroxide (both were supplied by Merck Ltd.), distilled water (obtained from Laboratory of Membrane Research Centre Diponegoro University), Methanol, and Glycerol (both were supplied by Indrasari Chemical Store).

2.2. Catalyst synthesis

Catalyst was synthesized according to Latourette *et al.* [11] method. Firstly, Solution A as the source of alumina was prepared. The solution contains 26.7 grams of Aluminum Sulfate ($Al_2(SO_4)_3.18H_2O$), 56 grams of 98% Sulfuric Acid (H_2SO_4), and 15 grams of distilled water. After that, Solution B as the source of silica was prepared. The solution contains 56 grams of Sodium Silicate and the same amount of 40% Sodium Hydroxide (NaOH). Subsequently, Solution A and Solution B were mixed. After that, the mixture was homogenized by a homogenizer at a speed of 12000 rpm. After obtaining a homogeneous mixture, crystallization is done by using an autoclave reactor at 200 °C in 6 and 8 hours. After the crystallization, ZSM-5 catalysts as products were washed with distilled water and then dried overnight.

The ZSM-5 products crystals were analyzed by using XRD to determine their degree of crystallinity. After that, ZSM-5 undergone calcination process at various temperatures of 500 °C, 600 °C, 700 °C and 800 °C in 5 and 7 hours. After this process, the product was analyzed using XRD to determine their crystallinity and their catalysts structure. SEM analysis was also conducted to find out the morphology of the catalysts.

2.3. Catalyst impregnation

ZSM-5 Catalyst was impregnated using a Nickel and Cobalt metal ions. 0.5 grams of Nitrate Cobalt ($Co(NO_3)_2.6H_2O$) and Nickel Nitrate ($Ni(NO_3)_2.6H_2O$) was dissolved in 25 ml of distilled water. Then, the solution was mixed with 5 grams of ZSM-5 catalyst, heated to 60 °C, and stirred with a magnetic stirrer at a speed of 600 rpm for 4 hours. After that, the mixture was left to stand for 24 hours, and then filtered and dried by using an oven at 110 °C for 1 hour. Finally, the mixture was undergone calcination process at 600 °C for 6 hours.

2.4. Catalyst characteristic

The ZSM-5 catalyst products were characterized with morphology and crystallinity analysis. The analysis of crystallography conducted by using XRD-7000S Shimadzu with Copper X-ray tube target, 30 kV voltage, 30 mA current, and Kα radiation. XRD data was analyzed with PCXRD software. Morphology of catalyst data was analyzed with JEOL PC Scanning Electron Microscope (PCSEM) model JSM-6510LA with x5000 magnification. These analysis processes were conducted in Center of Research and Service Diponegoro University (CORES DU).

3. Results and discussions

3.1. The effect of crystallization time

In the production of the ZSM-5 catalyst variations of crystallization time of 6 and 8 hours was performed. Crystallization time is one parameter that plays an important role on the properties of the catalyst since the catalyst products were harvested at different times [12]. During the variation of crystallization time, the crystallization takes place at temperatures of 200 °C.

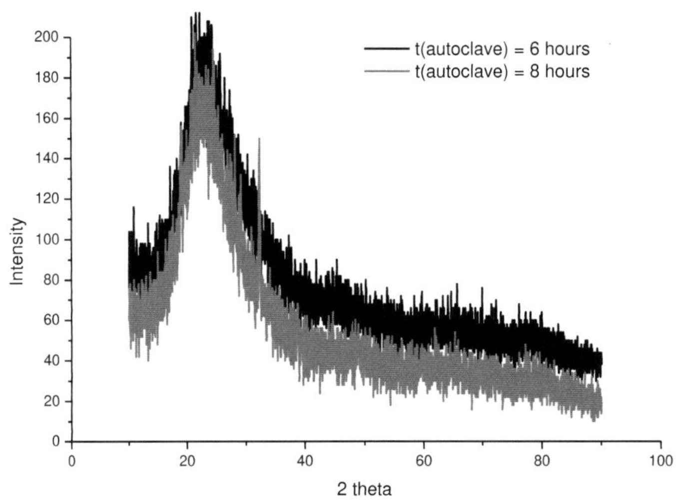

Figure 1.XRD pattern of ZSM-5 Synthesis on difference crystallization times.

Figure 1 shows the results of the XRD analysis of the ZSM-5 catalyst in the autoclave for 6 and 8 hours. The graph shows that the longer the crystallization time, the more compact the catalyst structure is. This fact can be seen from the intensity at 2 theta degree 21 and 23. Furthermore, the long crystallization time also affects the degree of crystallinity.

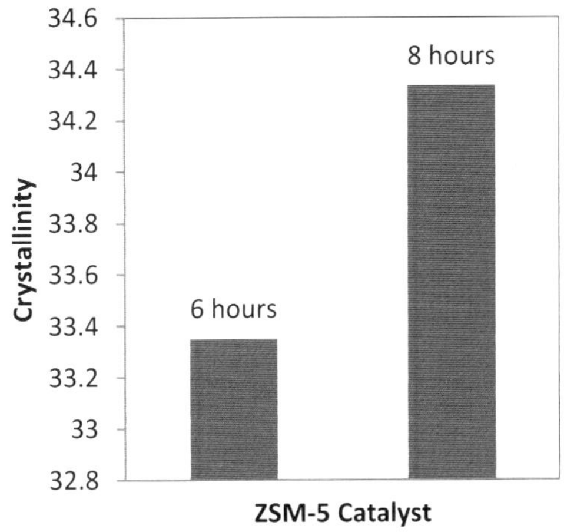

Figure 2.ZSM-5 crystallinity on difference crystallization time.

Figure 2 shows that the longer time of crystallization, the higher the catalyst crystallinity is. At 6 hours crystallization time, the ZSM-5 catalyst has a crystallinity of 33.35% whilst at 8 hours it has a crystallinity of 34.33%. Nevertheless, their degree of crystallinity is still low, consequently after the crystallization process is done the calcination process is conducted.

3.2. The effect of calcination temperature

To improve their crystallinity, the ZSM-5 catalysts were undergone calcination process. Calcination process was carried out at various temperatures of 500 °C, 600 °C, 700 °C and 800 °C for 5 hours. After calcination process, ZSM-5 catalysts were analyzed by using XRD to determine their degree of crystallinity. Then, the structure was being compared with the commercial one (standard).

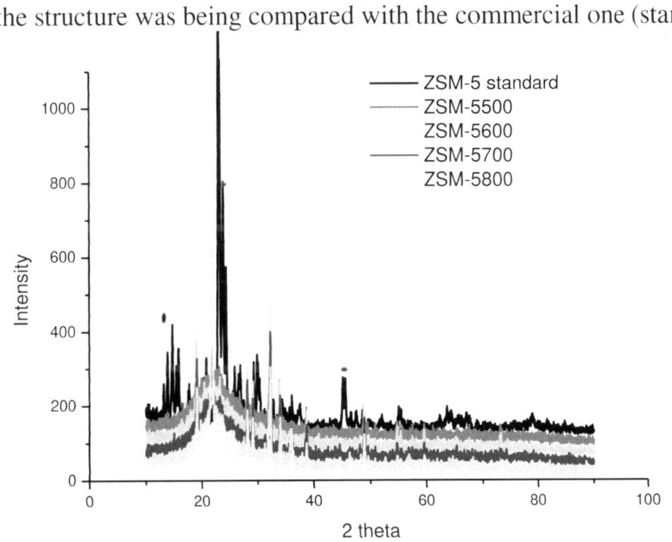

Figure 3.XRD pattern of synthesized ZSM-5 at different temperatures and the comparison to standard ZSM-5.

Figure 3 shows synthesized and standard ZSM-5 diffractograms which were undergone calcination at various temperatures. In general, all of the diffractograms shows a similar result. The presence of aluminum and silica in a commercial ZSM-5 is shown in 2 theta degrees 13°, 23° and 45° [10, 13]. However, synthesized ZSM-5 that has the peak at 2 theta degrees 13°, 23° and 45°, has a lower intensity compared to commercial ZSM-5. Furthermore, synthesized ZSM-5 also has other peaks. This is because the degree of crystallinity of synthesized ZSM-5 is quite low which made the catalyst has an amorphous structure [14]. The SEM analysis results in Figure 4 reveals that the morphology of synthesized ZSM-5 is still fairly amorphous. This is shown by the surface structure of the catalyst which clumped each other, whereas on commercial ZSM-5 has a separated structure due to its fairly high crystallinity [15].

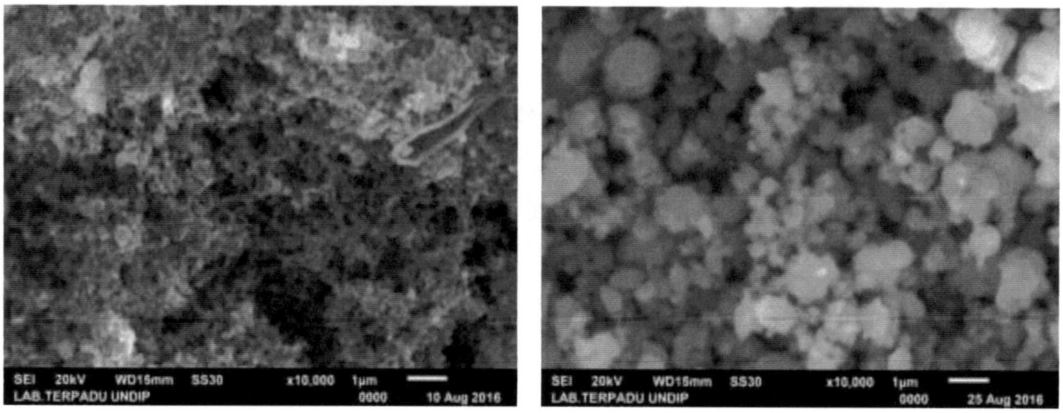

(a) **(b)**

Figure 4. Photographs of ZSM-5 catalyst (a. Synthesized; b. Standard ZSM-5).

Compared to the crystallinity of commercial ZSM-5, synthesized ZSM-5 has a fairly low crystallinity. Figure 5 shows the crystallinity of synthesized ZSM-5 calcination are 39.2% at 500 °C, 44.1% at 600 °C, 42.9% at 700 °C and 44.9% at 800 °C. In the other hand, commercial ZSM-5 has a crystallinity of 76.1%.

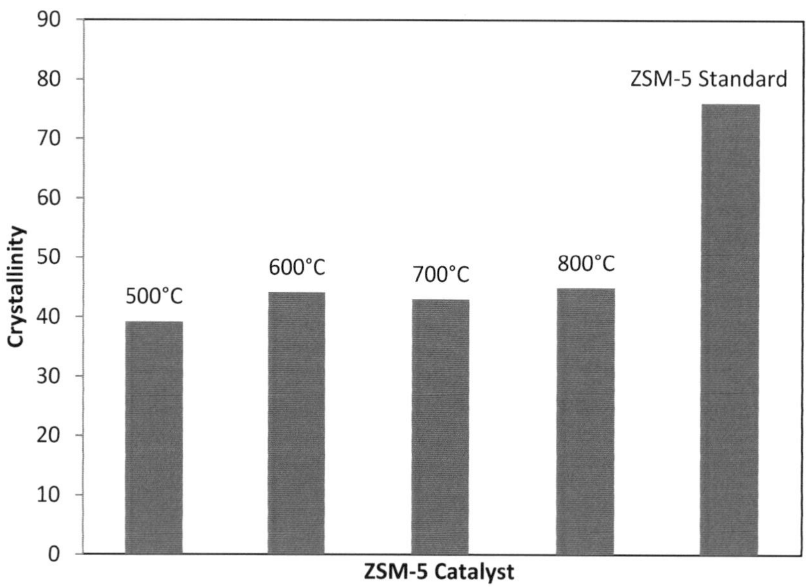

Figure 5.ZSM-5 Crystallinity on difference calcination temperature.

3.3. The effect of calcination time

ZSM-5 catalyst products were not only undergone calcination at various temperatures but also in various calcination times. The variations of calcination times were 5 and 7 hours with calcination temperature of 700 °C. Then, the synthesized ZSM-5 was analyzed by using XRD to determine the degree of crystallinity and its suitability to the standard ZSM-5. Figure 6 shows that the synthesized ZSM-5 with calcination time of 5 hours has a better intensity than the 7 hours. This also can be seen from the degree of crystallinity provided in Figure 7. The synthesized ZSM-5 with calcination time of

5 hours has a crystallinity of 42.9%, while synthesized ZSM-5 with calcination time of 7 hours has a crystallinity of 40.2%. It can be concluded that the optimal calcination time is 5 hours. In conclusions, the variables that affect the properties of the catalyst are the crystallization time and temperature of calcination.

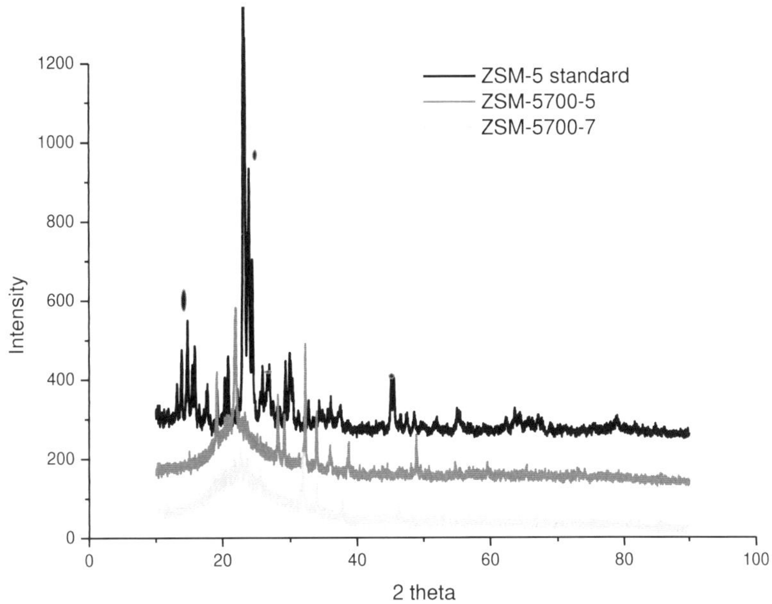

Figure 6.XRD pattern of synthesized ZSM-5 in various calcination times and standard ZSM-5.

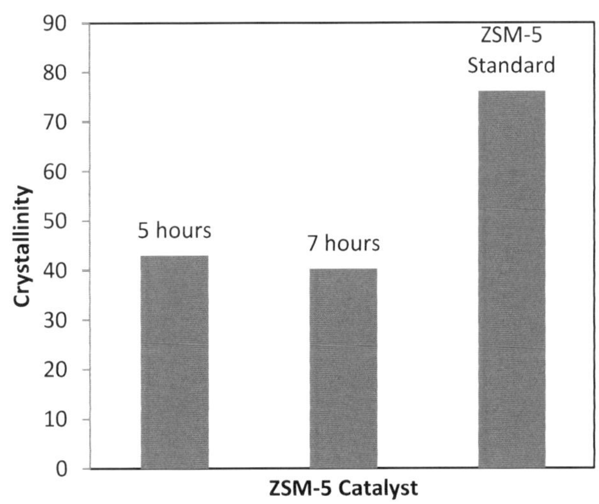

Figure 7.ZSM-5 crystallinity in difference calcination times.

4. Conclusions

It can be concluded that crystallization time affects the physical properties of ZSM-5 catalyst. The longer the time of crystallization, the more compact the physical properties are. This is because they have better crystallinity. In addition, the calcination temperature on ZSM-5 catalyst production also affects the crystallinity of the catalyst. The higher the temperature is, the better the catalyst's crystallinity. Besides, the morphological structure of the catalyst is also improved by the high calcination temperature.

5. Acknowledgements

This research was supported by Diponegoro University with Riset Pengembangan dan Penerapan (RPP) Grant with the letter of assignment No. 1051-1-7/UN.7.5.1/PG/2016. Authors also would like to thank Center of Research and Service Analysis Diponegoro University (CORES DU) for XRD analysis, Scanning Electron Microscope analysis, Gas Chromatography, and Mass Spectrophotometry analysis.

References

[1] Widayat W, Wicaksono AR,Firdaus LH and Okvitarini N 2016 *IOP Conference Series: Materials Science and Engineering***107** 012044
[2] Vinodh R, Babu CM, Abidov A, Ravikumar R, Peng MM, Palanichamy M, Cha WS and Jang HT 2016*International Journal of u and e- Service Science and Technology***9** **(5)**383-394
[3] Charusiri W and Vitidsant T 2005 *Energy & Fuels* **19** 1783-1789
[4] Goncalves, Lansoni M, Dimitrov LD, Jorda˜o MH, Wallau M and Urquieta-Gonza`lez EA 2008 *Catalysis Today,* **133-135** 69-79
[5] Abelló A, Bonilla A and Pérez-Ramírez 2009 *Applied Catalysis A: General* **364** 191-198
[6] Li J, Liu S, Zhang H, Lu E, Ren P and Ren J 2016 *Chinese Journal of Catalysis* **37** 308-315
[7] Wei Y, de Jongh PE, Bonati MLM, Law DJ, Sunley GJ and de Jong KP 2015 *Applied Catalysis A: General***504** 211-219
[8] Abbasian S and Taghizadeh M 2014 *International Journal Nanosci. Nanotech.* **10 (3)** 171-180
[9] Lu R, Tangho H, Wang Q and Xiang S 2003 *Journal of Natural Gas Chemistry* **12** 56-62
[10] Nezamzadeh-Ejhieh A and Shams-Ghahfarokhi Z 2013 *Journal of Chemistry* **2013** 1-11
[11] Latourette B, Le Raincy and Magnier C 1990 *US Patent 4,891,199*
[12] Kokotailo GT and Fyfe CA 1995 *The Rigaku Journal* **12 (1)**
[13] Tao L, Li GS, Yin SF, Ou-Yang Q, Luo SL, Zhou XP and Au CT 2011 *Reac. Kinet. Mech. Cat* **103** 191-207
[14] Phung TK, Radikapratama R, Garbarino G, Lagazzo A, Riani P and Busca G 2015 *Fuel Processing Technology* **137** 290-297
[15] Xu L, Wu S, Guan J, Wang H, Ma Y, Song K, Xu H, Xing H, Xu C, Wang Z and Kan Q 2008 *Catalysis Communications* **9** 1272-1276

IOP Conf. Series: Materials Science and Engineering **214** (2017) 012033 doi:10.1088/1757-899X/214/1/012033

Zinc Oxide/TiO_2 Bilayer Heterojunction as a Working Electrode in Quasi Solid Dye Sensitized Solar Cells

A Aprilia[1*], L Safriani[1], Wa Ode S Arsyad[2], N Syakir[1], T Susilawati[1], C Mulyana[1] Fitrilawati[1], R Hidayat[3]

[1] Department of Physics, Faculty of Mathematics and Natural Sciences, Universitas Padjadjaran, JL. Raya Jatinangor KM 21. Kabupaten Sumedang, 45363, West Java, Indonesia.

[2] Department of Physics, Faculty of Mathematics and Natural Sciences, Universitas Halu Oleo, Anduonohu 93232, Kendari, Sulawesi Tenggara, Indonesia.

[3] Department of Physics, Faculty of Mathematics and Natural Sciences, Institut Teknologi Bandung, Jl. Ganesha 10, Bandung 40132, West Java, Indonesia.

*a.aprilia@phys.unpad.ac.id

Abstract. Bilayer heterojunction of aluminium doped zinc oxide (AZO) and titanium dioxide (TiO_2) mesoporous has been successfully deposited on fluorine tin oxide (FTO) substrate as working electrode in dye sensitized solar cell. This layer was used as working electrode in quasi solid dye sensitized solar cell. The solar cell structure is FTO/ZnO/TiO_2/PGE/Pt/FTO using polymer gel electrolyte (PGE). In polymer gel electrolyte system, hybrid copolymer based on poly-TMSPMA (3-methoxysilyl propyl methacrylate) was used as a matrix to trap ionic liquid. An addition of aluminum as atom dopant also studied to observe the physical properties changes of photoanode related to solar cell performance. $AlCl_3$ was used as dopant material with the concentrations at 0.5 weight % and 1.0 weight% of zinc acetate dehydrate as raw material. Based on our previous result, the existence of Al dopant would decrease the surface roughness of ZnO layer, reduce the grain size of ZnO particles, transmittance at visible light increase and also change the charge carrier density. Nevertheless, the highest efficiency was achieved for undoped ZnO/TiO_2 photoanode (η=0.67%). Based on current-voltage measurement data analysis (using diode model equation) the ideality factor (n) of device using undoped ZnO was smaller (n=2.96) than AZO 0.5 wt% and 1.0 wt% (n=~4), indicate better quality of undoped ZnO/TiO_2 interfaces rather than AZO/TiO_2 .

1. Introduction

Zinc oxide is a II-IV semiconductor material which is widely used in variety of optoelectronics devices, including ultraviolet photodetector, gas sensing, field effect transistor, Schottky diodes, light emitting device arrays and also in emerging photovoltaic (PV) devices [1, 2]. This type metal oxide has a high energy exciton binding (~ 60 meV), direct wide band gap (~ 3.34 eV), exhibit near UV emission, high electron mobility (250 cm^2 V/s), and also shows a transparent conductivity and piezoelectricity [2,3]. Many researchers have been studied the effect of ZnO morphologies correlated with some applications, particularly in solar cell devices [4, 5]. One of the type emerging PV device which is used this kind of metal oxide is dye sensitized solar cell (dssc). Many morphology of ZnO structures have been developed for dssc application, for the examples are nanosphere, nanowire, nanorod, nanobelt, and nanocone, in order to enhance device performance [6,7].

In dye sensitized solar cell (dssc) system, photoanode plays an important role as an electron transport material to accept the photogenerated electrons from the excited state of the dye. Considering the role of photoanode, several requirements need to be possessed by metal oxides (as photoanodes on dssc) i.e. highly transparent, possess high surface area to maximize adsorption of dye molecules, has high electron mobility for efficient electron transport and be inert to the redox electrolyte to reduce electron recombination rate [8]. Zinc oxide (ZnO) and titanium dioxide (TiO_2) are commonly - used as photoanode materials in solar cell devices. S.A Mozaffari et.al. have been investigated the use both of ZnO nanoparticles and TiO_2 as entity of photoanode system. ZnO nanostructure could be acted as carrier trap and suppress the recombination process of the injected photoelectrons that produced by space charge on individual TiO_2 nanoparticle [9]. The main factor of reducing the solar cell performance by utilizing the TiO_2 particle as individual photoanodes, is charge recombination of photogenerated excitons. Therefore an addition of ZnO nanostructures was expected could be anticipate charge recombination process due to ZnO high electron mobility. Besides that, employing ZnO nanostructure as photoanode generally produce high short circuit current (I_{sc}) regards to an efficiently in charge transport [10].

The series resistance characteristic on dye sensitized solar cell was attributed by many factors such as transport resistance (R_t), ohmic series resistance (R_{OS}), diffusion resistance (R_d), and charge transfer resistances (R_{ct}). R_{os} is correspond to the resistance of FTO sheet and redox electrolyte solution. R_d is the resistance related to electrolyte properties and R_{ct} is the resistance that induced from interfacing between photoanode and electrolyte [11]. The series resistance can be simply calculated using diode model equation for solar cell as shown in equation (1) [12]:

$$I_{cell} = I_{ph} - I_0 . \left[exp\left(\frac{V+I_{cell}R_S}{nV_t}\right) - 1\right] - \left[\frac{V+I_{cell}R_S}{R_{sh}}\right] \quad (1)$$

where I_{ph} is photogenerated current, I_0 is dark saturation current, n is ideality factor and V_t is voltage of Boltzmann factor at 298 K ($V_t = \frac{k_BT}{q} \cong 25.85\ mV$). R_{sh} is shunt resistance correlated to leakage current across the cell edge, and the presence or the impurities in the junction between dye and photoanode system. The ideality factor (n) can be estimated from the diode model equation (1) for open circuit condition ($V = V_{oc}$ and $I_{cell} = 0$),

$$V_{oc} = nV_t \ln\left(\frac{I_{ph}+I_0}{I_0}\right) \cong nV_t \ln\left(\frac{I_{ph}}{I_0}\right) \quad (2)$$

using $I_{ph} + I_0 \approx I_{ph}$ because of $I_0 \ll I_{ph}$.

In this work, combination between ZnO (with/without aluminum doped) and TiO_2 bilayer heterojunction was used as working electrode (photoanode) in quasi solid dssc. Poly-TMSPMA (3-methoxysilyl propyl methacrylate) was used as gel matrix for ionic liquid as polymer gel electrolyte system (PGE). The influence of different aluminum content was studied by investigating the absorption spectra of photoanode, surfaces morphology and measuring current density-voltage curve (*J-V*) of the cell. An ideality factor (n), series resistance (R_s) and shunt resistance (R_{sh}) also derived from J-V curve using diode model equation (equation 2).

2. Experimental Methods

2.1. Photoanode preparation and DSSC *assembly*

In this work, ZnO layer was prepared by sol-gel method with zinc acetate dehydrate ($Zn(CH_3COO)_2.2H_2O$) (Merck) as raw material. 2-methoxyethanol anhydrous, aluminum chloride ($AlCl_3$) and diethanolamine were used as solvent, dopant material and stabilizer respectively. All materials were purchased from Sigma-Aldrich without further purifications. The aluminum doped ZnO (AZO layer) preparation procedure was followed our previous research [13]. $AlCl_3$ (0.5 weight % and 1.0 weight % respect to the 0.5 M of ZnO precursor) was added into precursor solution and stirred until clear solution was achieved. The precursor solution was spin coated onto transparent conductive oxide substrate (FTO ~ 7 Ohm/square). The sample was subsequently subjected to heat treatment up to 500°C using a hotplate. Afterward, one layer of TiO_2 mesoporous (Ti-Nanoxide T/SP; with anatase particle size about 15-20 nm purchased from Solaronix-SA) was deposited by screen printing technique and gradually annealed until 500°C for 30 minutes on hotplate. The photoanodes was immersed in dye solution containing of 535-bisTBA (Solaronix-SA) and chenodeoxolydacid (1:10) overnight. Before used, the FTO/AZO/TiO_2/Ru-dye was washed several times by acetonitrile and ethanol to remove the residual of dye molecules.

The dye sensitized solar cell was built by employing the FTO/AZO/TiO_2/Ru-dye as active layer and drilled platinum coated FTO substrate as counter electrode (CE). The CE was sandwiched with FTO/AZO/TiO_2/Ru-dye separated by hot melt surylin film (thickness of 25 μm). The polymer gel electrolyte (PGE) was prepared from Poly-TMSPMA gel mixed with 0.2 ml of Mosalyte. The PGE was injected to the hole of Pt/FTO side, and sealed with transparent tape to avoid leaking. The device with AZO/TiO_2 photoanode of 0.5 wt% and 1 wt% aluminum dopant are assigned as AZO 0.5 and AZO 1.0, respectively.

2.2. Sample characterizations

The atomic force spectroscopy (AFM) was used to identify the surface morphology of ZnO with different Al content. Ultra-violet visible spectroscopy (T70+ PG Instrument) was carried out for each layer which is built the photoanode system. Current density voltage (J-V) characteristic was investigated using halogen lamp (power input 30 mW/cm^2) integrated with Yokogawa GS 200 DC voltage-current source and Yokogawa digital multimeter 7555.

3. Results and Discussions

The effect of aluminum doped into ZnO morphology was observed by atomic force microscopy (AFM). Undoped ZnO layer (without aluminum dopant) shows a smallest surface roughness compared to others. Undoped ZnO layer morphology has a mountain like surface structure with approximate height about 13.9 nm (figure 1(a)). An addition of Al dopant increases the surface roughness that attributed to the increment of average particle height (28.78 nm for AZO 0.5 and 24.79 nm for AZO 1.0). The average of particle size also reduce as appeared in figure 1. It was already reported, that aluminum as dopant atom could be reduced the ZnO grain size due to the grain packing density behavior [13, 14].

Figure 2 shows the absorbance spectra of each layer that used as photoanode and also for the heterojunction active layer (AZO/TiO_2/Ru-dye). There is no significant difference in absorbance spectra between ZnO and AZO 0.5. The blue shift on absorbance peak (368 nm to 345 nm) is related to the small change on energy gap ~3.23 eV to ~3.19 eV (determined by Tauc plot). When titanium dioxide was coated onto AZO layer, the absorbance spectra was increased at ultra-violet region attributed to TiO_2 absorbances. The active layer of AZO/TiO_2/Ru-dye absorb at visible light region due to the existence of ruthenium particle (as sensitizer). It was already reported that TiO_2 has large surface area and can enhance dye loading inside the cell [8].

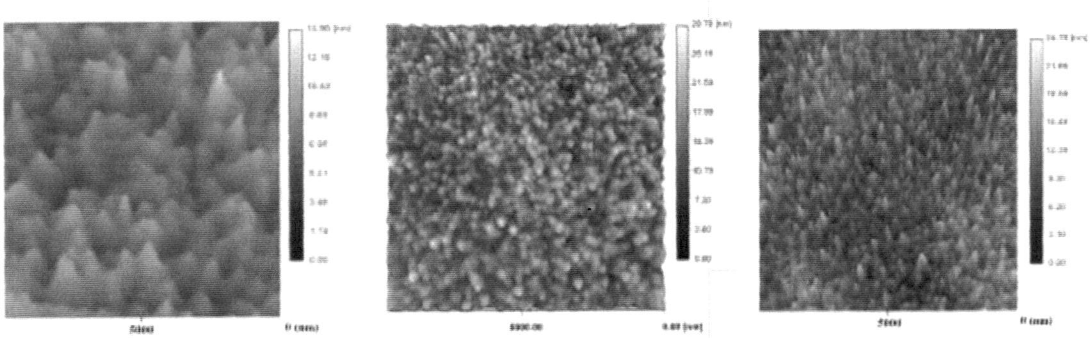

(a) (b) (c)

Figure 1. Atomic force microscopy images of ZnO layer deposited on glass substrate (a) undoped ZnO, (b) 0.5 wt% AZO and (c) 1.0 wt% AZO.

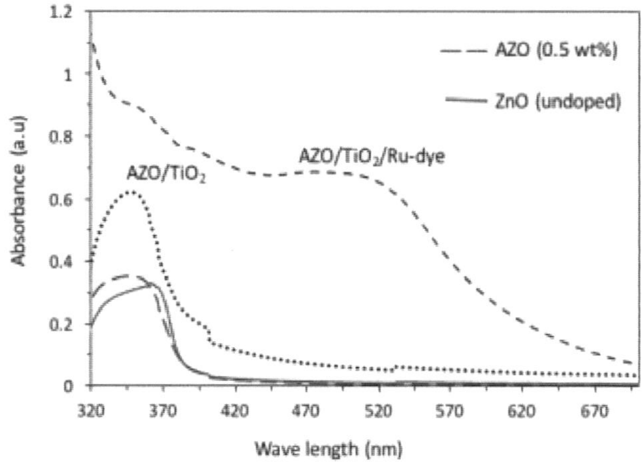

Figure 2. UV-Vis spectra of each layers; thin layer of aluminum doped ZnO (AZO-0.5 wt%) and ZnO, thin film AZO/TiO_2 photoanode before and after immersed in Ru-dye solution.

Figure 3 shows the current density-voltage (*J-V*) characteristic under irradiation by halogen lamp (30 mW/cm^2). The J-V characteristic confirm that utilizing AZO in photoanode system (particularly for 0.5 and 1.0 wt % of Al content) do not show a better result compared with undoped ZnO. The efficiency (η) of the device with AZO 0.5 and AZO 1.0 is 0.38% and 0.47% respectively, while undoped ZnO is 0.68%. Actually, this result was not expected before, considering from our previous research that an addition of small aluminum concentration as doping agent can enhance device efficiency [13]. We assume that lower efficiency when using AZO layer is correlated by AZO/TiO_2 interface quality. Figure 4 represent the illustration of heterojunction between ZnO and TiO_2 based on AFM images.

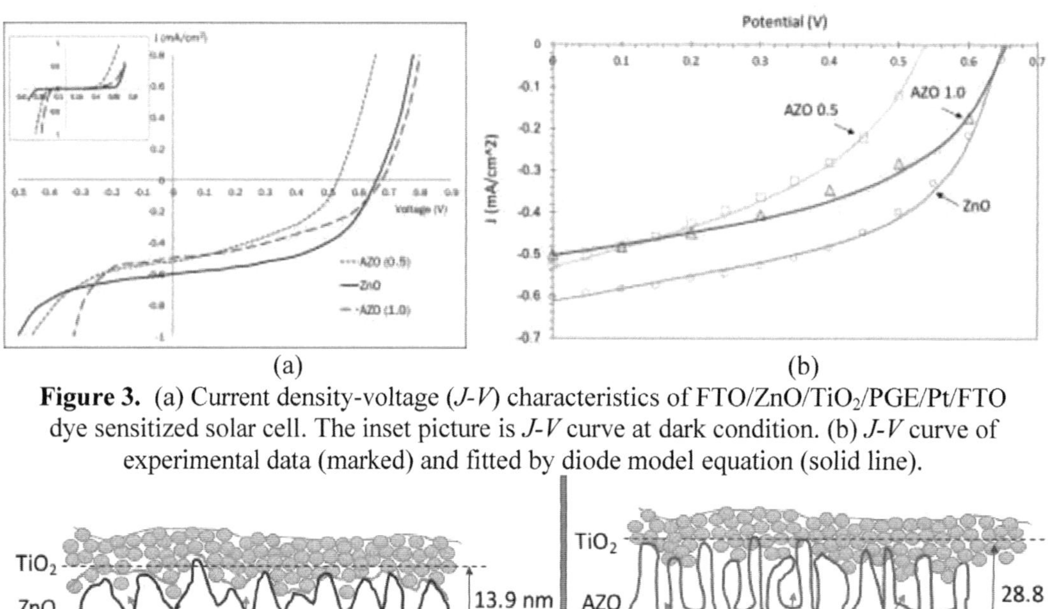

(a) (b)

Figure 3. (a) Current density-voltage (*J-V*) characteristics of FTO/ZnO/TiO_2/PGE/Pt/FTO dye sensitized solar cell. The inset picture is *J-V* curve at dark condition. (b) *J-V* curve of experimental data (marked) and fitted by diode model equation (solid line).

Figure 4. Illustration of ZnO and AZO/TiO_2 heterojunction in photoanode system in this work. AZO layer has relatively small grain size with high altitude (mountain like surface) produces more trap sites at interface area with TiO_2.

Table 1. Parameters determined from current-voltage characteristics

Photoanode	J_0 (mA/cm^2)	J_{sc} (mA/cm^2)	V_{oc} (volt)	n	R_s (Ohm)	R_{sh} (Ohm)	FF (%)	Efficiency (%)
ZnO–Undoped (0%)	1.25 x 10^{-4}	0.61	0.65	2.96	120	8.5 K	51	0.67
AZO (0.5 %)	1.67 x 10^{-3}	0.53	0.53	4	130	19.5 K	41	0.38
AZO (1.0%)	8.3 x 10^{-2}	0.5	0.69	4.17	150	33.5 K	41	0.47

Device parameters were observed and calculated from *J-V* curves are shown in table 1. An ideality factor (n) which is represent the back transfer process (recombination) inside the cell was calculated using equation (2). The series resistance (R_s) and shunt resistance (R_{sh}) were estimated using equation (1). The relation of cell voltage (V_{PE}) under applied bias (V) is [11]

$$V_{PE} = V + I_{cell}.R_s \tag{3}$$

therefore, the equation (1) becomes

$$I_{cell} = I_{ph} - I_0.\left[exp\left(\frac{V_{PE}}{nV_t}\right) - 1\right] - \left[\frac{V_{PE}}{R_{sh}}\right]. \tag{4}$$

The device with undoped ZnO has an ideality factor of n = 2.96, while AZO 0.5 and AZO 1.0 has n = 4 and n = 4.17 respectively. The ideality factor value is representation of junction quality and recombination process inside the cell. In ideally junction, the recombination process represented by the recombination of minority carriers in the quasi-neutral regions, thus the n equal to 1. If the n has a value about 2, it means that other recombination mechanism was occur [11,12]. A high n value not only affect the *FF* (fill factor) but also produce a low V_{oc} in consequently with decreasing device efficiency. Besides that, the *FF* also influenced by incident light power density. However, in this work all samples show a relatively small *FF* perchance contributed by device quality and low irradiation incident light (30 mW/cm^2). H.S Hafez et.al. have reported the effect of illumination intensity to the device parameters (J_{sc}, V_{oc} *and* P_{max}). The change of P_{max} to the power incident light was practically linear, hence the value of internal resistance would not show a significant difference when using the low incident light compared with standard irradiation (100 mW/cm^2) [12]. The series resistance (R_s) for all samples in this work do not show a substantial variance. Nevertheless, the shunt resistance (R_{sh}) of ZnO photoanode has a smallest value than others. According to the diode model equation (equation 4), higher R_{sh} provide a lower leakage photogenerated current and can increase the *FF* value. R_{sh} value between ZnO and AZO seems cannot be comparable because ideality factor (n) between them is quite large. Device with AZO photoanode suspected has a more variety recombination process in comparison with undoped ZnO photoanode. This situation probably correlated with the junction quality in AZO/TiO_2 photoanode. More investigation are needed, in order to study the type recombination inside the cell.

4. Conclusions

Combination between ZnO and (AZO) with titanium dioxide (TiO_2) was investigated, particularly for 0.5 wt% and 1.0 wt% of aluminum respect to ZnO precursor. From AFM measurement, the ZnO and AZO layer have a mountain like surface structure. The existences of aluminum dopant would be decrease the ZnO grain size, meanwhile the height of mountain like structure was increased. UV-vis spectroscopy reveal a blue shift on AZO layer absorbance spectrum due to band gap change and particle size reduction. Moreover, the AZO/TiO_2 interfacing area produce more trap sites leading to bad contact in consequences with high ideality factor (*n*) value and small efficiency. This work emphasizes that the small concentration of Al ($\leq$ 1 wt%) on ZnO layer (relative flat surface) is not proper to use as photoanode in dssc (in combination with TiO_2 mesoporous), however the large concentration (> 1 wt%) of Al dopant still needs to be investigated.

5. References

[1] Gyu-Chul Y, Chunrui Wand Won Il P 2005 *Semicond. Sci. Technol.* **20** S22–S34
[2] Schmidt-Mende L and MacManus-Driscoll J L 2007 *Mater. Today* **10** No. 5 p40-48
[3] Sheng X and Zhong L W 2011 *Nano Res.* **4** issue 11 pp 1013–98
[4] Zi M, Zhu M, Ling C, Haoming W, Xiaopeng Y and Bingqiang C 2014 *Ceram. Int.* **40** 7965-70
[5] Dae-Yon S, Jeong-Hyeok I, Hui-Seon K and Nam-Gyu P 2014 *J. Phys.Chem. C.* **118** 16567-73
[6] Seung H K, Daeho L, Hyun W K, Koo H N, Joon Y Y, Suk J H, Costas P G and Hyung Jin S 2011 *Nano Lett.* **11** 666–71
[7] Lori E G, Matt L, Dawud H T, Max M, Josh G, Gabor S and Peidong Y 2005 *Nano Lett.* **5** 7 1231-36
[8] Aravind K C, Mojtaba A-J, Mohammad K N and Michael G 2014 *ACS Nano* **8** (3) 2261-68
[9] Mozaffari S A, Ranjbar M, Kouhestanian E, Salar Amoli H and Armanmehr M H 2015 *Mater. Sci. Semicond. Process.* **40** 285-92
[10] Eskandari M and Ahmadi V 2015 *Electrocim. Acta* **165** 239-46
[11] Subrata S, Hyun Woo S and Dong Min K 2014 *J. Power Sources* **248** 739-744
[12] Hoda S H, Yahia I S, Sakr G B, Abdel-Mottaleb M S A and Yakuphanoglu F 2012 *Adv. Mat. Corr.* **1** 8-13
[13] Annisa A, Priastuti W, Veinardi S, Herman, Rahmat H, Akihiko F, and Masanori O 2013 *Sol. Energ. Mat. Sol. Cells* **111** 181-88
[14] Jin-Hong L and Byung-Ok P 2003 *Thin Solid Film* **426** 94-99

Acknowledgement

The authors would like to acknowledgment for the financial support from research PUPT (Program Unggulan Perguruan Tinggi), Padjadjaran University, with contract No. 431/UN6.3.1/PL/2016.

Thermophysical Parameters of Organic PCM Coconut Oil from T-History Method and Its Potential as Thermal Energy Storage in Indonesia

Alfriska O. Silalahi,[1] Nissa Sukmawati,[1] I.M. Sutjahja,[1] D. Kurnia,[1] S. Wonorahardjo[2]

[1] Dept. of Physics, Faculty of Mathematics and Natural Sciences, Institut Teknologi Bandung
[2] Building Technology Research Division, School of Architecture, Planning and Policy Development, Institut Teknologi Bandung
Jl. Ganesha No. 10, Bandung 40132

*) inge@fi.itb.ac.id

Abstract. The thermophysical parameters of organic phase change material (PCM) of coconut oil (co_oil) have been studied by analyzing the temperature vs time data during liquid-solid phase transition (solidification process) based on T-history method, adopting the original version and its modified form to extract the values of mean specific heats of the solid and liquid co_oil and the heat of fusion related to phase transition of co_oil. We found that the liquid-solid phase transition occurs rather gradually, which might be due to the fact that co_oil consists of many kinds of fatty acids with the largest amount of lauric acid (about 50%), with relatively small supercooling degree. For this reason, the end of phase transition region become smeared out, although the inflection point in the temperature derivative is clearly observed signifying the drastic temperature variation between the phase transition and solid phase periods. The data have led to the values of mean specific heat of the solid and liquid co_oil that are comparable to the pure lauric acid, while the value for heat of fusion is resemble to those of the DSC result, both from references data. The advantage of co_oil as the potential sensible and latent TES for room-temperature conditioning application in Indonesia is discussed in terms of its rather broad working temperature range due to its mixture composition characteristic.

Keywords: Phase Change Material (PCM), coconut oil, T-history method, thermophysical parameters.

1. Introduction

Thermal energy storage (TES) is a technology that saves thermal energy by heating or cooling a storage medium so that the stored energy can be used at a later time for heating and cooling applications [1-2]. For passive system, TES are used particularly in buildings to keep the thermal comfort of occupants [3-4]. For this application, TES system can help balance energy demand and supply on a daily, weekly and even seasonal basis. They can also reduce electrical energy consumption and its peak demand, as well as to reduce CO_2 emissions as the side effect of the use of airconditioning (AC) system. It will also increase overall efficiency of energy systems.

IOP Conf. Series: Materials Science and Engineering **214** (2017) 012034 doi:10.1088/1757-899X/214/1/012034

There are two kinds of physical TES systems, namely: 1) sensible heat storage that is based on storing thermal energy by heating or cooling a liquid or solid storage medium (e.g. water, bricks, rocks, etc.); 2) latent heat storage using phase change materials or PCMs (e.g. it involved the material that undergoes solid to liquid phase transition, or vice versa). Sensible heat storage is relatively inexpensive compared to PCM material and is applicable to domestic and district systems. However, in general sensible heat storage requires large volumes because of its low energy density and proper design to discharge thermal energy at constant temperatures. Thus, thermal energy storage system based on phase change material (PCM) is one of the feasible ways for energy conservation in the future because it can store relatively large heat at temperature around its phase transition (solid-liquid) region, and releasing the absorbed heat back to the environment when the ambient temperature is lower than the temperature of the material.

Mettawee et.al. in 2012 and 2013 have shown that coconut oil (co_oil) can be used as the medium of active and passive air conditioning system, namely as the agent that influence the initial temperature of the heat transfer fluid in air conditioning machine [5] and as the medium of heat exchanger in ceiling system [6]. These applications also in agreement with the result of DSC study that shown a big storing capacity of co_oil in the temperature range of 12 to 35°C [7]. Hence, co_oil as one of organic PCM have a capability to stores the latent heat and also sensible heat when the environment temperature is higher than its temperature and release the stored heats later to keep the room temperature relatively constant or conditioned. To get a reliable information for the application of co_oil, we present in this paper the thermophysical parameters of co_oil by means of T-history method proposed by Yinping et al. [8] and its modification by Hong et al. [9] to get the values of mean specific heat of the liquid and solid co_oil ($c_{p,l}$ and $c_{p,s}$) as well as the heat of fusion (h_m) for the solid-liquid transition of co_oil [1-3]. In order to get a reasonable value of solid specific heat, we proposed to adopt a combination of Yinping and Hong method, in particular for the determination of phase transition region, which might be related with the gradual nature for the phase transition of co_oil.

2. Methods

The coconut oil (co_oil) as PCM was purchased from local supermarket in Bandung. The setup of measurement adopting the T-history method is represented schematically in Figure 1 that shows the test tube in cooling bath. Each tube is equipped with the temperature sensors (T-type thermocouple with diameter about 1mm) that integrated to the data logger (Applent AT4508A from Applent Instruments Inc.) and computer. To characterize co_oil by this method, about 20gr reference sample (water) and 20g co_oil sample were placed in 150mm long glass test tube with 14mm internal diameter and 1mm wall thickness. These tube dimensions were selected to ensure that the heat transfer is occur one-dimension along the length of the tube, as measured by the Biot number defined by $Bi = hR/2\kappa \leq 0.1$, where h is the natural convective heat-transfer coefficient, R is radius of the tube and κ is the thermal conductivity of the tube material [8]. In this case the temperature distribution in the sample can be regarded as uniform and the lumped capacitance method can be used [10].

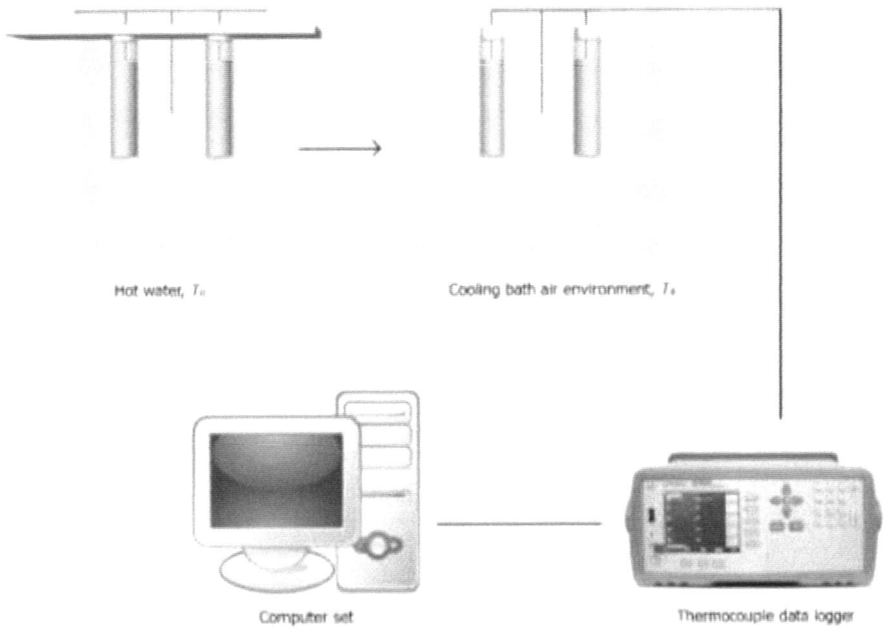

Figure 1. Schematic diagram of T-history method: co_oil sample and water as reference in two different tubes were heated at temperature T_0. The tubes were then suddenly exposed into cool air environment (temperature T_a), and the temperature-time data were recorded automatically by using data logger and computer during the decrease of temperature by time.

Prior to the measurement, water and liquid co_oil contained in two different tubes were heated to a high temperature (T_0) above the melting point (T_m) and stabilized for a few minutes to ensure the temperature homogeneity in all part of the sample. The tubes were then subsequently exposed to a cooler environment (T_a) provided by a cooling bath (with diameter 100 × 50 × 45cm). During the cooling process, the temperature of the water and co_oil were recorded and plotted against time until the co_oil solidified. The experiments were repeated several times to ensure the accuracy and reproducibility of the data.

3. Results and Discussion

When the temperature of PCM (co_oil) and water in the tube are suddenly exposed to cool air atmosphere whose temperature is T_a (which might be time dependent), the temperature vs time curve, i.e. the T-history curves are shown in Fig. 1. From this figure, while water in this temperature region experience only the temperature change without phase transition, co_oil material experience the temperature change and liquid to solid phase transition, with supercooling temperature and solidification temperature are each denoted by T_s and T_m. The values of T_s and T_m as well as the difference between them as defined as the supercooling degree ($\Delta T_m = T_s - T_m$) for 5 set of data are shown in Table 1. From this table one can see that the variation of solidification temperature for different set of experiment is about 18–21°C, while the supercooling degree is about 2-3°C. We note that the resulted solidification temperature (heat release) is lower than the reported values [7] of melting temperature (heat absorption) based on exothermic mode of DSC measurement, which commonly occur in other kinds of material.

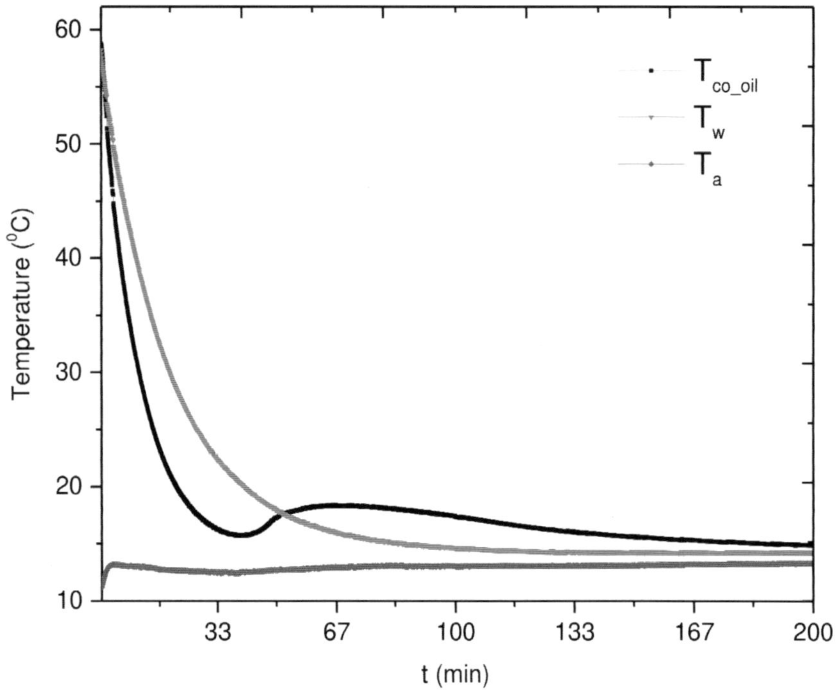

Figure 2. The temperature-time curves of PCM (co_oil) and water during the cooling process in cool air environment (T_a).

Table 1. Supercooling Temperature (T_s), solidification temperature (T_m), and supercooling degree (ΔT_m) of PCM (co_oil) for 5 set of data experiment.

No. of measurement	T_s(°C)	T_m(°C)	$\Delta T_m = T_m - T_s$ (°C)
1	15.69	18.37	2.68
2	16.09	18.48	2.39
3	18.48	20.65	2.17
4	16.37	19.13	2.76
5	15.06	18.28	3.22
Average values	16.34	18.98	2.64

Following the original T-history method by Yinping [8] and its modification by Hong [9], the energy equations are described in Eqs. (1)-(3) for PCM, each of them for liquid state, phase change, and solid state periods.

$$\left(m_{t,p}c_{p,t} + m_p c_{p,l}\right)\left(T_0 - T_s\right) = hA_c A_1 \quad (1)$$

$$\left(m_{t,p}c_{p,t} + m_p \frac{c_{p,l} + c_{p,s}}{2}\right)\left(T_m - T_i\right) + m_p h_m = hA_c A_2 \quad (2)$$

$$\left(m_{t,p}c_{p,t} + m_p c_{p,s}\right)\left(T_i - T_f\right) = hA_c A_3 \quad (3)$$

We note that the energy equations for water as the reference material with the similar periods are described in Eqs. (4)-(6),

$$\left(m_{t,w}c_{p,t} + m_w c_{p,w}\right)\left(T_0 - T_s\right) = hA_c A_1' \quad (4)$$

$$\left(m_{t,w}c_{p,t}+m_{w}c_{p,w}\right)\left(T_{m}-T_{i}\right)=hA_{c}A_{2}' \tag{5}$$

$$\left(m_{t,w}c_{p,t}+m_{w}c_{p,w}\right)\left(T_{i}-T_{f}\right)=hA_{c}A_{3}' \tag{6}$$

In those above equations, m_p, m_w and $m_{t,p}$, $m_{t,w}$ are the masses of the PCM, water and tube used for PCM and water; $c_{p,l}$ and $c_{p,s}$ are the mean specific heats of liquid and solid PCM; $c_{p,w}$ is the mean specific heat of water; $c_{p,t}$ is the mean specific heat of tube material and h_m is the heat of fusion of PCM, while A_c is the convective heat-transfer area of the tube. In addition, set values of $\{A_1, A_2, A_3\}$ correspond to the area below the curve of PCM' temperature towards air environment, each of them for liquid, liquid-solid and solid periods, while the corresponding set values for water are $\{A_1', A_2', A_3'\}$, each of them for the area below the curve of water' temperature towards air environment, with the boundaries follow those of the PCM's temperature [9], see Table 2. We note, however, that in the modified T-history method [9], the equation (2) that describe the liquid to solid phase transition has been corrected by the terms related to temperature change belongs to the tube and PCM, the later to take into account the possible release of sensible heat in the liquid and solid phases from the original formulation [8]. Besides that Hong et al. also proposed the end of phase transition process to be determined from the inflection point, where the first derivative of T-history curve for PCM becomes minimum [9]. This is rely on the fact that the temperature keeps constant in the latent heat release, but decreases exponentially in cooling process accompanying sensible heat alone [8].

Table 2. Area of each period used for analysis of T-history curves of PCM and water as reference material (T_p and T_w are temperature of co_oil and water, respectively)

	Period	Temperature Range	Range of time	Area
	liquid	T_0–T_s	t_0-t_1	$A_1=\int_{t_0}^{t_1}(T_P-T_a)dt$
co_oil	phase change	$T_s-T_m-T_s'$	t_1-t_2	$A_2=\int_{t_1}^{t_2}(T_P-T_a)dt$
	solid	$T_s'-T_f$	t_2-t_3	$A_3=\int_{t_2}^{t_3}(T_P-T_a)dt$
	liquid	T_0–T_s	$t_0'-t_2'$	$A_1'=\int_{t_0'}^{t_2'}(T_w-T_a)dt$
Water	liquid	T_m–T_i	$t_1'-t_3'$	$A_2'=\int_{t_1'}^{t_3'}(T_w-T_a)dt$
	liquid	T_i–T_f	$t_3'-t_4'$	$A_3'=\int_{t_3'}^{t_4'}(T_w-T_a)dt$

The result for analysis the T-history curves for one set of experiment data of co_oil and water are shown graphically in Fig. 3.

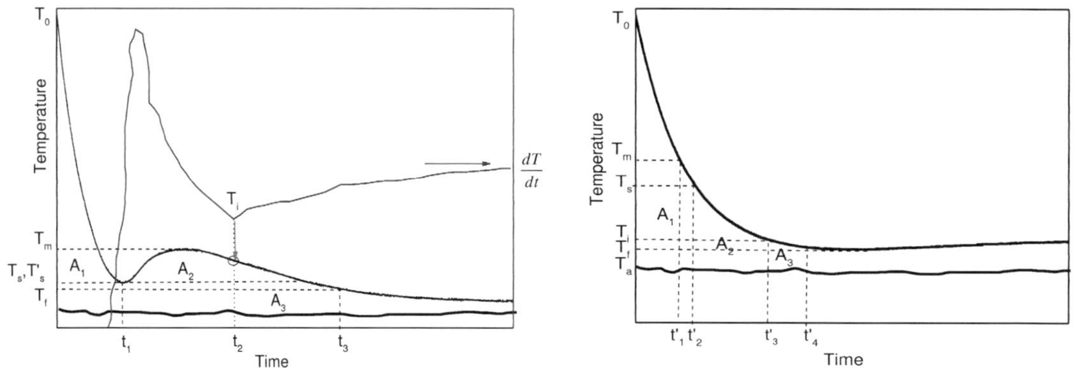

(a) (b)

Figure 3 (a) Illustration for the analysis of temperature-time curve for PCM (co_oil) and its derivative; T_0 is initialization temperature, T_m is melting temperature, T_s is supercooling temperature, T_f is final temperature, T_a is cool air temperature and T_i is inflection point at the first derivative of co_oil' temperature. (b) Illustration for the analysis of the corresponding temperature-time curve of water.

From this figure (Fig. 3(a)) one can see a clear inflection point in the first derivative of co_oil's temperature. We found, however, that for the co_oil as the organic PCM that is investigated in this paper, the liquid to solid phase transition occurs rather gradually in the temperature regime of T_m and T_s, and therefore the meaning of inflection to the phase transition of co_oil has become smeared out. We note that the gradual nature for the phase transition of co_oil might be related to the fact that it consists of many kinds of saturated and unsaturated fatty acids with the largest amount of lauric acid (about 50%) [7]. However, the inflection point is still needed to measure the corresponding area of the water as the reference material, in order to obtain a reliable value of solid specific heat as discussed in the following paragraph.

Finally, the thermophysical parameters of PCM consist of the mean specific heats of the liquid and solid co_oil ($c_{p,l}$ and $c_{p,s}$) and the heat of fusion (h_m) derived from this method are expressed as,

$$c_{p,l} = \frac{m_{t,w}c_{p,t} + m_w c_{p,w}}{m_p}\frac{A_1}{A_1'} - \frac{m_{t,p}}{m_p}c_{p,t} \tag{7}$$

$$c_{p,s} = \frac{m_{t,w}c_{p,t} + m_w c_{p,w}}{m_p}\frac{A_3}{A_3'} - \frac{m_{t,p}}{m_p}c_{p,t} \tag{8}$$

$$h_m = \frac{m_{t,w}c_{p,t} + m_w c_{p,w}}{m_p}\frac{A_2}{A_2'}\left(T_m - T_i\right) - \left(\frac{m_{t,p}}{m_p}c_{p,t} + \frac{c_{p,l} + c_{p,s}}{2}\right)\left(T_m - T_i\right) \tag{9}$$

The results of analysis for 5 set data of co_oil are shown in Table 3, together with the average and standard deviation values for each parameters, to be compared with the same or similar materials from references using various experiment methods (Table 4).

Table 3. The values of liquid and solid specific heats of co_oil and heat of fusion related to phase transition of co_oil taken from the analysis of 5 set of data and their average values.

Sampel	$c_{p,l}$ [kJ/kg.K]	$c_{p,s}$ [kJ/kg.K]	h_m [kJ/kg]
1	2.83	3.26	88
2	2.25	3.29	72
3	2.20	3.14	78
4	2.31	3.24	83
5	2.14	3.13	90
Average values	2.35	3.23	82
Standard deviation values	0.28	0.07	7

Table 4. The thermophysical parameter values of co_oil, pure lauric acid, or mixture between lauric and capric acids derived from DSC and T-history measurements from references.

Ref.	Material	Method	$c_{p,l}$ [kJ/kg.K]	$c_{p,s}$ [kJ/kg.K]	h_m [kJ/kg]
[7]	Coconut oil	DCS	-	-	103
Ref.	Material	Method	$c_{p,l}$ [kJ/kg.K]	$c_{p,s}$ [kJ/kg.K]	h_m [kJ/kg]
[8]	Lauric acid	T-history	1.60-1.75	-	160-179
[9]	Lauric acid	T-history	2.14	2.81	186
[11]	Capric-lauric acids	T-history	2.12	1.80	179

From these two tables one can see that the mean values of liquid and solid specific heats of co_oil close to the corresponding values of pure lauric acid ($T_m \cong 45.9–50.3$°C [12]), while the heat of fusion resemble the value obtained from DSC [7]. We note hence, that the potential usage of co_oil as TES for room-temperature conditioning application in tropical climate area was supported by its composition that consists of many kinds of fatty acids where each kind of fatty acid have their own melting temperature, in which case it is commonly occur that the longer fatty acid chain imply the higher its melting temperature, which in turn give an implication to its rather broad working temperature range.

4. Conclusion

We have described in this paper the thermophysical parameters of coconut oil (co_oil) as one of organic PCM, based on the temperature vs data during solidification (T-history method) with water as the reference material. The data show that liquid to solid phase transition occur rather gradually with relatively small supercooling degree. The data have been further analyzed based on the lumped capacitance method to obtain the mean specific heat of the liquid and solid co_oil ($c_{p,l}$ and $c_{p,s}$) and the heat of fusion (h_m) related to the phase transition of co_oil. In different with the previous analyses, the period for latent heat release was determined based on the assumption that the process occur at temperature regime between the supercooling temperature (T_s) and solidification temperature (T_m). Although the first derivative of T-history curve of co_oil show a clear inflection point as the boundary between phase change and solid-state periods, its importance is to specify the related period of water in order to obtain the reliable value of mean specific heat of the solid co_oil. The obtained values of mean specific heats of the liquid and solid co_oil are $c_{p,l} \cong 2.35$ kJ/kg.K and $c_{p,s} \cong 3.23$ kJ/kg.K, i.e. they are close to the values belong to pure lauric acid, while the heat of fusion value ($h_m \cong 82$ kJ/kg) is close to the result from DSC. In addition, the co_oil that consists of many kinds of fatty acids has rather broad working temperature range so that it is potential as the sensible and latent TES in Indonesia.

5. References

[1] I. Dincer and M. Rosen 2010 *Energy Storage System and Aplications* (Ontario Canada: A John Wiley and Sons, Ltd., Publication)

[2] Mehling, Harald and L. F. Cabeza 2008 *Heat and Cold Storage with PCM An Up to Date Introduction Into Basics and Applications* (Berlin, Germany: Springer)

[3] Vineet Veer Tyagi, D. Buddhi 2007 PCM thermal storage in buildings: A state of art, *Renewable and Sustainable Energy Reviews* **11** 1146–1166

[4] E. Ostermana, V.V. Tyagib, V. Butalaa, N.A. Rahimb, U. Stritih 2012 Review of PCM based cooling technologies for buildings, *Energy and Buildings* **49** 37–49

[5] Eman Bellah Sayed Mettawee, Eldesouki Ibrahim Eid and Sherif Amin Mohammed Amin 2012 Experimental study on space cooling with PCM thermal Storage *Journal of Applied Sciences Research*, **8** (7): 3424-3432 ISSN 1819-544X

[6] E.S. Mettaweea, A.I. Eadb 2013 Energy saving in building with latent heat storage *Int. J. of Thermal & Environmental Engineering* Volume **5**, No.1 21-30

[7] Tatsawan Tipvarakarnkoon, Reinhard Blochwitz and Bernhard Senge 2008 Rheological properties and phase change behaviors of coconut fats and oils, *Annual Transactions of the Nordic Rheology Society* **16**

[8] Zhang Yinping, Jiang Yi and Jiang Yi, A simple method 1999 The T–History method, of determining the heat of fusion, specific heat and thermal conductivity of phase-change materials *Meas. Sci. Technol.* **10** 201–205

[9] Hiki Hong, Sun Kuk Kimb, Yong-Shik Kim 2004 Accuracy improvement of T-history method for measuring heat of fusion of various materials *International Journal of Refrigeration* **27** 360–366

[10] Incropera F P and Dewitt D P 1996 *Fundamentals of Heat and Mass Transfer* 4th edn (New York: Wiley) pp 212–17

[11] A. Hasan, S.J. McCormack, M.J. Huang, B. Norton 2014 Characterization of phase change materials for thermal control of photovoltaics using Differential Scanning Calorimetry and Temperature History Method, *Energy Conversion and Management* **81** 322–329

[12] R. D. O'Brien, 2004, *Fats and Oils Formulating and Processing for Applications* 2nd *ed.*, (USA: CRC Press LLC)

Acknowledgements

The results in this research are funded by Desentralisasi DIKTI 2016 research program under contract number: 5830/I1.C01/PL/2016.

Fabrication and Characterization of Planar Spring Based on FR4-PCB for Electrodynamics Vibration Energy Harvesting Application

Gandi Sugandi[1]*, Grace A Mambu[1], Dadang Mulyadi[1], and Edi Mulyana[2]

[1]Research Center for Electronics and Telecommunication – LIPI, Bandung Indonesia
[2]Department of Electrical Engineering, UIN Sunan Gunung Jati, Bandung- Indonesia

*Email: gandi511@yahoo.com

Abstract. Planar spring as a mechanical resonator is very important in designing an electrodynamic vibration energy harvesting application (EVEH) to generate output power with high efficiency. Generally, component of the mechanical resonator is a cantilever beam that is designed using one cantilever with an inertial mass placed cantilever tip. In this study, a planar spring which has four arms cantilever beam was designed and fabricated using an extra-thin FR4-PCB material with a total thickness of 130 μm. There are four types of planar spring that were designed and fabricated in this research to produce resonant frequencies at about 30, 40, 50 and 60 Hz with 1 mm width cantilever arm and various length of 13.5, 11.2, 9.8 and 8.7 mm, respectively. FR4 resonator is fabricated using technology LASER-cutting in order to obtain results precisely. The resonant frequency generated by the mechanical resonator is characterized using vibrator system with certain acceleration. The resonant frequency of the planar spring was obtained at a frequency where the maximum induced voltage occurs. The resonant frequency generated by each type of planar spring was obtained at 24.81, 34.24, 40.2, and 46.8 Hz with three conditions of acceleration of 0.02, 0.06, and 0,1g ($g=9.8\ m/s^2$).

1. Introduction

Advances in technology microfabrication of integrated circuits (ICs) and sensor micro electro-mechanical system (MEMS sensor) has led to the springing up of wireless sensor node (Wireless Sensor Node, WSN) that has a small size, low cost and low power consumption. Most WSN is equipped with one or more sensors, a microcontroller, a wireless transceiver and resources. At present, the battery is still a major resource to maintain the operation of many wireless sensor nodes. However, batteries have limited energy, limited life and the need for the regular replacement when it runs out of energy, so this will increase the cost of maintenance [1]. To resolve this problem, energy harvesting (energy harvesting) that derived from the surrounding environment has become a source of alternative energy for the battery.

Vibrating kinetic energy sources is one of the many sources of energy is converted into electrical energy to provide electrical energy for WSN, among the sources of energy that vibrate includes a step or human motion, vibration of industrial machinery, bridges, vehicles, and so forth. Some literature has reported research on harvesting the kinetic energy into electrical energy using the method of harvesting, which is piezoelectric, electrostatic and electromagnetic or electrodynamics [2-5]. The third of the kinetic energy harvesting techniques, the method of electrodynamics energy harvesting is a technique that is very interesting and even become the focus of this study. Converting the energy

IOP Conf. Series: Materials Science and Engineering **214** (2017) 012035 doi:10.1088/1757-899X/214/1/012035

produced by the electrodynamics energy conversion is greater than piezoelectric or electrostatic method. The working principle of harvesting mechanical energy into electrical energy is using the concept of magnetic induction, Faraday's law.

There are many considerations have to be fulfilled in designing the energy harvesting module, that derived from the mechanical energy source. One of the most significant issue concerning the frequency of vibration of the energy sources is, for example, in many industrial applications of industrial machines having a frequency of a vibration in the range of tens to 100 Hz, while the frequency of vibration is raised by motion or step of the human being in the range below 10 Hz. This paper will discuss the fabrication and characterization of a miniaturized energy harvester's module using mechanical resonator. When the resonator has the same resonant frequency with the vibration source, the energy that will be generated will be maximum. Mechanical resonators used in this study was designed to have a resonant frequency of 30, 40, 50 and 60 Hz using extra thin PCB-FR4 material.

2. Principle of Electrodynamic Vibration Energy Harvesters

The Electrodynamic Vibration Energy Harvesters (EVEH) converts kinetic energy into electrical energy by electromagnetic transduction mechanism. A general spring-mass model was built to describe the harvester system by C.B. Williams and R.B. Yates [6], as shown in Figure 1. This system can be modeled analytically as a mass-spring-damper consisting of a total damping coefficient, *ct*, spring constant, *k*, and the seismic mass, m.

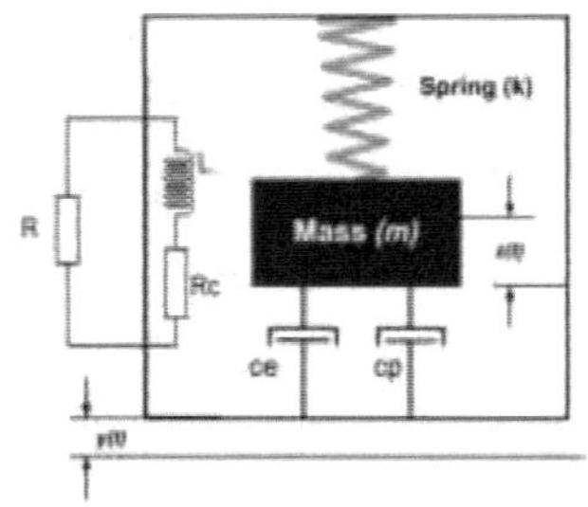

Figure 1. The schematic electrodynamic vibration energy harvesting (EVEH).

Voltage generated in the wire coil, that is supported by a spring mechanism is caused by the presence of magnetic induction that moves along the z -axis. When the module system is vibrated by the displacement, *y(t)*; mass transfer, *z(t)*, produced by the relative mass which is attached to a spring. Energy losses in the system (consisting of the parasitic losses, *cp*, and the reduction of electrical energy generated by the mechanical transduction, *ce*), are represented by the total attenuation coefficient, *ct*. From these analytic modeling, the equations of motion can be derived as follows:

$$m\frac{d^2z}{dt^2} + c_t\frac{dz}{dt} + kz(t) + F_{mn}[z(t)] = F_d(t) \qquad (1)$$

where F_{mn} is the magnetic force acting on the coil and $F_{d,}$ is the external driving force due to harmonic vibrations. The maximum energy that can be harvested by this system occurs when the external frequency is equal to that of the resonant frequency of the system, ω_n^2 = k/m. Williams and Yates showed that the power dissipated in the damper (i.e, extracted by both transduction mechanisms and parasitic damping mechanism) is given by:

$$P_{av} = \frac{m\xi_t Y^2\left(\frac{\omega}{\omega_n}\right)^3 \omega}{\left[1-\left(\frac{\omega}{\omega_n}\right)^2\right]^2 + \left[2\xi_t\left(\frac{\omega}{\omega_n}\right)\right]^2} \qquad (2)$$

where m is the mass of moving parts, *Y* and ω are the amplitude and angular frequency of the motion harmonic, respectively, ω_n is the resonant angular frequency of the mass-spring oscillator, $\xi_t = c_t / 2m\omega_n$ is total damping ratio, and *ct* is the total damping factor. Equation (1) assumes that the generator

IOP Conf. Series: Materials Science and Engineering **214** (2017) 012035 doi:10.1088/1757-899X/214/1/012035

is driven by a harmonic movement, hence in steady state; P_{av} is equal to the kinetic energy supplied per second of from the existing vibrations. The maximum power dissipation occurs at resonance, which is given as

$$P_{av} = \frac{mY^2\omega_n{}^3}{4\xi_t} \qquad (3)$$

Mechanical resonator is a very important component of the EVEH as previously described. Mechanical resonators are used to increase the amplitude of mechanical vibrations that are transferred from the surrounding environment. They are designed to match the frequency characteristics of the application environment. A mechanical resonator is characterized by the resonant frequency.

$$f_n = \frac{1}{2\pi}\sqrt{\frac{n\cdot E\cdot w\cdot t^3}{4\cdot L^3\cdot m}} \qquad (4)$$

3. Design and Fabrication of Device

Figure 2 shows the 3D structure of the Electrodynamic Vibration Energy Harvesting (EVEH) and a shape type of PCB planar spring or mechanical resonator. Generally, EVEH structure (Figure 2a) consists of a planar spring, coil and the permanent magnet and supporting components for the assembly of the system. Copper coils are placed at the bottom of the frame, and extra thin-FR4 PCB planar spring having four cantilever beam and a platform for attachment NdFeB permanent magnets is in the center of the platform. PCB planar spring and a permanent magnet is used as part of the move, meaning the coil frame is mounted and used as the static part of the generator. In the work principle of harvesting energy, the electrical energy is produced subjected to external vibration; the mass of evidence is that a permanent magnet mounted induction is responds to deflection. According to Faraday's law of electromagnetic induction, the degree of deflection induces a change in the terminal voltage of the coil output.

Figure 2b shows the design layout of the mechanical resonator that has four cantilevered arms while the platform in the center of the planar spring is where the inertial mass is placed. In this experiment, the structure of planar spring is designed and fabricated with four type of various length and fixed width of cantilever arm. Each type of planar spring is designed for resonant frequency within 30, 40, 50 and 60 Hz with various cantilever arms of 13.5, 11.2, 9.8 and 8.7 mm length respectively with a 1 mm of fixed width. Planar spring is fabricated using double-sided cooper layer of extra thin FR4-PCB (1 oz.) with a size of 2.5 x 2.5 mm^2 and a total thickness PCB about 0.13 mm.

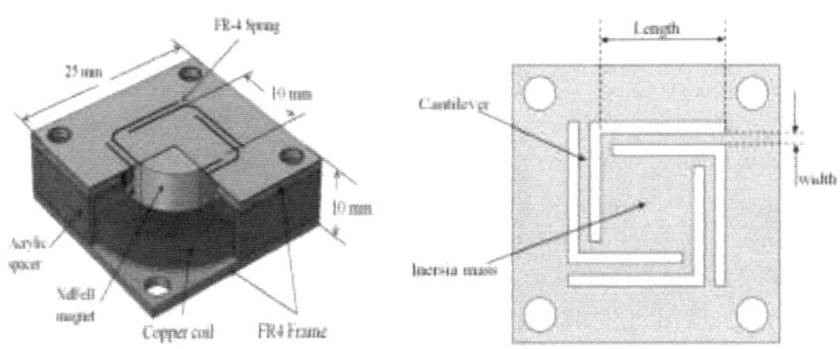

Figure 2. (a) Schematic structure of electrodynamics vibration energy harvesting and (b) a shape type of PCB planar spring or mechanical resonator.

A simple fabrication process for the mechanical resonator components and assembling components into one system energy harvester is shown in Figure 3. The process steps include design layout of the planar spring (Fig. 3a), followed by the FR4 PCB substrate cleaning process and cutting with a laser cutter in accordance to the design layout (Fig. 3b). The etching of copper layers (Fig. 3c) on both sides of the PCB is the next process before finally cleans off the FR4 substrate. After the mechanical resonator fabrication, the next step is to characterize the resonant frequency of the resonator components. The components are assembled into a vibration energy harvesting system as shown in figure 3d.

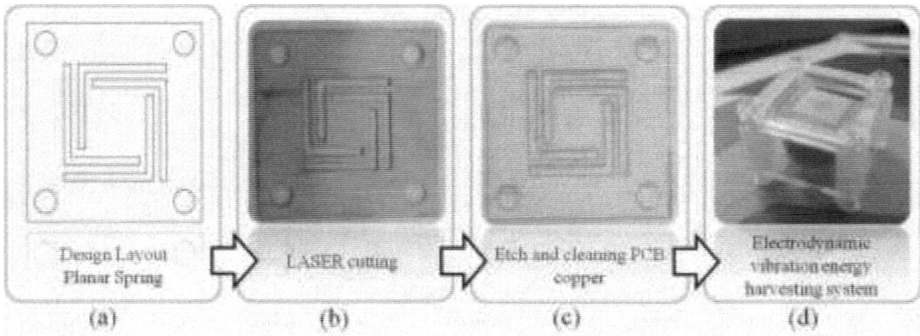

Figure 3. (a-c) Step fabrication of planar spring and (d) assembling into EVEH system.

4. Experiment Results and Discussion

Figure 4 shows the custom vibration system equipment diagram of experiment setup for the characterization of mechanical resonant frequency of PCB planar spring. The subwoofer loudspeaker with good performance capabilities is well suited for low frequency vibration input source [7]. The input signal from the tone generator software is delivered to audio amplifier for subwoofer vibration. The EVEH generator system as shown in Fig 3d is attached on the jig of the vibrator. A digital multimeter is then connected to the coil of generator to measure the open-circuit voltage of the generator. When the frequency of the input vibration is equal to the natural frequency of the mechanical resonator structure, the spring-mass system operates at resonance; at that point, the permanent magnet has the maximum of the amplitude, and the coil produces the maximum voltage.

Figure 4. The experiment test setup.

Figure 5 shows the output voltage response of the EVEH with changes in the input frequency. The relationship of the results obtained from measurements of the peak output voltage to frequency changes show the response of the resonant frequency of the planar spring. From the measurement results for the fourth resonant frequency planar type of spring with the vibration sources of 0.02g, 0.06g and 0.1g brackish obtained for each of these conditions are type-1: 23.61, 25 and 24.81 Hz (Fig. 5a), and for type-2: 30.42, 33.12 and 34.24 Hz (Fig. 5b), whereas for type-3: 35.51, 38.91 and 40.2 Hz (Fig. 5c), and the latter type-4 as shown in Figure 5d responded to the resonant frequency at 41.05, 44.4 and 46.8 Hz.

Based on the result of design and obtained measurements, a significant difference can be found in the value where the differences in the resonant frequency that occurs in the range of 5 Hz to 10 Hz. From the measurement results shown, the relationship of the resonance frequency of a planar spring is clearly affected by the acceleration of its vibration source. We can establish the relationship as when the speed is high, the response of the resonant frequency of the spring is higher too. Meanwhile, the longer arm of the cantilever planar spring resonance frequency becomes lower.

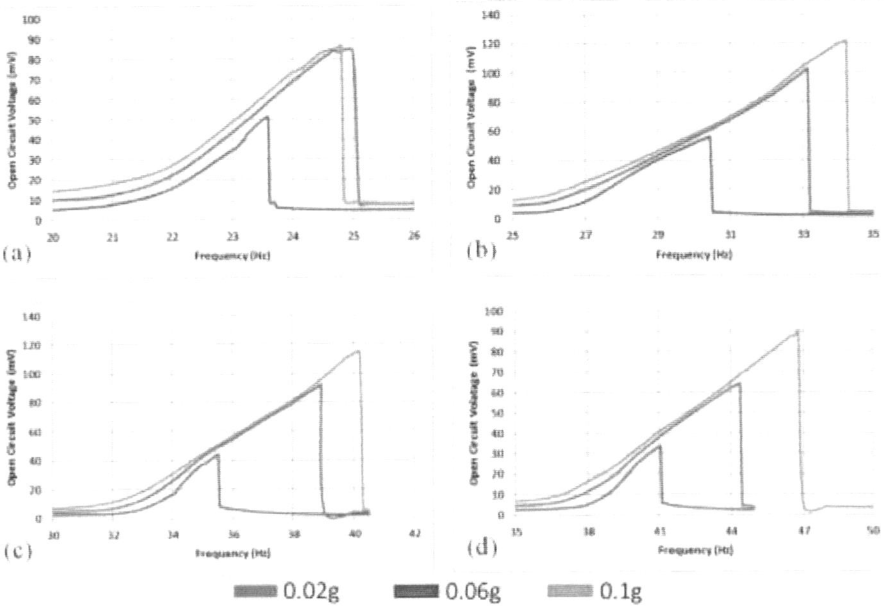

Figure 5. The output voltage response of the EVEH versus input frequency. Resonant frequency of PCB FR4 planar spring for (a) type-1, (b) type-2, (c) type-3 and (d) type-4 with vibration source acceleration in 0.02g (gray color), 0.06g (red color) and 0.1g (blue color) respectively.

The generator demonstrated nonlinear response which produced a different of resonance frequency when acceleration vibration source changed. And also the induced output voltage depends on the vibration frequency source. The maximum output voltage is generated in the resonant frequency, and drops significantly out of the frequency range.

5. Conclusions

This paper presents the design, fabrication and characterization of the planar springs as the mechanical resonator that made from FR4-PCB for electrodynamics vibration energy harvesting (EVEH) applications. The use of a very thin material PCB-FR4 can be integrated with electronic drive components on a one chip PCB for the energy harvesting systems. Based on the results of measurements, the FR4 used as the planar spring material is very suitable for the sources of energy vibrating. It has a very low acceleration, just below 0.1 g with vibration frequencies below 100 Hz. In this study, we have designed a mechanical resonator with variations of resonant frequencies of planar springs at 30, 40, 50 and 60 Hz. The measurement results showed the differences of the resonant frequencies, in the range of 5 to 10 Hz with a vibration source acceleration condition at 0.02, 0.06 and 0.1g.

6. References

[1] G Hatipoglu & Rey, H. U 2010 *Fr4-Based Electromagnetic Energy Harvester for Wireless Sensor Nodes*. IOP Publishing 19 (Smart Materials and Structures): 11

[2] Byung-Chul Lee & Chung, G.-S 2013 Design and Fabrication of Low Frequency Driven Energy Harvester Using Electromagnetic Conversion, *Transaction on Electrical and Electronic Materials* **14** (3) 5

[3] Mitcheson P. D., Miao P., Stark B. H., Yeatman E. M., Holmes A. S. & C., G. T 2004 Mems Electrostatic Micro-Power Generator for Low Frequency Operation, *Sensors Actuators A* **115** (7)
[4] Neil N. H. Ching, Gordon M. H. Chan, Wen J. Li, Hiu Yung Wong & Leong, P. H. W. *Pcb Integrated Micro-Generator for Wireless Systems*
[5] Renaud M., Fiorini P., Shaijk R. & C., H 2009 Harvesting Energy from the Motion of Human Limbs: The Design and Analysis of an Impact-Based Piezoelectric Generator Smart Mater. *Struct.* **18**(16)
[6] S.P. Beeby, R.N. Torah & Tudor, M. J 2008 Kinetic Energy Harvesting. ACT Workshop on Innovative Concepts
[7] Litwhiler, D. H 2011 A Custom Vibration Test Fixture Using a Subwoofer. The 2011 IAJC-ASEE International Conference

IOP Conf. Series: Materials Science and Engineering **214** (2017) 012036 doi:10.1088/1757-899X/214/1/012036

PTFE Additive and Re-annealing Effect on Thermoluminescence Response of $CaSO_4$:Dy Derived from Co-precipitation Method

Nunung Nuraeni[1,2], Ricky Dwi Septianto[1], Ferry Iskandar[1,3], Freddy Haryanto[1], Abdul Waris[1] and Eri Hiswara[2]

[1]Department of Physics, Faculty of Mathematics and Natural Sciences, Institut Teknologi Bandung, Jl. Ganesha 10 Bandung 40132, Indonesia
[2]Work Safety and Dosimetry Division, Centre for Safety Technology and Radiation Metrology
National Nuclear Energy Agency (BATAN)
Jl. Lebak Bulus Raya Kav. 49, Jakarta Selatan 12440, Indonesia
[3]Research Center for Nanosciences and Nanotechnology (RCNN), Institut Teknologi Bandung, Jl. Ganesha 10 Bandung 40132, Indonesia

Corresponding author: ferry@fi.itb.ac.id

Abstract. Effect of re-annealing treatment in thermoluminescence response of thermoluminescent dosimeter (TLD) $CaSO_4$:Dy and $CaSO_4$:Dy with PTFE (Polytetrafluoroethylene) addition was investigated. $CaSO_4$:Dy was prepared by a co-precipitation method. The PTFE was added before re-annealing treatment which the mass ratio of $CaSO_4$:Dy and PTFE was fixed to 2:3. The re-annealing treatments of the samples were done at temperature 700°C for 1 hr. The obtained samples were characterized using a Fourier-transform infrared (FTIR) and X-ray diffraction (XRD) to observe the molecule bonding in sample and crystal properties, respectively. From the experimental results, it was observed that the thermoluminescence intensity of $CaSO_4$:Dy, $CaSO_4$:Dy re-annealed at 700 °C, and $CaSO_4$:Dy + PTFE re-annealed at 700 °C are 57.03, 75.15, and 1191.11 nC, respectively. The intensity of 700 °C-re-annealed $CaSO_4$:Dy increased significantly after PTFE addition.

1. Introduction

The utilization of thermoluminescence dosimeter (TLD) as a radiation dosimeter has been widely used for individual radiation monitor and in vivo measurement. In the radio diagnostic field, TLDs were used for monitoring patient's radiation dose regularly and will be reported to the United Nations Scientific Committee on the Effects of Atomic Radiation (UNSCEAR)[1]. Meanwhile in the radiotherapy field, TLDs were used for patients by means of in vivo dosimetry, both in quality assurance activities and dosage monitoring in special cases such as in cases with complex geometry, the critical organ dose, total body irradiation (TBI), brachytherapy, verification techniques treatment, dosimetry audit by the IAEA and WHO in dose audit program using the TLD post, and inter comparison between hospitals [2, 3]. TLDs are also used for environment dosimetry applications. However, the performance criteria for TLD in this application are different from requirements for personnel monitoring.

The synthesis methods of TLD have been widely developed as the development of material fabrication technology. The preparation method is very important because it can control the final properties of material. The common methods that used to fabricate TLD are precipitation, evaporation, precipitation from solution or molten phase (flow method), chemical vapour deposition (CVD), spray-pyrolysis, and sol-gel [4].

$CaSO_4$:Dy is well known TLD which has a good sensitivity. The presence of Dy^{3+} will increase the luminescence of TLD $CaSO_4$[5]. PTFE was a good additive for material, moreover, when it is applied for cohesive TLD, a new crystal properties will occur [6]. However, there is a lack information about the effect of heat treatment of $CaSO_4$:Dy and $CaSO_4$:Dy with PTFE addition on its thermoluminescence properties. Therefore, in this research the effect of re-annealing in thermoluminescence properties of $CaSO_4$:Dy and $CaSO_4$:Dy with PTFE addition prepared by co-precipitation method was investigated.

2. Experimental Method

$CaSO_4$:Dy was prepared by a co-precipitation method [7] with a reaction as shown in Eq. 1.

$$(CH_3COOH)_2Ca + (NH_4)_2SO_4 + Dy_2(SO_4)_3 \rightarrow CaSO_4 : Dy \downarrow + 2\,CH_3COONH_4 \quad (1)$$

Dysprosium (Dy) is used as a dopant which has concentration of 0.1 mol%. In a brief, Calcium acetate (MERCK) was dissolved in double distilled water and subsequently was mixed with the mixture of dysprosium sulphate and sulfuric ammonium (MERCK) in the presence of ethanol. The precipitate was then washed and dried in an oven for 4 hrs. After the drying process, the sample was annealed at a fixed temperature for 1 hr in a furnace. In order to synthesis $CaSO_4$:Dy/PTFE, the prepared samples were mixed with PTFE by a fixed mass ratio ($CaSO_4$:Dy/PTFE = 2:3). The composite samples were re-annealed at 700°C for 1 hr. The illustration of $CaSO_4$:Dy and $CaSO_4$:Dy with PTFE addition preparation is shown in Fig. 1.

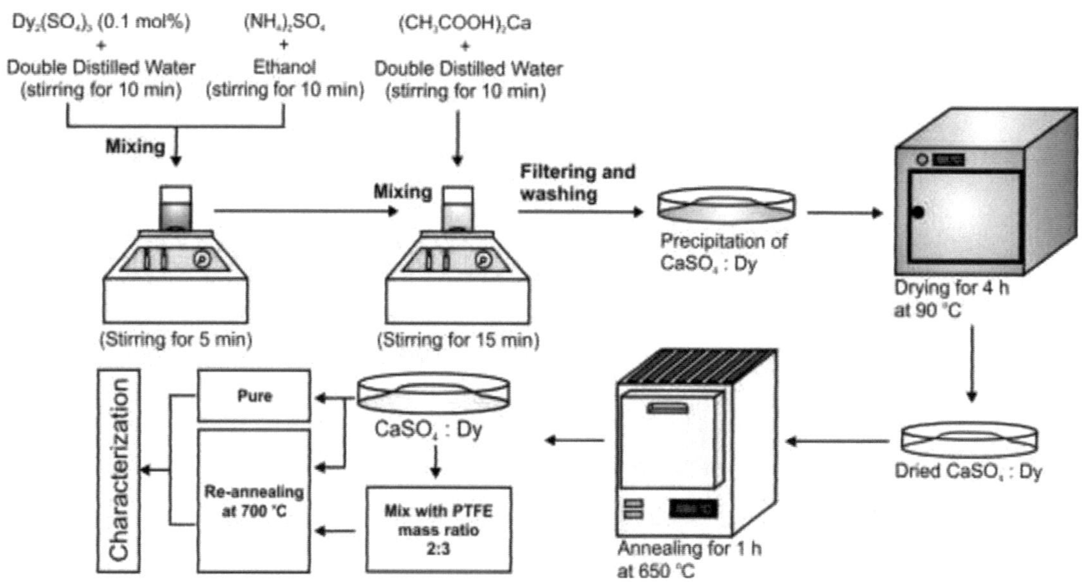

Figure 1. Schematics of preparation method of $CaSO_4$:Dy and $CaSO_4$:Dy with PTFE addition.

The prepared samples were characterized by Fourier-transform infrared (FTIR, Omron Bucker Alpha FTIR spectrometer) and X-Ray diffractometer (XRD, Philip Analytical PW 1710 BASED). The average crystal size of samples was determined by using Scherer's formula based in XRD result as shown in Eq. 2.

$$d = \frac{0.9\lambda}{\beta \cos \theta} \tag{2}$$

Furthermore, TL response was observed by Harshaw 3500 TLD Reader with a maximum temperature of 260 °C after a given radiation source Strontium-90 (Sr-90).

3. Result and Discussion

Figure 2 shows FTIR spectra of pure PTFE, $CaSO_4$:Dy, re-annealed $CaSO_4$:Dy, and re-annealed $CaSO_4$:Dy with PTFE addition. The results show that all samples have identical appearance peaks. The appearance peaks at wavenumber of 1000 - 1200 cm^{-1} show S-O bonding that indicates the presence of $CaSO_4$[8]. The additional of PTFE in sample decrease FTIR spectrum intensity of $CaSO_4$. After the re-annealing process at 700 °C, a sharp peak of pure PTFE in $CaSO_4$:Dy/PTFE could not be observed, it means that PTFE degraded and $CaSO_4$ bonding dominated the sample. Meanwhile, the re-annealing process itself did not affect the bonding of $CaSO_4$:Dy.

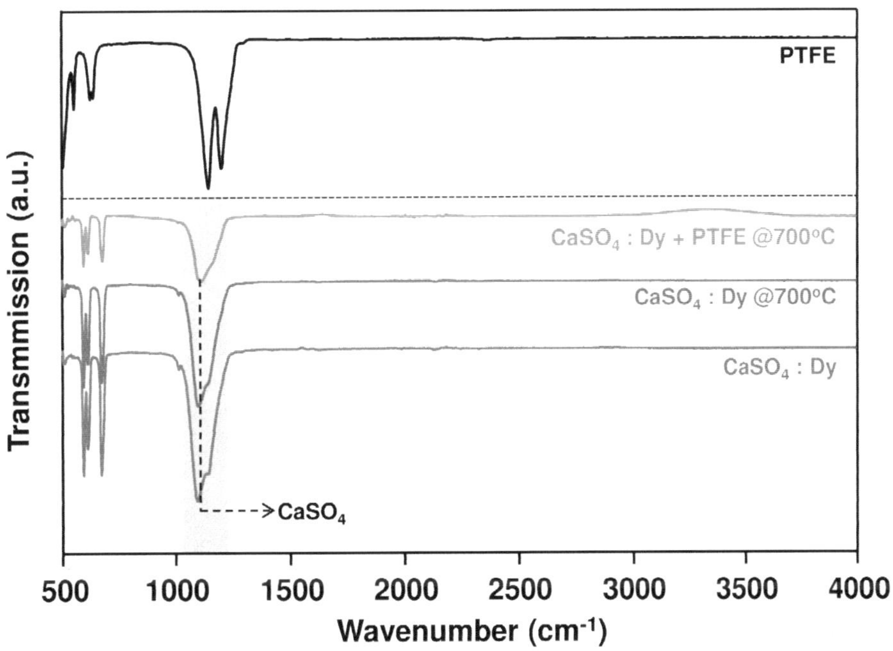

Figure 2. IR spectra of pure PTFE, $CaSO_4$:Dy, $CaSO_4$:Dy after re-annealing at temperature of 700 °C, and $CaSO_4$:Dy with PTFE addition after re-annealing at temperature of 700 °C.

Figure 3 shows XRD result of $CaSO_4$:Dy, re-annealed $CaSO_4$:Dy, and re-annealed $CaSO_4$:Dy with PTFE addition. According to XRD result, there are peaks that indicates the appearance of $CaSO_4$ crystal[7]. Meanwhile, there is no difference between XRD spectra of $CaSO_4$:Dy on both with and without re-annealing treatment. In the other hand, in sample of $CaSO_4$:Dy with PTFE addition, the intensity was decreasing. The decreasing of $CaSO_4$:Dy intensity might be caused by the interaction or reaction between $CaSO_4$:Dy surface and PTFE during the annealing process. The other possibility is the PTFE substance may cover up the surface of $CaSO_4$, causing the decreasing of the signal of X-ray for entering the $CaSO_4$ crystal layer. In the result, we also observed an unknown peak appeared around 28°. This peak occurred due to several possibilities such as change of $CaSO_4$:Dy crystal structure during re-annealing process due to additional of PTFE or the presence of remained PTFE component after re-annealing process. Furthermore, this phenomenon needs to be more investigated.

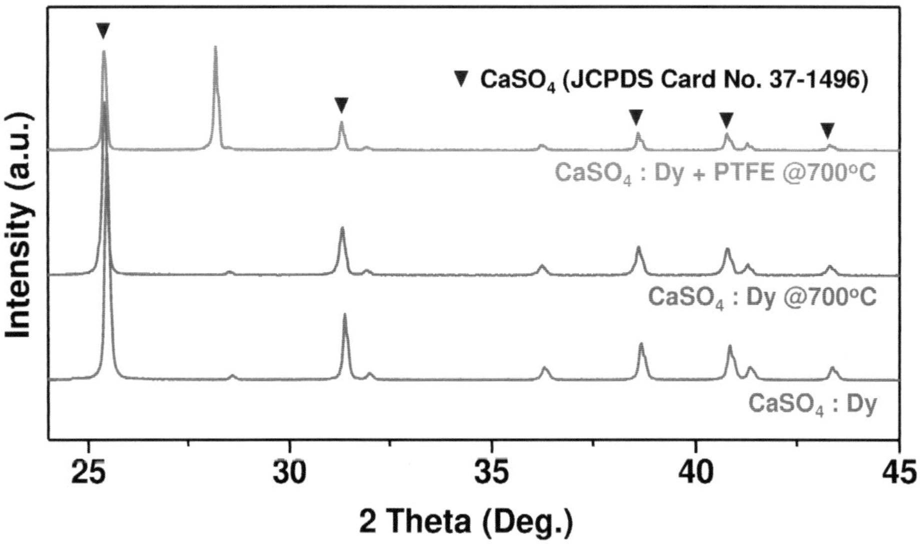

Figure 3. XRD result of pure $CaSO_4$:Dy,$CaSO_4$:Dy after re-annealing at temperature of 700 °C, and $CaSO_4$:Dy with PTFE addition after re-annealing at temperature of 700 °C.

Using Eq. 2 and XRD result, the average crystal size are 54.09, 50.91, and 63.58 nm of pure $CaSO_4$:Dy, $CaSO_4$:Dy with re-annealing treatment at 700 °C, and $CaSO_4$:Dy with PTFE addition re-annealed at 700 °C, respectively.

To evaluate the performance of TLD $CaSO_4$ as dosimeter, these materials were irradiated use Sr-90 source. The response of TLD $CaSO_4$:Dy before and after re-annealing are 57.03 and 75.15 nC, respectively. After $CaSO_4$:Dy mixed with PTFE, the response of TLD became greater than $CaSO_4$:Dy without PTFE addition which has 1191.11 nC of TL response. The addition of PTFE affect the TL response increase up to 15 times than without PTFE after re-annealing. This phenomena might due to the structure change of $CaSO_4$:Dy crystal as shown in Fig. 3. This result shows that re-annealing treatment can increase the response of TLD material, especially after PTFE addition.

4. Conclusion

$CaSO_4$:Dy was successfully prepared by co-precipitation method; and re-annealing treatment was investigated on properties of $CaSO_4$:Dy and $CaSO_4$:Dy with PTFE addition. Re-annealing treatment did not significantly change crystal size of $CaSO_4$:Dy without PTFE addition (54.09 mm become 50.91), meanwhile the presence of PTFE in sample increased the crystal size of $CaSO_4$:Dy after re-annealing treatment (63.58 nm). The thermoluminescence response of $CaSO_4$:Dy after re-annealing at 700 °C was increased. However, with additional of PTFE, the TL response was increased without PTFE after re-annealing.

5. References

[1] United Nations Scientific Committee on The Effect of Atomic Radiaiton, Source and Effect of Ionizing Radiation, Report to General Assembly with Scientific Annexes, (UN Publications, New York 2010)
[2] T.Kron 1995 Australian Physical & Engineering Science in Medicine **18** 1-25
[3] F.M Khan 1987 The Physics of Radiation Therapy, (MD: Lippincott, Williams and Wilkins, Baltimore 144-148
[4] Juan Azorin 2014 Applied Radiation and Isotopes **83** C 187-191
[5] J.A.Wani, M.S.Atone, N.S. Dhoble, S.J. Dhoble 2013 Journal of Luminescence **134** 640-648

IOP Conf. Series: Materials Science and Engineering **214** (2017) 012036 doi:10.1088/1757-899X/214/1/012036

[6] R.A.P.O. d'Amorim, M.I. Teixeira, S.O. Souza, J.M. Sasaki, L.V.E. Caldas 2012 Journal of Luminescence **132**(2) 266-269
[7] Numan Salah, P.D. Sahare, S.P. Lochab, Pratik Kumar 2006 Radiation Measurements, **41**(1) 40-47
[8] Wang Shuqin, Zhao Yi, Zhang Peipei, Liu Yandong 2011 Chemical Engineering Research and Design **89**(7) 1061-1066

Acknowledgements
The research was funded by the research grant of Riset Inovasi ITB Fiscal Year 2016. The authors thank to Keselamatan Kerja dan Dosimetri Division, Environment lab Pusat Teknologi Keselamatan dan Metrologi Radiasi, Badan Tenaga Nuklir Nasional and all colleagues in Energy and Environmental Material Laboratory, Department of Physics, ITB.

IOP Conf. Series: Materials Science and Engineering **214** (2017) 012037 doi:10.1088/1757-899X/214/1/012037

Influence of Incident Angle of Electron on Transmittance and Tunneling Current in Heterostructures with Bias Voltage by Considering Spin Polarization Effect

Fatimah A. Noor*, Ezra Nabila, Euis Sustini, Khairurrijal

Physics of Electronic Materials Research Division, Faculty of Mathematics and Natural Sciences, Institut Teknologi Bandung, Bandung 40132, Indonesia

*fatimah@fi.itb.ac.id

Abstract. In this work, an analytical expression is presented of electron transmittance through a potential barrier by applying a bias voltage with spin polarization consideration. A zinc-blende material was employed for the barrier in the heterostructure to calculate the transmittance, which depends on the spin states indicated as "up" and "down". The obtained transmittance was then employed to compute the tunneling current. It was shown that the transmittances are different for each state and asymmetric with incident angle. The polarization is positive for a positive incident angle of and negative for a negative incident angle. It was also shown that the tunneling current did not reach its highest value at an incident angle of 0° (z-direction).

1. Introduction

It is known that the electron-tunneling phenomenon depends on spin polarization. It has received great attention in the study of charge transport in heterostructures due to its application in spintronic devices [1-3]. Since the invention of diluted magnetic semiconductors (DMS) by Ohno many researchers have reported that spin polarization in semiconductor heterostructures can be achieved by using a ferromagmetic semiconductor material [4-6]. On the other hand, Voskoboynikov et al. have proposed that the Rashba spin-orbit coupling can cause a non-magnetic semiconductor material to become a spin filter [7-9]. Furthermore, by using only nonmagnetic materials that exploit the unique characteristics of bulk inversion asymmetry in (110)-oriented semiconductor heterostructures, a spin transistor can be obtained, as proposed by Hall et al. [10-11]. Perel et al. have reported that spin polarization is dependent on the electron tunneling through a zinc-blende semiconductor. Moreover, the current interface enhancement in the case of a thin barrier that arises due to the Dresselhaus effect occurring in the zinc-blende [12]. Previously, Suryamas et al. have calculated the transmittance of electron tunneling through a nanometer-thick zinc-blende semiconductor square barrier with spin consideration. However, they did not consider the effect of bias voltage applied to the barrier in calculating the tunneling current [13, 14].

In this paper, we report our study on electron tunneling through a potential barrier with spin polarization consideration by applying a bias voltage to the barrier (called a trapezoidal barrier) following the method from Ref. [13]. The electron transmittance, which depends on spin orientation, was obtained by combining the Dresselhaus term to the Hamiltonian and solving the Schrödinger equation. The obtained transmittance was then utilized to calculate the tunneling current. The effect of the incident angle of the electron on the transmittance and tunneling current are discussed thoroughly.

2. Theoretical Model

Figure 1 depicts the potential profile of heterostructures with a voltage applied to the barrier. We consider that the material in Region 1 is the same as that in Region 3. The zinc-blende semiconductor in Region 2 acts as a potential barrier with width and height d and Φ_0, respectively. The potential profile is expressed as:

$$V(z)=\begin{cases} 0 & z<0 \\ \Phi_0 - e\dfrac{V_b}{d}z & 0\le z<d \\ -eV_b & z\ge d. \end{cases} \tag{1}$$

where V_b is the oxide voltage and e is the electronic charge.

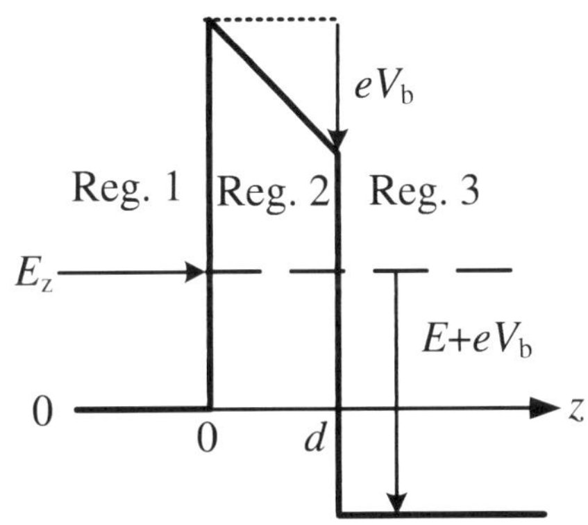

Figure 1. Potential profile of heterostructures with bias voltage applied to the barrier.

The electron tunnels through the potential barrier from the Regions 1 to 3 with initial wave propagation vector are given by

$$\vec{k} = k_\rho \hat{\rho} + k_z \hat{z} \tag{2}$$

where k_ρ and k_z are the momentum in parallel and normal directions, respectively. $\hat{\rho}$ and $\hat{z}$ are the unit vectors in parallel and normal directions to the interfaces between two regions, respectively. Furthermore, the incident angle, θ, is expressed as:

$$\theta = Arc\tan(k_\rho / k_z) \tag{3}$$

An electron in a semiconductor heterostructure with spin polarization is described by the Hamiltonian equation, which is given by:

$$H\Psi = E\Psi \tag{4}$$

where E is the total electron energy and Ψ is the electron wavefunction. H is the Hamiltonian, which is composed of H_O for the heterostructure without spin polarization and H_D for considering the spin polarization (Dresselhauss term), which is given by [15]:

$$H_D = \gamma(\alpha_x k_x - \alpha_y k_y)(\partial^2/\partial z^2) \tag{5}$$

where γ is the Dresselhauss constant and α_i are the Pauli matrices.
By solving the Schrödinger equation and using the boundary condition at each interface [16] it is easy to find the transmission coefficient, which is expressed as:

$$t_{\pm} = -2i\frac{k_{1\pm}\delta_{1\pm}}{m_1}\exp(-ik_{3\pm}d)\times\left\{\left(\frac{2eV_b}{m_2{}^2\hbar^2 d}\right)^{\frac{1}{3}}\delta_{2\pm} + i\left(\frac{k_{1\pm}\delta_{3\pm}k_{3\pm}\delta_{4\pm}}{m_1 m_3}\right) - \left(\frac{k_{1\pm}k_{3\pm}}{m_1 m_3}\right)\left(\frac{2eV_b}{m_2{}^2\hbar^2 d}\right)^{-\frac{1}{3}}\delta_{5\pm}\right\} \tag{6}$$

with

$$k_{1\pm} = \left(\frac{2m_1 E_{z+}}{\hbar^2}\right)^{\frac{1}{2}}\left(1 \pm \frac{2\gamma m_1 k_\rho}{\hbar^2}\right)^{-\frac{1}{2}}, \tag{7}$$

$$k_{3\pm} = \left(\frac{2m_1}{\hbar^2}(E_z + eV_b)\right)^{\frac{1}{2}}, \tag{8}$$

$$\eta_{\pm}(z) = \left(\frac{2m_2 eV_b}{\hbar^2 d}(E_z + eV_b)\right)^{\frac{1}{3}}\left\{(\Phi_0 - E_z)\frac{d}{eV_b} - z\right\}, \tag{9}$$

$$\delta_{1\pm} = Ai'(\eta(d))Bi(\eta(d)) - Ai(\eta(d))Bi'(\eta(d)), \tag{10}$$

$$\delta_{2\pm} = Ai'(\eta(d))Bi'(\eta(0)) - Ai'(\eta(0))Bi'(\eta(d)), \tag{11}$$

$$\delta_{3\pm} = Ai(\eta(0))Bi'(\eta(d)) - Ai'(\eta(d))Bi(\eta(0)), \tag{12}$$

$$\delta_{4\pm} = Ai(\eta(d))Bi'(\eta(0)) - Ai'(\eta(0))Bi(\eta(d)), \tag{13}$$

$$\delta_{5\pm} = Ai(\eta(0))Bi(\eta(d)) - Ai(\eta(d))Bi'(\eta(0)). \tag{14}$$

The transmittance is then obtained as follows:

$$T_{\pm} = t_{\pm}{}^{*} t_{\pm}, \tag{15}$$

where $t_{\pm}{}^{*}$ is the conjugate of $t_{\pm}$.

The polarization of spin P is written as:

$$P = \frac{|t_+|^2 - |t_-|^2}{|t_+|^2 + |t_-|^2} \times 100\%. \tag{16}$$

Furthermore, the tunneling current is calculated by using the following equation [17]:

$$J_z = \frac{em_1 kT}{2\pi^2\hbar^3} \times \int_0^\infty T(E_z)\ln\left\{\frac{(1+\exp[(E_F - E_z)/kT])}{1+\exp[(E_F - E_z - eV_b)/kT]}\right\}dE_z, \tag{17}$$

where k is the Boltzmann constant, T is the temperature, E_F is the Fermi energy of metal, and $T(E_z)=T_+(E_z)-T_-(E_z)$ is the total transmittance. The tunneling current in Equation (17) is easily evaluated by using the Gauss-Laguerre Quadrature method [18].

3. Calculated Results and Discussion

In order to study the effects of the electron incident angle, we use metal/GaSb/metal structures. The following parameters were used to calculate the electron transmittance and tunneling current: $\gamma_1 = 0$, $\gamma_2 = 187$ eV/Å3, $\Phi_0 = 0.2$ eV, $E = 0.1$ eV, $V_b = 0.1$ V. Since the Dresselhauss constant, γ, in Region 1 is the same as that in Region 3, the wave vectors are also the same.

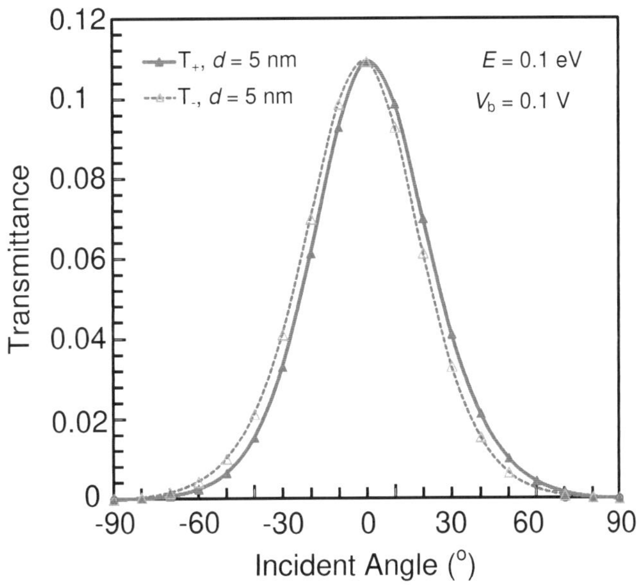

Figure 2. Transmittance versus incident angle of electron with barrier width at 5 nm.

Figure 2 illustrates the electron transmittance as a function of the electron incident angle with the barrier width at 5 nm. The transmittance had the highest value at an incident angle, θ, of about +5° and -5° for spin "up" and "down" polarization respectively. It decreased as the incident angle increased. The result is the same as that obtained without considering the bias voltage [13, 14] in which the transmittance for each state is not the same and asymmetric with the incident angle because of the bulk inversion asymmetry properties of the zinc-blende.

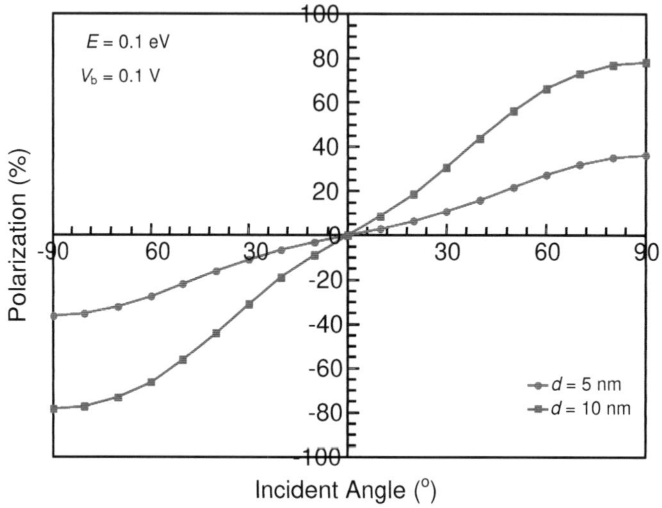

Figure 3. Transmittance versus incident angle of electron with barrier width at 5 nm.

The effect of incident angle on polarization is shown in Figure 3. It can be seen that the polarization increases as the incident angle of electron increases and has zero value at an incident angle of 0° (normal direction). It is also shown that the polarization gives a negative value for $\theta < 0^{\circ}$ and a positive value for $\theta > 0^{\circ}$ because the negative and positive states are dominant for $\theta < 0^{\circ}$ and $\theta > 0^{\circ}$, respectively. Moreover, the polarization shows the highest value at an incident angle in the parallel direction.

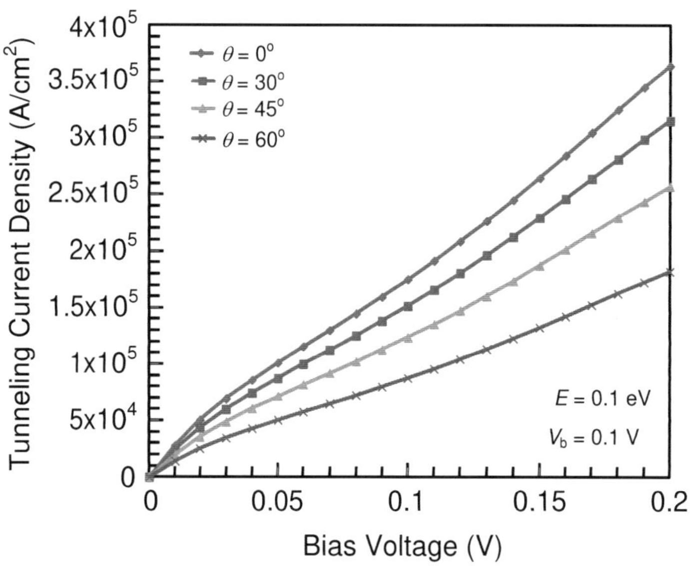

Figure 4. Tunneling current as a function of bias voltage for a variety of incident angles of the electron.

The electron tunneling current in metal/GaSb/metal heterostructures as a function of the oxide voltage for a variety of electron incident angles is displayed in Figure 4. It can be seen that as the incident angle decreases, the tunneling current increases and reaches the highest value at an incident angle of 0°. However, the variation of the incident angle is not significant for an oxide voltage less than 0.01 V. It can also be seen that the tunneling current increases with increasing bias voltage.

4. Conclusions

The electron transmittance and tunneling current in metal/GaSb/metal heterostructures with bias voltage under spin polarization consideration were calculated by taking into account the Dresselhauss term. A zinc-blende material for the barrier in the heterostructure was employed to calculate the transmittance. The transmittance was evaluated for "up" and "down" states. It was found that the transmittance for each state is different and asymmetric with the incident angle. It was also shown that the polarization was optimum at the incident angle in the parallel direction. On the other hand, the tunneling current reached its highest value at an incident angle in the normal direction.

5. References

[1] Zutic I, Fabian J, Das Sarma S 2004 *Rev. Mod. Phys.* **76** 323.
[2] Fabian J, Matos-Abiague A, Ertler C, Stano P, Zutic I 2007 *Acta Phys. Slov.* **57** 566.
[3] Lu J W, Chen E, Kabir M, Stan, M R, Wolf S A 2016 *Int. Mater. Rev.* **61** 456.
[4] Dakhlaoui H, Jaziri S 2005 *Physica B* **355** 401.
[5] Lebedeva N, Kuivalainen P 2003 *J. Appl. Phys.* **93** 9845.
[6] Luo M, Shen Y H 2016 *Optik* **127** 8951.
[7] Voskoboynikov A, Liu S S, Lee C P 1998 *Phys. Rev. B* **58** 15397.
[8] Boda A, Boyacioglu B, Erkaslan U, Chatterjee A 2016 *Physica B Condens Matter* **498** 43.
[9] Tutunculer H, Koc R, Olgar E 2004 *J. Phys. A* **37** 11431.
[10] Hall K C, Lau W H, Gundogdu K, Flatte M E *2003Appl. Phys. Lett.* **83** 2937.
[11] Flatte M E, Byers J M, Lau W H 2002 Spin Dynamics in Semiconductors *Semiconductor Spintronics and Quantum Computation* (Berlin: Springer-Verlag).
[12] Perel' V I, Tarasenko S A, Yassievich I N, Ganichev S D, Bel'kov V V, Prettl W 2003 *Phys. Rev. B.* **67** 201304(R).
[13] Suryamas A B, Abdullah M, Khairurrijal 2006 *Indonesian J. Phys.* **17** 43.
[14] Suryamas A B, Abdullah M, Khairurrijal 2006 Transmittance Coefficient of Electron Tunneling through a Nanometer Thick Square Barrier with Spin Polarization Consideration *International Conference on Mathematics and Natural Sciences (ICMNS)* (Bandung, 29-30 November 2006) pp. 941.
[15] Dresselhauss G 1955 *Phys. Rev.* **100** 580.
[16] BenDaniel D J, Duke C B 1966 *Phys. Rev.* **152** 683.
[17] Noor F A, Abdullah M, Sukirno, Khairurrijal, Ohta A, Miyazaki S 2010 *J. Appl. Phys.* **108** 093711.
[18] De Vries P L 1993 *A first course in computational physics* (New York : John Wiley & Sons).

Acknowledgments

This research was supported by "Riset Desentralisasi" Research Grants in the fiscal year of 2017.

IOP Conf. Series: Materials Science and Engineering **214** (2017) 012038 doi:10.1088/1757-899X/214/1/012038

Comparison of Electron Transmittance and Tunneling Current through a Trapezoidal Potential Barrier with Spin Polarization Consideration by using Analytical and Numerical Approaches

Ezra Nabila, Fatimah A. Noor* and Khairurrijal

Physics of Electronic Materials Research Division, Faculty of Mathematics and Natural Sciences, Institut Teknologi Bandung, Bandung 40132, Indonesia

*Email: fatimah@fi.itb.ac.id

Abstract. In this study, we report an analytical calculation of electron transmittance and polarized tunneling current in a single barrier heterostructure of a metal-GaSb-metal by considering the Dresselhaus spin orbit effect. Exponential function, WKB method and Airy function were used in calculating the electron transmittance and tunneling current. A Transfer Matrix Method, as a numerical method, was utilized as the benchmark to evaluate the analytical calculation. It was found that the transmittances calculated under exponential function and Airy function is the same as that calculated under TMM method at low electron energy. However, at high electron energy only the transmittance calculated under Airy function approach is the same as that calculated under TMM method. It was also shown that the transmittances both of spin-up and spin-down conditions increase as the electron energy increases for low energies. Furthermore, the tunneling current decreases with increasing the barrier width.

1. Introduction

Modern technology relies hugely on semiconductor transistors, in which data processing and computing take place. For decades, the growth and improvement of microelectronic devices was mainly driven by attempts to miniaturize device dimensions. This miniaturization is reflected in Moore's Law [1], which predicts that the number of transistors per area doubles every 18 months. However, this progressive trend will come into conflict with fundamental physical limitations because of which the size of transistors cannot be further reduced. Earlier this decade indications of a decline in the growth of miniaturization have emerged. A semiconductor technology only utilizes electron charge and completely ignores the associated spin state of the electron. However, integrating spin polarized currents and using electron spin as a new degree of freedom could not only further enhance and improve this technology but could also unveil a new kind of technology.

This is enabled by the invention of diluted magnetic semiconductors (DMS) [2], in which the spin polarization in semiconductors can be obtained by inducing a ferromagnetic material to become a semiconductor material. Hideo Ohno et al. in 1998 were the first to determine the ferromagnetic properties of semiconductor materials doped with transition metals [2]. Since then, research regarding DMS has become one of the core research areas in the magnetic semiconductor field. Additionally, Voskoboynikov et al. [3] proposed the idea of using Rashba spin orbit coupling in nonmagnetic

IOP Conf. Series: Materials Science and Engineering **214** (2017) 012038 doi:10.1088/1757-899X/214/1/012038

semiconductors as a spin filter [4]. Meanwhile, Hall et al. [5,6] suggested the idea of using nonmagnetic semiconductors as a spin transistor, which exploits bulk inversion asymmetry (BIA) in the (110)-oriented semiconductor heterostructure as the spin polarizator. Perel et al. [7] in their report showed that spin polarization affects the process of electron tunnelling through zinc blende material. Wang et. al [8] reported that the Dresselhaus effect caused by BIA occurs in zinc blende material, which causes tunnelling current enhancement in the case of a thin barrier. Here, the study of electron tunnelling through a nanometer-thick trapezoidal barrier with spin polarization consideration is reported.

2. Theoretical model

The potential profile of the heterostructure composed by metal-GaSb-metal grown along the z-axis is shown in Fig. 1, where q is an electron charge, V_0 is the GaSb height, V_b is the bias voltage applied to the GaSb, and d is the GaSb width. The mathematical representation of the potential profile is given by:

$$V(z) = \begin{cases} 0, & z \leq 0 \\ V_0 - \frac{qV_b}{d} z, & 0 < z \leq d \\ -qV_b, & z > d \end{cases} \tag{1}$$

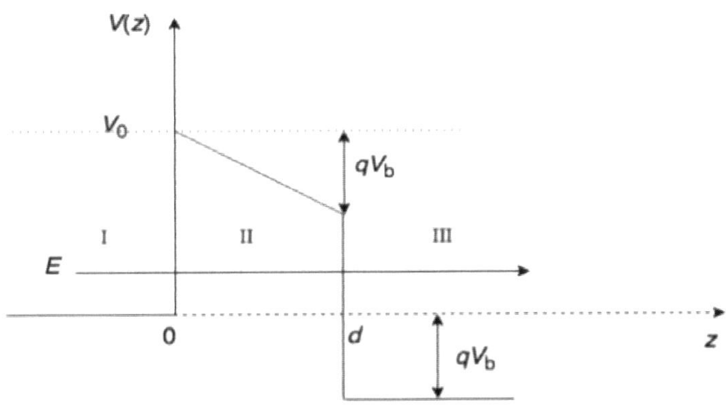

Figure 1. Potential energy profile by applying bias voltage to the barrier.

The alignment of the electron spin was caused by the inversion asymmetry effect, namely bulk inversion asymmetry (BIA). The asymmetry effect is represented by the Dresselhaus Hamiltonian which is given by [6,9]:

$$H_D = \gamma\left(\sigma_x k_x - \sigma_y k_y\right)\frac{\partial^2}{\partial z^2} \tag{2}$$

here γ is the Dresselhaus constant, for the material grown only along the z axis the (k) wave number represented for every axis is $k_x^2 = k_y^2 = 0$ and $k_z = \frac{-i\partial}{\partial z}$. By calculating the eigenvalue and eigen function we can calculate the total Hamiltonian system ($H = H_0 + H_D$), which can be represented by [10,11]:

$$H = -\frac{\hbar^2}{2m_\pm^*}\frac{\partial^2}{\partial z^2} + V(z) + \frac{\hbar^2 k_p^2}{2m^*} \tag{3}$$

here $m_\pm^*$ is the effective mass of a spin dependant electron, which can be represented as:

$$m_{\pm}^{*} = m^{*}\left(1 \pm 2\frac{\gamma m^{*} k_p}{\hbar^2}\right) \tag{4}$$

with ϕ as the smallest angle between k_y and k_x vector, k_p can be written as:

$$\vec{k}_p = k_x\hat{x} + k_y\hat{y} \tag{5}$$
$$k_y = k_p \cos\phi \tag{6}$$
$$k_x = k_p \sin\phi \tag{7}$$

2.1. *Exponential wavefunction-approach*

Using the exponential function, the solution for every region is represented by:

$$\varphi(z) = \begin{cases} A\exp(ik_1 z) + B\exp(-ik_2 z), & z \le 0 \\ C\exp\left(\int_0^z k_2(z')dz'\right) + D\exp\left(-\int_0^z k_2(z')dz'\right), & 0 < z \le d \\ F\exp(ik_3 z), & z > d \end{cases}, \tag{8}$$

with the wavenumber as follows:

$$k_1 = \left(\frac{2m_1 E_z}{\hbar^2}\right)^{\frac{1}{2}} \tag{9}$$
$$k_2 = \left(\frac{2m_2 E_z}{\hbar^2}\left(V_0 - \frac{qV_b z}{d}\right)\right)^{\frac{1}{2}} \tag{10}$$
$$k_3 = \left(\frac{2m_1 E_z}{\hbar^2}(E_z + qV_b)\right)^{\frac{1}{2}} \tag{11}$$

The transmission coefficient then can be derived as:

$$\left(\frac{F}{A}\right) = 2k_1\left(\frac{k_2(d)}{k_2(0)}\right)\exp(-ik_3 d) \times \frac{\left[\left(k_3 + \left(k_1\frac{k_2(d)}{k_2(0)}\right)\right)\cosh(a)\right] + i\left[\left(\frac{m_2 k_1 k_3}{m_1 k_2(0)} - \frac{m_1 k_2(d)}{m_2}\right)\sinh(a)\right]}{\left[\left(k_3 + \left(k_1\frac{k_2(d)}{k_2(0)}\right)\right)\cosh(a)\right]^2 + \left[\left(\frac{m_2 k_1 k_3}{m_1 k_2(0)} - \frac{m_1 k_2(d)}{m_2}\right)\sinh(a)\right]^2} \tag{12}$$

with

$$a = \int_0^d k_2 z \tag{13}$$
$$a = \left(\frac{2m_2}{\hbar^2}\right)^{\frac{1}{2}}\frac{2d}{3qV_b}\left((V_0 - E)^{\frac{3}{2}} - (V_0 - E - qV_b)^{\frac{3}{2}}\right) \tag{14}$$

2.2. *Airy wavefunction-approach*

By defining a function η as:

$$\eta(z) = \left(\frac{2m}{\hbar^2}\frac{qV_b}{d}\right)^{\frac{1}{3}}\left[(V_0 - E_z)\frac{d}{qV_b} - z\right], \tag{15}$$

the solution for every region can be represented as:

$$\varphi(z) = \begin{cases} A\exp(ik_1 z) + B\exp(-ik_1 z), & z \le 0 \\ C\,Ai(\eta(z)) + D\,Bi(\eta(z)), & 0 < z \le d \\ F\exp(ik_3 z), & z > d \end{cases}, \tag{16}$$

with the wavenumber as follows:

$$k_1 = \left(\frac{2m_1 E_z}{\hbar^2}\right)^{\frac{1}{2}} \tag{17}$$

$$k_3 = \left(\frac{2m_1}{\hbar^2}(E_z - qV_b)\right)^{\frac{1}{2}} \tag{18}$$

The transmission-coefficient then can be derived as:

$$\left(\frac{F}{A}\right) = -2i\frac{k_1 a_1}{m_1}\exp(-ik_3 d) \times \left[\left(\frac{2}{m_2^2\hbar^2}\frac{qV_b}{d}\right)^{\frac{1}{3}} a_2 + i\left(\frac{k_1 a_3}{m_1}\frac{k_3 a_4}{m_3}\right) - \frac{k_1 k_3}{m_1 m_3}\left(\frac{2}{m_2^2\hbar^2}\frac{qV_b}{d}\right)^{-\frac{1}{3}} a_5\right] \tag{19}$$

with the Airy function as follows:

$$a_1 = Ai'(\eta(d))Bi(\eta(d)) - Ai(\eta(d))Bi'(\eta(d)) \tag{20}$$
$$a_2 = Ai'(\eta(d))Bi'(\eta(0)) - Ai'(\eta(0))Bi'(\eta(d)) \tag{21}$$
$$a_3 = Ai(\eta(0))Bi'(\eta(d)) - Ai'(\eta(d))Bi(\eta(0)) \tag{22}$$
$$a_4 = Ai(\eta(d))Bi'(\eta(0)) - Ai'(\eta(0))Bi(\eta(d)) \tag{23}$$
$$a_5 = Ai(\eta(0))Bi(\eta(d)) - Ai(\eta(d))Bi'(\eta(0)) \tag{24}$$

2.3. Matrix transfer method

Using the matrix transfer method, the GaSb region is divided into N regions, with n = 2, 3, …. , N-1. The solution can be represented as:

$$\varphi_1 = A_1\exp(ik_1 z_1) + B_1\exp(-ik_1 z_1), \quad z < 0 \tag{25}$$
$$\varphi_n = A_n\exp(ik_n z_n) + B_1\exp(-ik_n z_n), \quad 0 \le z < z_n \tag{26}$$
$$\varphi_n = A_{n+1}\exp(ik_{n+1} z_n) + B_{n+1}\exp(-ik_{n+1} z_n), \quad 0 < z \le z_n \tag{27}$$
$$\varphi_N = A_N\exp(ik_N z_{N-1}), \quad z \le z_N \tag{28}$$

By applying the boundary conditions, the solution can then be written into matrix

$$\begin{pmatrix}1\\B_1\end{pmatrix} = \begin{pmatrix}a_{11} & a_{12}\\a_{21} & a_{22}\end{pmatrix}\begin{pmatrix}A_N\\0\end{pmatrix}, \tag{29}$$

and the transmission coefficient can then be defined as:

$$t = A_N = \frac{1}{a_{11}} \tag{30}$$

2.4. WKB approximation

The solution of the Schrödinger equation is written as:

$$\varphi(z) = R(z)exp\left(i\frac{S(z)}{\hbar}\right) \tag{31}$$

and the transmission coefficient can then be written as:

$$T = \left|\frac{\varphi_{trans}}{\varphi_{inc}}\right|^2 = exp\left(\int_{z_2}^{z_1}(2m(V(z) - E)^{\frac{1}{2}}(\hbar dz)^{-1}\right) \tag{32}$$

2.5. Spin dependent transmittance

The effective Schrödinger equation for each region can be written as:

$$\frac{d^2\varphi_\pm}{dz^2}+\frac{2m_1}{\hbar^2}\left(1\pm\frac{2\gamma m_1 k_p}{\hbar^2}\right)^{-1}E_{z\pm}\varphi_\pm=0,\ z\leq 0 \tag{33}$$

$$\frac{d^2\varphi_\pm}{dz^2}+\frac{2m_2}{\hbar^2}\left(1\pm\frac{2\gamma m_2 k_p}{\hbar^2}\right)^{-1}\left[[V_0-E_{z\pm}]\frac{d}{qV_b}-z\right]\varphi_\pm=0,\ 0<z\leq d \tag{34}$$

$$\frac{d^2\varphi_\pm}{dz^2}+\frac{2m_1}{\hbar^2}\left(1\pm\frac{2\gamma m_1 k_p}{\hbar^2}\right)^{-1}(E_{z\pm}+qV_b)\varphi_\pm=0,\ z>d \tag{35}$$

Using the Airy function, the solution then can be written as:

$$\varphi_\pm(z)=\begin{cases}A_\pm\exp(ik_{z1\pm}z)+B_\pm\exp(-ik_{z1\pm}z), & z\leq 0\\ C_\pm Ai\left(\eta_\pm(z)\right)+D_\pm\,Bi\left(\eta_\pm(z)\right), & 0<z\leq d\\ F_\pm\exp(ik_{z3\pm}z), & z>d\end{cases} \tag{36}$$

with

$$k_{z1\pm}=\left(\frac{2m_1E_{z\pm}}{\hbar^2}\right)^{\frac{1}{2}}\left(1\pm\frac{2\gamma m_1 k_p}{\hbar^2}\right)^{-\frac{1}{2}} \tag{37}$$

$$k_{z3\pm}=\left(\frac{2m_1}{\hbar^2}(E_z+qV_b)\right)^{\frac{1}{2}} \tag{38}$$

$$\eta_\pm(z)=\left(\frac{2m_{2\pm}}{\hbar^2}\frac{qV_b}{d}\right)^{\frac{1}{3}}\left[(V_0-E)\frac{d}{qV_b}-z\right] \tag{39}$$

the transmission coefficient can then be derived as:

$$\left(\frac{F_\pm}{A_\pm}\right)=-2i\frac{k_{z1\pm}a_1}{m_{1\pm}}\exp(-ik_{z3\pm}d)\times\left[\left(\frac{2}{m_{2\pm}^2\hbar^2}\frac{qV_b}{d}\right)^{\frac{1}{3}}a_2+i\left(\frac{k_{z1\pm}a_3}{m_{1\pm}}\frac{k_{z3\pm}a_4}{m_{3\pm}}\right)-\frac{k_{z1\pm}k_{z3\pm}}{m_{1\pm}m_{3\pm}}\left(\frac{2}{m_{2\pm}^2\hbar^2}\frac{qV_b}{d}\right)^{-\frac{1}{3}}a_5\right] \tag{40}$$

2.6. *Spin dependent tunneling current*

The tunneling current can be calculated by deriving the following equation [12,13]:

$$J=\int_0^\infty\frac{qm_1k_bT}{2\pi^2\hbar^3}T(z)\ln\left(\frac{1+\exp(E_f-E_z)}{1+\exp(E_f-E_z-qV_b)}\right)dE_z, \tag{41}$$

with

$$T(z)=T_+(z)+T_-(z) \tag{42}$$

In which *T*(z) is the total of the transmittance in spin up and spin down state, m_1 is the electron mass in the metal, k_b is the Boltzmann constant, *T* is the temperature in Kelvin, E_f is the Fermi energy, E_z is the electron energy.

3. Results and Discussion

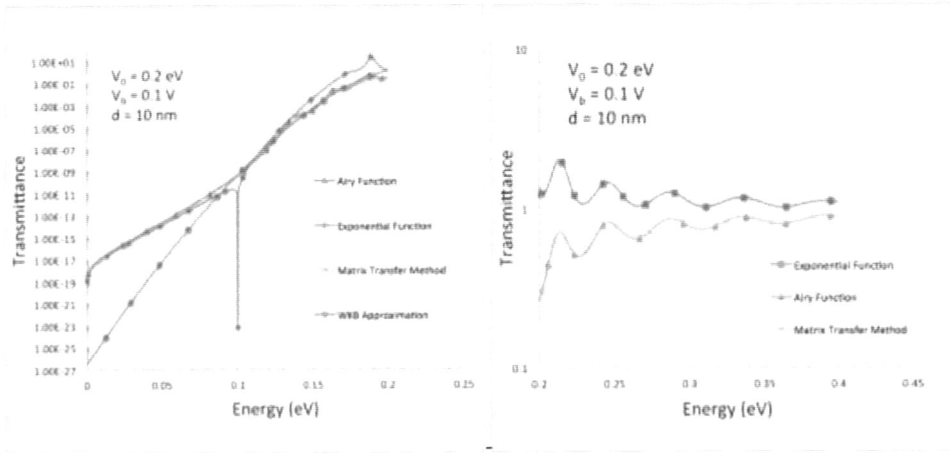

Figure 2. Comparison of methods with electron energy lower than potential barrier.

Figures 2 (a) and (b) show the transmittance versus electron energy at low and high energy regimes, respectively. The transmittances are calculated by using Airy- and Exponential-wavefunction approaches, matrix transfer method, and WKB approximation. It is shown that only the transmittances computed by Airy wavefunction-approach fit those calculated by TMM. In addition, the transmittances computed under Exponential wavefunction-approaches and WKB approximation show the deviation results from TMM. They indicate that the Airy wavefunction-approach is the best analytical method in calculating the transmittance in the spintronic devices. These results are the same as those obtained for MOSFETs device without spin polarization consideration [14].

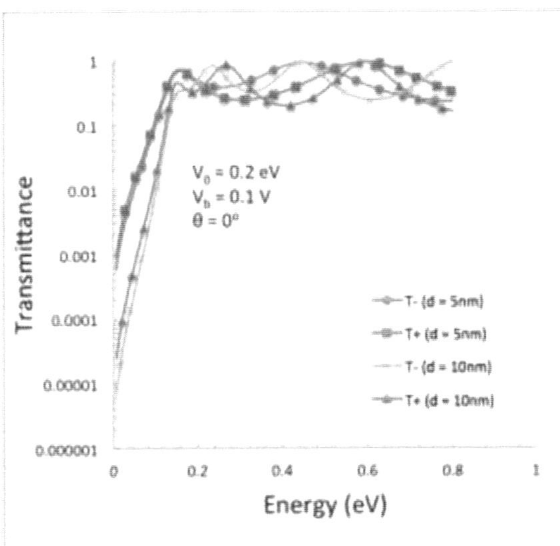

Figure 3. Transmittance versus energy with variation of the barrier width.

Fig. 4 shows transmittance as a function of energy for the barrier width of 5 and 10 nm. It is seen that the transmittances increase as energy increases for the energy lower than barrier height. It is also seen that the transmittances show the oscillatory behaviour for energy higher than barrier height. The transmittance of the spin up state is greater than the spin down state. Moreover, the transmittance increase with decreasing the barrier width.

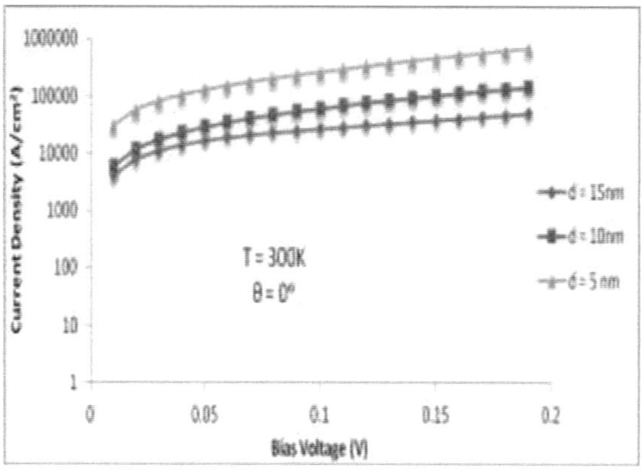

Figure 4. Tunneling current density as a function of bias voltage with variation of the barrier width.

Fig. 4 illustrates the tunnelling current density versus bias voltage plotted against external voltage with variation of the barrier width. It is seen that the width of the barrier is very influential on the tunnelling current density obtained: the smaller the width of the barrier, the greater the value of the tunnelling current density obtained. This is due to the narrower width of the barrier making it easier for the electrons to tunnel through the potential barrier, resulting in a higher electron transmittance and tunnelling current density.

4. Conclusion

We have derived the analytical expression of transmittance and tunnelling current through a trapezoidal potential barrier by including the spin polarization effect. It is shown that the transmittances calculated under the Airy wavefunction-approach match those computed under the matrix transfer method. It is also shown that the transmittances in the spin up state are higher than those in the spin down state. In addition, the tunnelling currents increase as the bias voltage increases and the barrier width decreases.

5. References

[1] G E Moore 1998 *Cramming More Components onto Integrated Circuits* P. IEEE **86** 1
[2] H Ohno 2002 *Ferromagnetic III–V Semiconductors and Their Heterostructures* in *Semiconductor Spintronics and Quantum Computation* ed D Awschalom, D Loss N Samarth Germany, Springer
[3] A Voskoboynikov, S S Liu, C P Lee 1998 *Spin-dependent electronic tunneling at zero magneticfield* Phys. Rev. B. **58** 15397
[4] H Tütüncüler, R Koç, E Olğar 2004 J. Phys. A: Math. Gen. **37** 11431
[5] K C Hall, W H Lau, K Gündoğdu, M E Flatté, T F Boggess 2003 Appl. Phys. Lett. **83** 2937 (2003).
[6] M E Flatte, J M Byers, W H Lau 2002 *Spin Dynamics in Semiconductors,* in *Semiconductor Spintronics and Quantum Computation*, ed D Awschalom, D Loss, N Samarth Germany, Springer
[7] V I Perel', S A Tarasenko, I N Yassievich, S D Ganichev, V V Bel'kov, W Prettl 2003 Phys. Rev. B. **67** 201304(R)
[8] L G Wang, W Yang, K Chang, K S Chan 2005 Phys. Rev. B. **72** 153314
[9] V Lechner 2012 *Bulk and Structure Inversion Asymmetry in Semiconductor Quantum Well Structure*, Dissertation, Universität Regensburg.
[10] A B Suryamas, M Abdullah, Khairurrijal 2006 *Spin Dependent Tunneling through a Nanometer-thick Square Barrier Based on Zinc-blende Structure Material*, Indonesian J. Phys. **17** 43-47.

[11] A B Suryamas, M Abdullah, Khairurrijal 2006 *Transmittance Coefficient of Electron Tunneling through a Nanometer Thick Square Barrier with Spin Polarization Consideration*, International Conference on Mathematics and Natural Sciences (ICMNS). Bandung 941-944

[12] F A Noor, M Abdullah, Sukirno, Khairurrijal 2010 *Comparison of electron transmittances and tunneling currents in an anisotropic Tinx/HfO2/Sio2/pSi(100) metal/oxide/semiconductor (MOS) capacitor calculated using exponential- and Airy-wavefunction approaches and a transfer matrix method*, J. Semicond. **31** 124002

[13] F A Noor, M Abdullah, Sukirno, Khairurrijal, A Ohta, S Miyazaki 2010 *Electron and hole components of tunneling currents through an interfacial oxide- high-k gate stack in metal-oxide-semiconductor capacitors*, J. Appl. Phys. **108** 093711

[14] J C Ranuarez, M J Deen, C H Chen 2006 *A review of gate tunneling current in MOS devices*, Microelectron. Reliab. **46** 1939 – 1956.

Institute of Physics
Dirac House, Temple Back
Bristol BS1 6BE UK

ISSN: 1757-8981
ISBN 978-1-5108-4416-2
9 781510 844162

CURRAN ASSOCIATES INC.
proceedings.com